AF826553

Information Technology Security and Risk Management

Information Technology Security and Risk Management: Inductive Cases for Information Security is a compilation of cases that examine recent developments and issues that are relevant to IT security managers, risk assessment and management, and the broader topic of IT security in the 21st century. As the title indicates, the cases are written and analyzed inductively, which is to say that the authors allowed the cases to speak for themselves, and lead where they would, rather than approach the cases with presuppositions or assumptions regarding what the case should be "about". In other words, the authors were given broad discretion to interpret a case in the most interesting and relevant manner possible; any given case may be "about" many things, depending on the perspective adopted by the reader, and many different lessons may be learned. The inductive approach of these cases reflects the design philosophy of the advanced IT Security and Risk Management course we teach on the topic here at the University of Canterbury, where all discussions begin with the analysis of a specific case of interest and follow the most interesting and salient aspects of the case in evidence. In our course, the presentation, analysis, and discussion of a case are followed by a brief lecture to address the conceptual, theoretical, and scholarly dimensions arising from the case. The inductive approach to teaching and learning also comes with a huge advantage – the students seem to love it, and often express their appreciation for a fresh and engaging approach to learning the sometimes-highly-technical content of an IT security course. As instructors, we are also grateful for the break in the typical scripted "chalk-and-talk" of a university lecture afforded by the spontaneity of the inductive approach.

We were motivated to prepare this text because there seems to be no other book of cases dedicated to the topic of IT security and risk management, and our own success and satisfaction with inductive teaching and learning. We believe this book would be useful either for an inductive, case-based course like our own or as a body of cases to be discussed in a more traditional course with a deductive approach. There are abstracts and keywords for each case, which would help instructors select cases for discussions on specific topics, and PowerPoint slides are available as a guide for discussion about a given case.

Information Technology Security and Risk Management

Inductive Cases for Information Security

Edited by
Stephen C. Wingreen
Amelia Samandari

CRC Press
Taylor & Francis Group
Boca Raton London New York

CRC Press is an imprint of the
Taylor & Francis Group, an **informa** business

First edition published 2024
by CRC Press
2385 NW Executive Center Drive, Suite 320, Boca Raton FL 33431

and by CRC Press
4 Park Square, Milton Park, Abingdon, Oxon, OX14 4RN

CRC Press is an imprint of Taylor & Francis Group, LLC

© 2024 selection and editorial matter, Stephen C. Wingreen and Amelia Samandari; individual chapters, the contributors

Reasonable efforts have been made to publish reliable data and information, but the author and publisher cannot assume responsibility for the validity of all materials or the consequences of their use. The authors and publishers have attempted to trace the copyright holders of all material reproduced in this publication and apologize to copyright holders if permission to publish in this form has not been obtained. If any copyright material has not been acknowledged please write and let us know so we may rectify in any future reprint.

Except as permitted under U.S. Copyright Law, no part of this book may be reprinted, reproduced, transmitted, or utilized in any form by any electronic, mechanical, or other means, now known or hereafter invented, including photocopying, microfilming, and recording, or in any information storage or retrieval system, without written permission from the publishers.

For permission to photocopy or use material electronically from this work, access www.copyright.com or contact the Copyright Clearance Center, Inc. (CCC), 222 Rosewood Drive, Danvers, MA 01923, 978-750-8400. For works that are not available on CCC please contact mpkbookspermissions@tandf.co.uk

Trademark notice: Product or corporate names may be trademarks or registered trademarks and are used only for identification and explanation without intent to infringe.

Library of Congress Cataloging-in-Publication Data
Names: Wingreen, Stephen C., editor. | Samandari, Amelia, editor.
Title: Information technology security and risk management :
inductive cases for information security / edited by Stephen C. Wingreen and
Amelia Samandari. Description: First edition. | Boca Raton, FL : CRC Press, 2024. |
Includes bibliographical references and index.
Identifiers: LCCN 2023051811 (print) | LCCN 2023051812 (ebook) |
ISBN 9781032206141 (hbk) | ISBN 9781032206158 (pbk) | ISBN 9781003264415 (ebk)
Subjects: LCSH: Computer security. | Information technology--Security measures.
Classification: LCC QA76.9.A25 I548 2024 (print) | LCC QA76.9.A25 (ebook) |
DDC 005.8--dc23/eng/20240304 LC record available at https://lccn.loc.gov/2023051811
LC ebook record available at https://lccn.loc.gov/2023051812

ISBN: 978-1-032-20614-1 (hbk)
ISBN: 978-1-032-20615-8 (pbk)
ISBN: 978-1-003-26441-5 (ebk)

DOI: 10.1201/9781003264415

Typeset in Times
by SPi Technologies India Pvt Ltd (Straive)

Access the Support Material: www.routledge.com/9781032206141

Contents

SECTION 1 TECHNICAL CASES

TEACHING CASES

RESEARCH CASES

SECTION 2 BEHAVIORAL AND SOCIAL CASES

TEACHING CASES

RESEARCH CASES

SECTION 3 PROCESS CASES

TEACHING CASES

RESEARCH CASES

Editors

Stephen C. Wingreen is an Associate Professor of Information Systems and Decision Sciences at the University of Canterbury in Christchurch, New Zealand. Prior to his academic career, he was a manager of networks and telecommunications whose responsibilities included network security. He has taught IT security and risk management over the span of his 23-year teaching career at multiple universities, both as a component of more comprehensive courses and as its own course. His current research agenda on the topic of emerging technologies in 21st-century commerce includes a project with multiple papers about information privacy, trust, and ethical concerns in the context of advanced emerging technologies.

Amelia Samandari is a PhD candidate at the University of Canterbury (UC) in Christchurch, New Zealand. She received a BSc(Hons) First Class in Computer Science, with a focus on IoT security. She has tutored undergraduate and postgraduate courses in information systems, networks, and security. Amelia represented New Zealand at the ITS World Congress Youth Leadership Program, was selected to be a member of the US Embassy's New Zealand Youth Council, and is a founder of ITS New Zealand's Young Transport Professionals Mentoring Program. Her work in these spaces has fostered a passion for engaging with students in meaningful ways that encourage self-motivated learning and development. Amelia is the author of several peer-reviewed papers on the topics of networks and security.

Contributors

Editors

Stephen C. Wingreen
Department of Accounting and Information Systems
University of Canterbury
Christchurch, New Zealand

Amelia Samandari
Department of Computer Science
University of Canterbury
Christchurch, New Zealand

Review Editor

Paula Wingreen
Christchurch, New Zealand

Authors

Areeb Amanulla
Kainga Ora – Homes and Communities
New Zealand

Alastair Blackett
M2X, Auckland, New Zealand
Chartered Accountants of New Zealand

Tridhawan Choudhary
AFFCO New Zealand Limited
Hamilton, New Zealand

Roydon Dominic Correya
MYOB
Christchurch, New Zealand

Alisha Gilbert
Canterbury District Health Board
Christchurch, New Zealand

Zhigang Gong
Department of Accounting and Information Systems
University of Canterbury
Christchurch, New Zealand

Priyanka Jagre
UNIFY Solutions
Wellington, New Zealand

Anjali Krishnan
Department of Accounting and Information Systems
University of Canterbury
Christchurch, New Zealand

Abhishek Kumar
Department of Accounting and Information Systems
University of Canterbury
Christchurch, New Zealand

Joe Mani Kuttikatt
Z Energy NZ
Wellington, New Zealand

Ying-Pin Lan
Hato Hone St. John
Auckland, New Zealand

Yuchen Lin
Department of Accounting and Information Systems
University of Canterbury
Christchurch, New Zealand

Rion Jovial Luis
TussockLine
New Plymouth, New Zealand

Cherie Chun Mei Mak
LivePerson
Singapore, Singapore

Vignesh Makkada
Ministry of Justice
Wellington, New Zealand

Saadik Niyaz
Waste Management NZ, Ltd
New Zealand

Satish Prabhu
Capgemini
Christchurch, New Zealand
Chartered Accountants of New Zealand

Sasha Pugh
Ryman Healthcare
Christchurch, New Zealand

Ishani Rai
Department of Accounting and Information Systems
University of Canterbury
Christchurch, New Zealand

Nishanth Chakkere Ramesh
Kapiti Coast District Council
Paraparaumu, New Zealand

Tushar Ashok Sagathia
ASB Bank
Auckland, New Zealand

Amelia Samandari
Department of Computer Science
University of Canterbury
Christchurch, New Zealand

Juliet Samandari
Department of Computer Science
University of Canterbury
Christchurch, New Zealand

Xuejie Tai
Department of Accounting and Information Systems
University of Canterbury
Christchurch, New Zealand

Zuejie Tai
University of Canterbury
Christchurch, New Zealand

Shameer Thottoli
Acclaim Group Limited
Auckland, New Zealand

Hannah Yeo
Department of Accounting and Information Systems
University of Canterbury
Christchurch, New Zealand

Huixin Zhang
Department of Accounting and Information Systems
University of Canterbury
Christchurch, New Zealand

Zhehen Zheng
Department of Accounting and Information Systems
University of Canterbury
Christchurch, New Zealand

Yang Zhou
Department of Accounting and Information Systems
University of Canterbury
Christchurch, New Zealand

Section 1

Technical Cases

Teaching Cases

1.1 Data Breach at Nintendo Co. Ltd.

300,000 Nintendo Users Hacked

Areeb Amanulla
Kainga Ora – Homes and Communities, New Zealand

Saadik Niyaz
Waste Management NZ, Ltd, New Zealand

CASE SUMMARY

Nintendo Co., Ltd. is a Japanese consumer electronic video game company headquartered in Kyoto. Founded in 1889 by Fusajiro Yamauchi, the company has produced some of the most successful video game consoles in the industry such as Gameboy, Nintendo Wii, and Nintendo Switch. On April 21, 2020, a gaming reporter issued a warning stating that Nintendo Switch accounts were being hacked. Three days later, the Japanese giant initially confirmed that 160,000 Nintendo Network ID (NNID) accounts had been compromised through unauthorized logins. However, the event became even more concerning as Nintendo later confirmed that the actual number of users who had their accounts illegally accessed by hackers was 300,000 (Winder, 2020).

The company stated that the Nintendo Switch Console handheld device impacted users' nicknames, date of birth, email address, and regions and have been viewed by a third party. Furthermore, the hackers have accessed payment services linked to these accounts which include PayPal accounts and credit cards to purchase on Nintendo platforms (Warren, 2020). According to Nintendo, there has been no evidence of how the data breach occurred. However, the company has stated, "In order to ensure similar events do not occur, we will strive to further strengthen security and safety".

Unfortunately, the above statement by Nintendo does not provide precise information on how the accounts were hacked. However, the company has found a trace stating that through 'spoofing', the login IDs and passwords were obtained illegally by some means other than the company's service. Nintendo has predicted that one of the following three alternatives had been used to hack into the accounts.

DOI: 10.1201/9781003264415-3

The first alternative is through credential stuffing. This is a cyberattack method in which attackers breach into a system by using a list of compromised user credentials. Based on the assumption that many gamers reuse passwords between accounts, bots will be used in the attack for automation and scale. Thus, investigation and general knowledge of user habits identified that many users have used the same logins and passwords across both NNID and Nintendo accounts. Nintendo predicted that phishing is another method that attackers may have used to hack into the accounts. Attackers trick the users into disclosing their login details by sending an email from a legitimate company asking to provide sensitive information. Brute force is the final alternative that may have been used to hack into the users' accounts. This involves an attacker submitting many passwords and passphrases with the hope of eventually guessing the correct one. Nintendo users who have used weak and common passwords fall victim to this attack (Winder, 2020).

Nintendo has taken steps in response to the breach, firstly by discontinuing the option for switch console users to log into their Nintendo accounts via NNID. Nintendo has encouraged all users to check the purchase history in their accounts for any unauthorized transaction and request refunds. Furthermore, password resets have been sent to the 300,000 accounts that were impacted by the hacking attempts via email from the company. In addition, Nintendo has recommended that gamers and all switch console users enable two-step verification to secure their accounts.

In April 2020, Nintendo made an announcement on the company's official website by issuing the following statement.

> We would like to reassure our users that there is currently no evidence pointing towards a data breach of Nintendo's database, servers or services and we will continue to investigate more into this matter.

CASE ANALYSIS

Nintendo is a company that can be categorized into producing technological systems. One of their latest video gaming consoles which was released back in March 2017, Nintendo Switch, enables the users with the functionality to create an account using an email address as well as to link it to their NNID which was used for older Nintendo 3DS and Wii consoles. According to the company, the data breach was linked to its NNID login system and it issued a warning to the gamers via email stating that credit card/PayPal may be illegally used at My Nintendo Store or eShop if the same password has been used by the users to log into NNID and Nintendo accounts (Liao, 2020).

Similar to the data breach at Nintendo, Sony PlayStation suffered a massive data breach back in 2011. According to the article by Baker & Finkle (2011), an unauthorized person illegally stole names, addresses, logins, birth dates, and possible credit card details which belonged to 77 million user accounts through the video game online network. This led to Sony shutting down the PlayStation network immediately, preventing users of Sony from buying and downloading games through their network. Alan Paller, the research director of the SANS Institute, stated that Sony did not provide higher priority in the development of their software which runs the network. The system Administrator has the authority to access sensitive information about Sony customers and the Japanese giant has suspected that the hackers have entered the network by sending the administrator an email that contains a piece of malicious software which helps to download information to the hackers' PC. This deception brought about one of the largest internet security break-ins.

Therefore, in view of the data breach at Sony, a possible, similar assumption can be made that the hackers may have attacked the users of Nintendo to access their personal information.

The situation at Nintendo can be analyzed from the managerial point of view as well as the users' point of view.

When analyzing the data breach from the Nintendo users' perspective, Whittaker (2020) mentions that Nintendo has implied that hackers have cracked the weak passwords used by the gamers to access their accounts. This method of cyberattack is known as 'brute force'. The reality of this method's use is reinforced by the recent massive brute-force attack on Alibaba which affected 21 million accounts. According to the news reporter Seals (2016), the security expert and director Dave Martin stated that by the brute force password attacks of hackers, many of the weak passwords were defeated and the accounts were stolen. The hackers used the username and passwords of a 99 million database and managed to compromise 20.6 million accounts.

Winder (2020) has implied that the Nintendo Switch gamers who were attacked are victims of credential stuffing. Nintendo has indicated that many users have used the same credentials across different accounts. In the case of Nintendo, many users have used the same login ID and passwords across both NNID and Nintendo accounts.

This enables the attackers to steal the credentials of users from a low-profile website which will eventually increase the chance of working on services that hold more sensitive data. Akamai Technologies, a cybersecurity and cloud service company, found 61 billion credential stuffing attacks between January 2018 and June 2019 (Sjouwerman, 2019). Furthermore, a recent study has found that the gaming industry has become the most popular target of credential-stuffing attacks (Barth, 2019).

Hence, credential stuffing is a very popular and common method used by hackers targetting the gaming industry; this factor leads to a higher chance that the hackers may have used this method to attack Nintendo Switch users' accounts.

According to Winder (2020), Nintendo has predicted that 300,000 users were tricked into disclosing their login details by redirecting them to a cloned site. This type of cyberattack is known as 'phishing', where the attackers disguise the email as a weapon to trick the users into believing the message is from a legitimate company, bank, etc. An example of a phishing attack is the scam that took place at Pathé. The European cinema chain lost approximately 19 million pounds through an email scam; the two top-level executives were targeted. For over nearly a month, the hacker impersonated the company's CEO and got the executives to transfer multiple payments (DutchNews.NL, 2018).

Therefore, the Nintendo users who are not aware of this type of cyberattack are likely to be easy victims.

An analysis can be provided from the managerial or company's perspective as well. With regards to the hacking, Nintendo has stated that in order to prevent similar events, the company is planning to further strengthen its security and ensure the safety of its users. However, Nintendo was not aware of the breach until a gaming reporter and the hacked users discovered it. Prior to Nintendo's statement, the reporter published a warning and the users turned to social media to complain that hackers were accessing their Nintendo accounts (Winder, 2020).

Furthermore, the company has reassured that there has been no breach in their database and servers and suggested the attackers have used spoofed logins to gain access to the accounts. However, considering the incident at Sony, Alan Paller, the director of SANS Institute stated that Sony was in a rush to get out innovative products, and due to this, the company did not provide enough attention to security in the development of the software (Baker & Finkle, 2011). Therefore, this was one of the major reasons for the data breach. Thus, taking this incident into consideration, Nintendo should give more priority to risks, security, and safety concerns when it comes to producing future gaming consoles.

The Nintendo case is identified as one of the recent major data breaches with regard to the data privacy of the users. When a user registers to the Nintendo Network, Nintendo holds all the personal information of the user which includes username, email address, date of birth, country of residence, and credit card details. Securing this confidential information of the user is the company's responsibility. However, Nintendo has failed to secure the data of their users, and the personal data was exposed to an unsecure environment. According to Privacy.org.nz (2013), organizations can lose personal information of users in four ways: through complacency, due to inadequate security, poor procedures, or by accident. Once the personal information of the users are compromised, both the users and the company will be affected. According to Janakiraman et al. (2017), customers feel vulnerable about their information privacy and if these data are exposed to an unsecure environment this uncertainty will have a huge impact on their personal life. Furthermore, after a data breach crisis, customers' trust in the company significantly reduces.

LESSONS LEARNED

In general, being free from danger is known as security. To be secure is to be protected from damage, risk of loss, unwanted modification, or other hazards (Whitman & Mattord, 2017). The Nintendo case highlights the importance of information security and how achieving the appropriate level of security is vital for any organization.

One of the important concepts or theories to consider is that every organization should focus on Issue-specific security policies (ISSP). The case of Nintendo, viewed as being a brute force and credential stuffing attack, is a reminder that the users should create both a unique and strong password for every online account. This further highlights that individuals should also take responsibility for their own information security. Therefore, companies can implement and issue specific security policies such as restricting users from using the same password across multiple accounts. Furthermore, creating security functionalities for the system to automatically detect if the new user registered password is weak and notify the user.

Also, the Nintendo case and its literature show the importance for companies to follow theories and frameworks in order to mitigate security vulnerabilities. Nintendo Switch is one of Nintendo's recent gaming consoles whose users were victims of data

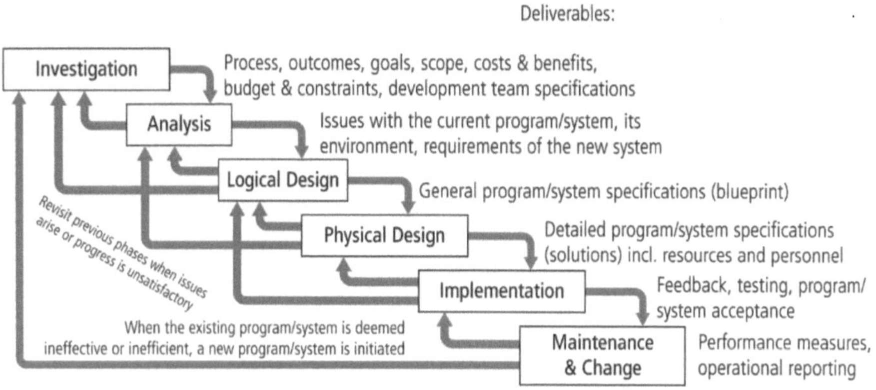

FIGURE 1.1.1 Security system development lifecycle.

(Source: Management of Information Security 6th Edition – Whitman & Mattord, 2017).

breaches, and in the future, the companies should follow frameworks and theories to further strengthen their security and safety in the development of new consoles. The Sony case study can be taken as a good example, as the company provided less priority to security in the development of the software that runs the network which led to the data breach (Baker & Finkle, 2011). Therefore, in order to mitigate this issue, companies can follow a framework such as the Security System Development Lifecycle (SecSDLC) with the waterfall model (Figure 1.1.1).

Following the SecSDLC will help to identify specific threats and the risk that they represent as well as the subsequent design and implementation of specific controls to counter those threats and manage the risk. A few examples of how the SecSDLC will work are writing security requirements while collecting the functional requirement or during the design phase of the process and also performing an architecture risk analysis. A security SDLC framework will bring primary advantages to the company such as better, secure software, early security detection flaws in the system, and stakeholder consideration in security awareness.

DISCUSSION QUESTIONS

1. Is the data breach at Nintendo a credential stuffing attack, a phishing attack, or a brute force attack? Explain your answer.
2. Does Nintendo have poor information security policies or has it not provided proper attention to security when developing their network in consoles?
3. How can Nintendo strengthen its information security to avoid or overcome future data breaches and data privacy issues?

REFERENCES

Baker, L. B., & Finkle, J. (2011). Reutus: Sony PlayStation Suffers Massive Data Breach. Retrieved from Reutus: https://www.reuters.com/article/us-sony-stolendata-idUSTRE73P6WB20110427

Barth, B. (2019). SC Media: Gaming Industry Has Becaome Popular Target of Credential Stuffing Attacks: Study. Retrieved from SC Media: https://www.scmagazine.com/home/security-news/gaming-industry-has-become-popular-target-of-credential-stuffing-attacks-study/#:~:text=Gaming%20industry%20has%20become%20popular%20target%20of%20credentail%20stuffing%20attacks%3A%20study,-Bradley%20Barth&te

DuthcNews.NL (2018). Retrieved from Dutch News: Internet Con Men Ripped Off Pathe NL for €19m in Sophisticated Fraud: https://www.dutchnews.nl/news/2018/11/internet-con-men-ripped-off-pathe-nl-for-e19m-in-sophisticated-fraud/

Janakiraman, R., Lim, J., & Rishika, R. (2017). The Effect of Data Breach on Customer Behavior: Evidence from a Multichannel Retailer. *Journal of Marketing*, 82. https://doi.org/10.1509/jm.16.0124

Liao, S. (2020). CNN: Business Said a Total of 300,000 Accounts Have Been Hacked. Retrieved from CNN Business: https://edition.cnn.com/2020/06/09/tech/nintendo-300000-accounts-hacked/index.html

Privacy.org.nz. (2013). Retrieved from Privacy Commissioner: Privacy Breaches: Privacy-for-agencies

Seals, T. (2016). Infosecurity Magazine: Massive Brute-Force Attack on Alibaba Affect Millions. Retrieved from Inforsecurity Magazine: https://www.infosecurity-magazine.com/news/massive-bruteforce-attack-on/

Sjouwerman, S. (2019). KnowBe4: 18 Months, 61 Billion Credential Stuffing Attacks. Retrieved from KnowBe4: https://blog.knowbe4.com/18-months-61-billion-credential-stuffing-attacks

Warren, T. (2020). The Verge: Nintendo confirms 160,000 Nintendo Accounts accessed in hacking attempts. Retrieved from The Verge: https://www.theverge.com/2020/4/24/21234205/nintendo-account-hack-nnid-breach-security-hacking-attempt-attempt

Whitman, & Mattord. (2017). *Management of Information Security*. CENGAGE Learning.

Whittaker, Z. (2020). Nintendo now says 300,000 accounts breached by hackers. Retrieved from TechCrunch: https://techcrunch.com/2020/06/09/nintendo-accounts-affected-breach/

Winder, D. (2020). Forbes: 300,000 Nintendo Users Hacked: What Gamers Need to Know. Retrieved from Forbes: https://www.forbes.com/sites/daveywinder/2020/06/12/300000-nintendo-users-hacked-what-gamers-need-to-know-switch-gamers-account-passwords/#42b4e93bdcd0

1.2 Target Corporation

Xuejie Tai and Huixin Zhang
University of Canterbury, Christchurch, New Zealand

CASE SUMMARY

Before the breach, Target took some actions to protect sensitive data and overall cybersecurity. Security professionals were employed by Target to protect data and predict potential risks. In addition, Target passed an audit of the Payment Card Industry Data Security Standard (PCI-DSS) which is a security standard for companies to deal with payment card information in 2013. Target also invested $1.6 million in malware detection tools to detect any potential cyberattacks (Plachkinova & Maurer, 2018). Security operations centers of Target around the world were established to provide thorough monitoring of cyber threats. Target, seemingly, has done best practices to prevent information security from any possible threats.

The malware detection software notified security personnel in Target about the potential cyberattack in November 2013, but the personnel did not take any action. Another notification was sent to security professionals in December 2013, but again there were no actions taken. Target did not begin to investigate the breach until the Department of Justice in the United States noticed Target's potential data breach. After realizing illegal access to the data of payment cards, Target took several steps to minimize the impacts of unauthorized access and avoid other potential threats. At that time Target immediately informed authorities and financial institutions after the breach. Meanwhile, Target cooperated with a third-party forensics corporation to thoroughly investigate the incident. Additionally, the malware was eliminated for the safe network environment. After taking these public actions, Target sent emails to customers promising they will provide free credit monitoring for one year. However, these individuals who received the emails expressed that they never had businesses with Target, which aroused people's suspicions of how Target gained the information (Plachkinova & Maurer, 2018).

To conduct the investigation, security experts at Verizon, which is a telecommunication enterprise in the United States, were commissioned by Target. There were two fundamental objectives of this investigation. One was to explore the reasons for the data breach; another was to identify opportunities for Target to improve its security infrastructure. As of the beginning of the investigation, hijacked credentials might have been stolen from Fazio Mechanical Services, which is a supplier of Target, providing refrigeration equipment and service. Fazio Mechanical has access to and uses electronic billing via Target's system. Although there was no strong evidence to make sure that Fazio Mechanical violated business rules, Fazio Mechanical depended on the free Malwarebytes anti-malware software which does not protect businesses

DOI: 10.1201/9781003264415-4

in real-time. Once Fazio was attacked, accounts in Target would be stolen via Target systems. Security professionals figured that Citadel malware caused the data breach since Citadel malware could sneak login credentials from systems.

Due to the data breach, 70 million customers were adversely affected since their payment card accounts, PIN numbers, and their other personal information were stolen. Target also had a considerable financial loss and negative impact on its reputation. Most customers lost their trust and confidence in shopping from Target, which resulted in a decreasing volume of sales. In addition to the decreasing income, Target had to spend more than $200 million on insurance and security infrastructure.

A critical analysis of this case on two topics, incident response and data breach, is given in the following sections. Weaknesses of Target's security infrastructure will be discussed, and solutions to handle the data breach will be found out. Lessons learned from the case will be represented at the end of the chapter.

CRITICAL ANALYSIS

Incident Response

Through the study of Target's case, Target has some weaknesses in its security infrastructure. The first problem Target has poor incident response to potential cyberattacks. Security personnel in Target are not sensitive to incidents and potential security threats, which makes them take no further actions after they repeatedly receive alerts from malware detection tools. They received the first notification related to potential malicious activities in November 2013, but they did not conduct investigations until the Department of Justice intervened several weeks later. If Target quickly responded to the incident, its financial loss might not have been so significant, and the data breach which Target suffered would not have been one of the largest in history. Incident response plays a vital role in protecting information security for corporations.

Incident response can be regarded as an organized method to handle a data breach and other cyberattacks. It aims to decrease and avoid any negative impacts of cyberattacks on companies and customers. Each enterprise needs to have an Incident Response Plan (IRP) because an IRP helps companies encounter and solve incidents with confidence and efficiency. Security incidents usually happen with panic and confusion, but security personnel and other employees in a company with well-organized IRP will know what they should do after a cyberattack. The second advantage of an IRP is that it can alleviate any potential damages following security incidents. There is a clear outline of mitigation and recovery steps in an IRP, which helps companies minimize adverse influences that a security breach has on privacy, integrity, and availability of sensitive data. In addition, an IRP allows corporations to maintain their relationship with their customers, suppliers, and investors. If corporations do not quickly respond to security incidents and provide enough details of incidents to their customers and suppliers, they will lose more businesses and cooperation with their customers and partners. An IRP also promotes important communication between different departments in a company. Due to the lack of interaction between various departments, many companies fail to mitigate the negative impacts of security incidents. However, an IRP helps to address communication problems, unite employees, and promote collaboration (Gerberding, 2018).

After realizing the significance of incident response, this analysis considers how to handle a security incident for Target. According to Cichonski et al. (2012), there are four phases in the incident response process. The first phase is the preparation which allows Target to be ready to respond to any security incidents. Target not only prepares contact information and facilities but also prepares hardware and software to analyze incidents in advance. Contact information includes phone numbers and email addresses for employees, customers, and partners. These facilities refer to an issue-tracking system used to track the information and status of security incidents; the encryption software is used for interactions between employees. For the preparation of hardware and software, digital forensic workstations which save incident data and spare workstations which are used to restore backups, laptops, and evidence-gathering accessories should be prepared for potential cyberattacks. Some practices for preventing security incidents can be recommended. For example, regular risk assessments can help Target understand specific threats and identify crucial resources; malware prevention software is useful for detecting malware. In addition to malware prevention, it is crucial to provide users with training about security systems and applications and improve users' awareness of information security (Cichonski et al., 2012).

The second phase of the incident response process is detection and analysis. Some cyberattacks are easy to detect, but many of them do not have apparent symptoms. Therefore, detecting cybersecurity threats is the most challenging task for a company after a security incident, but the detection can assist the affected company in realizing its weaknesses. In terms of cyberattack detection, Target hired security professionals who work with Verizon to detect incidents and any potential cyber threats. Through the incident detection and analysis, Target found some problems or weaknesses which they have on the security infrastructure. Any malicious actors can easily access Target's network and devices because of inadequate network segmentation. At the same time, the problem of password policies in Target was found by security consultants of Verizon. For instance, a weak administration password was used in Microsoft SQL servers, and a default executive password was used in Apache Tomcat servers. Additionally, Target had problems which were associated with system maintenance and patching.

The next phrase is containment, eradication, and recovery. Containment is a prioritized consideration in the early course of incident handling since cyber attackers might do more damage to companies without containment. Elimination is necessary to remove malware and identify all vulnerabilities. Recovery involves some practices that are restoring backups, re-establishing systems, and changing passwords (Cichonski et al., 2012). Based on the case, Target eliminated malware at the beginning of the investigation and applied chip readers with PIN codes to increase the security for its customers. These initial responses of Target to the data breach are recommended. The last phrase of incident response is a post-incident activity which represents lessons learned from security incidents.

Data Breach

After analyzing Target's incident response, a data breach itself must be analyzed for how data breach occurs and how to solve data breach. A data breach might happen due to various factors. Initially, insider misuse is the biggest threat to companies as insiders take advantage of their access rights and privileges to steal sensitive business information and customer information. Precious documents of companies might be physically stolen. Crimeware is the most common method for attackers to conduct cyberattacks. Ransomware, SQL injection, and phishing attempts are involved in malicious software. Some data breaches are from web application attacks. Personal details will be stolen by attackers when individuals register web applications. Additionally, financial information of individuals could be taken by malicious actors who use skimming devices on credit card readers (Symanovich, 2020). Indeed, there are other forms of attacks which lead to data breaches. For the data breach which happened in Target, Target was attacked by Citadel malware which was specifically designed to steal personal and financial information.

A data breach is one of the biggest issues that many companies encounter since it results in multiple consequences. Financial loss is the worst consequence for enterprises when they suffer data breaches. According to Davis (2019), a data breach on average costs $7.91 million in the United States. Target also experienced enormous economic loss due to the data breach. The percentage of those shopping at Target in 2014 dropped by 10%, compared with the same month in 2013. Simultaneously, the net income decreased by 34% because of extra costs related to investigation and remediation. Enterprises affected by data breaches also have damage to their reputation. Due to the leakage of personal information, corporations could be surrounded by negative news, which makes current customers lose confidence to continue businesses and makes potential customers refuse to cooperate with these companies. According to the investigation of Target, more than 70 million customers were negatively affected by the data breach, some discontinuing their shopping at Target altogether. Besides, companies' operations will be disrupted by data breaches. Depending on the severity of the breach, affected companies spend different periods of time from investigation to recovery; a data breach could have a severe impact on business operations. For Target, seemingly, its business operation did not get seriously affected.

Suffering from the data breach, Target does need to recover from huge financial and reputational damage. There are some steps that could help Target return to its good reputation and competitive retail position. Stopping the breach as quickly as possible is the first action that breach-affected companies should take (Jelen, 2018). Methods of elimination of attacks differ due to the different types of attacks. Once Target noticed the data breach, they eliminated malware as soon as possible lest they suffer more serious damage. After that, sending a notification to those affected is fundamental, which allows enterprises to save their reputation and preserve their integrity. Target issued a public statement that they were attacked by a malicious actor and sent emails to inform their customers. In addition, a security audit is required to assess the current security systems of companies and help to prepare recovery plans. Despite whether there is a data breach, companies should perform security audits regularly. However, many companies do not since they believe that their IT security is enough. Target also performed security audits after the data breach happened, which is not sufficient for its IT security. Therefore, it is necessary for Target to conduct security audits on a regular basis.

According to Dutton (2018), 80% of data breaches result from human or process errors. There are several ways to protect IT security for companies. Initially, companies should provide training to their employees and impart to them cybersecurity details specific to their responsibilities. From the case, we know Target had a lack of employee training since its security personnel could not give a quick response to the alerts from the malware detection tool. The second useful method to protect information security is using strong passwords which include numbers, symbols, capital letters, and lower-case letters. According to the investigation of the data breach, Target made a mistake on its administration passwords. The use of weak and default passwords was extremely common in Target, which enabled attackers to easily steal business information and install malware. Patching vulnerabilities is the next crucial way of security protection. Out-of-date systems and software might have various weaknesses that are easy to be attacked by malicious actors, so these outdated software and vulnerabilities should be patched a in timely fashion (Jelen, 2018). Target had loopholes on patches and systems as well. Attackers utilized these vulnerabilities to get access to the network because Target lacked Microsoft patches and used outdated web servers. To make future protection and recovery plans, Target should gain precious lessons from this data breach.

LESSONS LEARNED

Although Target suffered one of the most significant data breaches in history, it learned some lessons from this data breach to improve its IT security infrastructure. After this data breach, Target invested approximately $1 billion in cybersecurity operations. Target established the first Cyber fusion center in 2015, which was used to prevent similar cyberattacks. A Cyber fusion center is an advanced model which combines detection, threat hunting, and response, which helps Target increase the speed of incident response and decrease organizational costs. In addition to the investment in cybersecurity infrastructure, Target also put money into top talents and developed technologies so that they can constantly evaluate their security practices. Moreover, professional security analysts will work to analyze and detect the latest technologies and prevent the company from any cyber threats. To increase the confidentiality of customer information and the security of chips, Target provided chip readers with PIN codes to its customers. In a word, Target realized the consequence of the data breach and enforced many new practices and improvements to its information security.

Through the analysis of the Target case, we can also learn some lessons about cybersecurity and effective practices of eliminating and avoiding cyberattacks.

INCIDENT RESPONSE LESSONS

Incident response is important for organizations that are affected by cyberattacks. An effective incident response helps affected organizations decrease financial loss and reputational damage. In terms of incident response, companies should make a detailed IRP based on four phases of the incident response process, including preparation, detection and analysis, containment, eradication and recovery, and post-incident activity. Contact information, facilities, hardware, and software which are associated with cyberattack detection are needed to be prepared. After that, companies should identify what kind of attacks they suffer and find out the root causes which lead to these attacks so that they are able to take practical actions to contain and eliminate these attacks. Assigning responsibilities is required, so employees can perform their functions when a cyberattack happens.

Data Breach Lessons

Investing money in security infrastructure, management, and security personnel is another lesson learned from the Target case. It is essential to install different hardware and software to detect cyber threats and protect companies from potential cyberattacks. According to various security purposes, organizations can implement different tools. For example, Network Security Monitoring Tools, such as Argus and Splunk, enable companies to monitor network security. Encryption Tools allow companies to encrypt sensitive information to avoid attacks from malicious actors. Web Vulnerability Scanning Tools, which include Burp Suite and Nikto, can be used to identify critical weaknesses. Apart from the investment in hardware and software, the investment in talents is indispensable. Enterprises need to regularly provide employee training about cybersecurity and relevant technologies so that employees can make a reasonable and effective response to any security incident and do not violate company security policies. If necessary, enterprises can employ security professionals to monitor and analyze their current security systems and identify any vulnerabilities which might pose threats to the IT security of companies.

In addition, companies should update software and patch vulnerabilities regularly. Companies might ignore the reminder of updating software for some reasons, but outdated software could lead to many vulnerabilities which are easily found and attacked by malicious actors. Outdated software will not be maintained and not have patches for vulnerabilities, so outdated software poses a great threat to information security. Updating software regularly is essential for organizations. After completing a needed software update, corporations should patch vulnerabilities regularly. To check and identify vulnerabilities, vulnerability scanners can be used, and companies can use patch management tools to patch vulnerabilities.

Finally, sending notifications to individuals who are adversely affected by data breaches and performing security audits regularly is crucial. After a data breach, companies should inform their customers and partners about the data breach rather than keep it secret, since sending the notification is a means to maintaining customer trust to some extent. Security audits help companies evaluate their current information systems and improve security infrastructure to protect companies from cyberattacks.

DISCUSSION QUESTIONS

1. How should organizations make appropriate incident response plans for any threats?
2. How could organizations reduce the negative impacts that data breaches have on organizations and their customers?
3. Which are the best tools for companies to detect and eliminate cyberattacks?

REFERENCES

Cichonski, P., Millar, T., Grance, T., & Scarfone, K. (2012). Computer security incident handling guide. *NIST Special Publication*, *800*(61), 1–147.

Davis, M. (2019, July 25). *4 damaging after-effects of a data breach.* Retrieved from Cybint: https://www.cybintsolutions.com/4-damaging-after-effects-of-a-data-breach/

Dutton, J. (2018, January 30). *4 of the 5 top causes of data breaches are caused by human or process error.* Retrieved from IT Governance: https://www.itgovernance.co.uk/blog/4-of-the-5-top-causes-of-data-breaches-are-because-of-human-or-process-error

Gerberding, K. (2018, March 13). *The 5 benefits of an Incident Response Plan (1/5).* Retrieved from Hitachi Systems Security: https://www.hitachi-systems-security.com/blog/benefits-incident-response-plan/

Jelen, S. (2018, November 27). *Top 5 ways to handle a data breach.* Retrieved from SecurityTrails: https://securitytrails.com/blog/top-5-ways-handle-data-breach

Plachkinova, M., & Maurer, C. (2018). Teaching case: security breach at Target. *Journal of Information Systems Education*, *29*(1), 11–20.

Symanovich, S. (2020). *What Is a data breach and how do I handle one?* Retrieved from Lifelock: https://www.lifelock.com/learn-data-breaches-data-breaches-need-to-know.html

1.3 Case Study: Cyber-attack on Ukrainian Power Grid

Vignesh Makkada
Ministry of Justice, Wellington, New Zealand

Ishani Rai
University of Canterbury, Christchurch, New Zealand

INTRODUCTION

The rapid increase in the use of the internet has led to the advances in different domains of technologies such as artificial intelligence, IoT, and e-commerce, which have improved human life; yet these technologies have also increased the concerns and need for efficient security measures against cyber-attacks. A cyber-attack is a measured effort by an individual, group, or organization to breach the information system of the other party. The attack can target a single system or multiple networks to steal, modify, or destroy the data and information system. There are various methods to achieve this which include malware attacks, phishing attacks, and denial of service. Cyber-attacks are increasing against critical infrastructure such as power grids.

In December 2015, a malfunction at the Ukrainian Kyivoblenergo, the regional electricity distribution entity for Kyiv and its surrounding area caused a power outage in three provinces of Ukraine, which lasted for around six hours and affected 225,000 consumers (Whitehead, Owens, Gammel, & Smith, 2017). The outage was a result of an illegal intrusion into the company's computer and SCADA (Supervisory Control and Data Acquisition) system. SCADA are systems used to monitor and control a plant or equipment by collecting and analyzing the real-time data. Seven 110 kV and 23 35 kV substations were disconnected for almost three hours starting at 3:30 p.m. local time. Operators were unable to regain the remote access to more than 50 substations because the attacker had a remote control over the SCADA distribution management system. Further, a telephone-based denial of service attack was launched using the automated systems to overload the phone lines which reduced the communication among the operators. There was an additional impact on various portions of the distribution grid which forced the operators to switch to manual mode. This attack used various techniques and the sophistication demonstrated that the intruder was highly resourceful and well-structured (Case & Center, 2016).

DOI: 10.1201/9781003264415-5

The investigation of this attack showed that it was strategically planned and began in March 2015, when the hacker gained access to the system due to a vulnerability in Microsoft Office rather than the code on the devices. A spear-phishing email with an attachment was sent to users, a popup screen asked the users to enable macros in the document which allowed the installation of BlackEnergy 3 (BE3) malware on the systems (Shehod, 2016). This malware created a backdoor communication channel for the hacker allowing them to control and collect information. The intruder remained within the system for about six months and using the collected credentials moved through the network to gain remote access to SCADA systems using the Virtual Private Network (VPN). To make the power recovery difficult for the operators, hackers reconfigured the backup power systems of two control centers. They further studied the grid systems and corrupted the original firmware on serial to Ethernet converters with the coded malicious firmware at substations. The converters were made vulnerable so that operators could be prevented from sending commands to re-close the breakers once the shutdown happened (Figure 1.3.1).

Further, the attackers changed the passwords for crucial systems at multiple instances so that the access during the recovery could be delayed. The power supplies were remotely turned off for the server room and telephones and the infiltrators also tampered with the hard drives. On December 23, 2015, at 3:30 p.m. local time, the hackers tapped into the SCADA network and sent commands to deactivate the power system. The communication between the regional office and the distribution system was cut off by targeting an internal telephone communication server. The power shutdown scheduled for the main data center was scheduled using a local UPS. The telephone denial of service was the next phase in which hackers crashed the system with excessive fake calls so that legitimate users could not report the power outage. KillDisk which is a piece of malware was launched in the system that caused systems

FIGURE 1.3.1 Ukraine attack: consolidated technical components.

to crash as it infected the master boot record (Case & Center, 2016). This entire attack caused an outage for almost seven hours and the restoration process was delayed due to the attack by the malware KillDisk.

ANALYSIS

The rapid growth of information systems has led to an increase in the integration of control and business networks of power plants. The excessive use of information systems has led to new ways of attacking the systems and networks. The severity and frequency of cyber-attack on critical infrastructures is rapidly increasing. The hijacking of power grids is increasing due to their vulnerability, causing a huge social and economic disruption because of the huge loss of electricity. The malware launched in the system provides hackers with operational access that allows them to run commands to alter or halt the electricity flow. Often communication protocols are standardized across different infrastructures, limiting the security. For instance, a slight modification in an industrial control system malware could attack a power grid (Kshetri & Voas, 2017).

There have been cyber-attacks on the grids all across the world and the Ukrainian power grid was targeted multiple times during the years 2015 and 2016 (Dunietz, 2017). These attacks depict that the hackers are learning more about the industrial control system (ICS) and targeting the interconnected infrastructure, leading to the disruption of power in various parts of the world. Hackers usually sell the information from data breaches or extort monetary values for unblocking the hacked systems. However, the cyber-attacks on power grids can be considered as a cyber-warfare method among different nations. Societies are greatly dependent on electricity and if companies suffer such cyber-attacks, then the scale of consequence could be very large. For instance, long power shutdowns at hospitals could lead to delays in treatments of patients which could potentially increase the loss of lives. The lack of relevant defense measures, existing vulnerabilities, lack of training, and obsolete patch levels of the system are a few of the reasons for such cyber-attacks.

In the Ukraine power grid cyber-attack incident, the hackers used a variety of techniques to infiltrate the system. The initial intrusion started with spear phishing which is a targeted phishing method. In phishing, attackers attempt to trick users into sharing personal information and credentials by disguising themselves as a trustworthy organization or individual. Phishing is not customized according to individuals and usually, the same message is sent to a group of people (Giandomenico, 2019). However, for the Ukrainian power grid attack the hackers used spear phishing in which users are sent messages from organizations that the user is familiar with and are specifically addressed to them. The spear phishing files appeared to be an official email from the Ukrainian Energy Ministry to the targeted individuals. The files contained the malware which was installed on the computers as soon as the users enabled the macros (Whitehead et al., 2017).

The second stage after the malware installation was to learn about the systems and the coding of the malicious firmware, which was consistent and uploaded within a small timeframe at multiple sites. The attackers may have tested the deployment but went unnoticed as there were no accurate security policies and measures to intercept the intrusion over the network. Using the remote access and VPN services they were able to directly interact with the network and modify the ICS components as they desired. Approximately 27 substations went offline when the attackers opened the circuit breakers in the SCADA environment using the human–machine interface or HMI. The corrupted firmware prevented the operators from gaining remote access to bring back the power. Further, there was telephonic denial of service which led to a high number of fake calls and prevented the customers from reporting the issue. The review of the incident showed that the jamming of telephone services was to increase the frustration of customers as they were unable to reach the customer support to report the power outage (Case & Center, 2016). The primary section of the attack was hacking the SCADA system to open the circuit breakers while the disconnecting of the UPS, flooding of the telephone lines, clearing out the disks using the KillDisk malware, and the corrupted firmware were all used to amplify and increase the duration of the outage (Case & Center, 2016).

The shutdown of power stations in Ukraine due to cyber-attacks shows that any organization could be a victim of hacking and other online attacks. One wrong click on the systems could release a malware which could lead to the shutdown of the power grid or loss of millions of users' sensitive data. Organizations need to assess the risk and develop contingency plans and strategies to build a cyber-resilience against the attacks. Risk assessment allows organizations to identify important risks and control them after measuring them accurately. Risk assessment helps to implement the security controls effectively once the vulnerabilities and threats are understood, and appropriate evaluation is applied to mitigate the risks. Research shows that the Ukrainian power grid cyber incident had a very low or no risk assessment in place to handle such intrusions (Whitehead et al., 2017).

Organizations need to understand and accurately identify their assets and their values, any threats and vulnerabilities associated with these assets, and calculate their impact on the business and the users. Vulnerability assessment is vital to the risk management function and regular checks need to be in place for its remediation and timely fixes. Threats could be the result of human actions, where opening a phishing email will deploy a malware into the system providing access and allowing the intruders to corrupt the systems (Whitehead et al., 2017). The probability of a cyber-attack is usually considered high when there are no fixes, and the count of threats and vulnerabilities is high.

LESSONS LEARNED

The companies managing critical infrastructure should formulate and follow appropriate risk assessment strategies so that they can mitigate the threats associated with the cyber-attacks. The initial intrusion began when employees opened the malicious emails and attachments, which show that the companies should implement training and programs which will educate the employees about the threats linked with the phishing messages (Kessler, 2016). The programs could help employees differentiate between the phishing and normal emails by sending test emails and if they open the phishing emails, these employees could be redirected to the training portals. These trainings should be scheduled on a regular basis so that employees stay updated with the new ways of phishing targeting the organizations. Inbound email should be monitored, and any suspected email should be blocked before it reaches the inbox of an employee. Further, a firewall should be in place which would bounce back any email that is being sent to groups or individuals that are not associated with the organization.

After installing the malware through the phishing attack, the intruders were able to access the control system networks by using the credentials obtained from compromising the system. This showed that the company needed a robust network security infrastructure. The fact that the intruders were able to infiltrate through the network for six months without leaving any trace of their presence meanwhile continuously accessing the system raises an alarm and questions the level of network security. The power company needs to install tools to maintain continuous monitoring of their network which would help them to track any excessive connections to the system along with the launch of the malware (Higgins, 2016). The continuous monitoring of network data will allow the organizations to set a baseline which can be used to compare and track any anomalies causing the unusual amount of data flow within the system at any given time.

Any critical infrastructure that manages sensitive user data, network, or financial flows should follow at least two-factor authentications and have limited remote access to the control systems. If remote access is mandatory for the operations, then it should have a time-out fixed which once reached would again prompt the employee or a user to enter the credentials. The remote access should be limited to certain individuals and log details should be set up to monitor the commands executed by these users. A notification email should be sent to security teams if any executed commands could alter the system. Companies could also set up their own remote desktop connections and allow only specific local machines to connect after appropriate authentication.

The installations of malicious firmware and online command interface for the UPS were the issues that delayed the restoration of power and showed the requirement for continuous updates and patching of firmware and running system checks for continuous maintenance and detection of any risks or vulnerabilities. Any abnormal behavior such as an upgrade outside the regular maintenance window should be monitored. Data should be captured and appropriate backups for the systems should be maintained and secured at a different location to improve the chances of system recovery during cyber-attacks (Case & Center, 2016). Organizations should schedule audits at internal and external levels to keep the systems, network, and data checked so that any anomalies or disruptions within the system can be identified and remediated at any early stage.

Organizations should maintain a disaster recovery plan and an incident response plan, which will document the details of any issue that occurred and will also test plans and maintain the appropriate steps to follow by the companies during such attacks. These plans need to engage the internal participants such as employees, IT teams, security teams, operations teams, and external members which would include the stakeholders and vendors. The cyber-attack on the Ukrainian power grid shows that power companies should test various complex scenarios so that their response during actual attacks is effective and timely (Kessler, 2016). Systems should be tested for network outages, and any unnecessary communications should be terminated so that the hackers are not able to use them to their advantage. Employees should be updated with any issue that was faced and the fixes that were applied so that they could handle any similar security breaches in the future. Regular documentation and update of the fixes and their issues are important so that real-time security issues can be handled effectively.

CONCLUSION

Industrial control systems are beneficial as they provide remote access and automation to the power companies; however, the lack of employee training and awareness, monitoring tools, and security measures could change these benefits into threats and vulnerabilities which could be exploited by cyber-attackers. The Ukraine power outage was an outcome of multiple vulnerabilities which were left unchecked causing the cyber-attack. This incident increased awareness among power companies around the world and made them question if they had appropriate security measures to handle such situations. They were also made to consider if there were any vulnerabilities that could be exploited and whether their employees were being trained and informed properly about the threats and theirpreventive measures. The incident shows that the development of technology will increase the risk of security breaches and the best any organization can do is to follow measures and protocols that would minimize the effect of such cyber-attacks.

DISCUSSION QUESTIONS

1. What are the security policies implemented to block any phishing emails?
2. Are the employees getting regular training and being made aware of the threat associated with opening attachments in phishing emails?
3. Is there a system in place to monitor the network logs?
4. Are the systems being patched regularly to fix any vulnerabilities and threats?
5. What is the level of logging access available to the users and who can remotely access the systems?

REFERENCES

Case, D. U. J. E. I. S., & Center, A. (2016). Analysis of the Cyber Attack on the Ukrainian Power Grid. 388. https://www.huntonprivacyblog.com/wp-content/uploads/sites/28/2016/03/Documents_E-ISAC_SANS_Ukraine_DUC_18Mar2016.pdf

Dunietz, J. (2017). Is the Power Grid Getting More Vulnerable to Cyber Attacks? Retrieved from https://www.scientificamerican.com/article/is-the-power-grid-getting-more-vulnerable-to-cyber-attacks/

Giandomenico, N. (2019). What is Spear-phishing? Defining and Differentiating Spear-phishing from Phishing. Retrieved from https://digitalguardian.com/blog/what-is-spear-phishing-defining-and-differentiating-spear-phishing-and-phishing

Higgins, K. J. (2016). Lessons From The Ukraine Electric Grid Hack. Retrieved from https://www.darkreading.com/vulnerabilities---threats/lessons-from-the-ukraine-electric-grid-hack/d/d-id/1324743

Kessler, M. (2016). Lessons Learned from the Power Outage in Ukraine and How the Electric Grid of the Future Will Reduce Cybersecurity Risk. Retrieved from https://cip.gmu.edu/2016/05/24/lessons-learned-power-outage-ukraine-electric-grid-future-will-reduce-cybersecurity-risk/

Kshetri, N., & Voas, J. J. C. (2017). Hacking Power Grids: A Current Problem. *Computer*, *50*(12), 91–95.

Shehod, A. (2016). Ukraine Power Grid Cyberattack and US Susceptibility: Cybersecurity Implications of Smart Grid Advancements in the US. In *Cybersecurity Interdisciplinary Systems Laboratory (CILL)*, Massachusetts. https://web.mit.edu/smadnick/www/wp/2016-22.pdf

Whitehead, D. E., Owens, K., Gammel, D., & Smith, J. (2017). Ukraine Cyber-induced Power Outage: Analysis and Practical Mitigation Strategies. Paper presented at the *2017 70th Annual Conference for Protective Relay Engineers (CPRE)*.

1.4 Capital One Data Breach

Vignesh Makkada
Ministry of Justice, Wellington, New Zealand

Ishani Rai
University of Canterbury, Christchurch, New Zealand

CASE SUMMARY

Capital One is among the top consumer banks in the United States with approximately 50,000 employees and 28 billion US dollars of revenue. It is one of the organizations that followed an early adoption of cloud computing technologies and one of the first banks in the world that invested in the migration of their physical data centers to a cloud computing environment (Novaes Neto et al., 2020). This helped them to reduce the physical locations which in turn, minimized infrastructure, and operating costs. The adoption of and investment in cloud computing also helped in the expansion of the micro services along with improved customer services by efficient utilization of time and money. Capital One used Amazon Web Services or AWS as their cloud vendor and focuses on agile management practices along with the majority of customer-facing applications and the operating system being cloud-based.

In July 2019, Capital One declared that an external individual had accessed sensitive customer data. The data was comprised of more than 100 million Capital One's customer account details and credit card applications which included 140,000 social security numbers, one million Canadian social insurance numbers, 80,000 bank account numbers, and an undisclosed number of names, addresses, credit scores, and other personal information (Capital One, 2019). The data leaked included the regular details collected from customers during the credit card applications which consisted of name, address, postcodes, email, date of birth, and self-reported income. The data breach had affected approximately 100 million people in the United States and 6 million in Canada, which included information from individuals to small enterprises.

On July 17, 2019, this incident was reported to Capital One through their Responsible Disclosure Program instead of the regular cyber security operations. Responsibledisclosure@capitalone.com received an email message stating, "There appears to be some leaked s3 data of yours on someone's GitHub/gist." The FBI presented a screenshot of the email received by Capital One in the Seattle court (Novaes Neto et al., 2020).

The filed report also explains that during the investigation, the FBI came across a script on the GitHub repository that was deployed on cloud servers to access the data from Capital One. The script file contained three commands which provided access

DOI: 10.1201/9781003264415-6

to the AWS servers for Capital One. The first command provided security credentials for folder access; the second command provided the list of folder names or data saved in storage space and the last one allowed the copy of data from this space. These scripts were able to access and execute at Capital One's server because there was a firewall misconfiguration. The company later checked its logs and confirmed the execution of these commands.

The hacker was able to mislead the firewall into relaying requests due to this misconfiguration of the open-source Web Application Firewall (WAF). This led to the assignation of full permissions on the files which in turn, provided read access to all the data contents in the listed files. This type of vulnerability is known as Server-Side Request Forgery or SSRF attack, in which a server is misled to execute a set of commands that should never be permitted to run.

The intruder tried to sell the stolen data on the internet which helped the FBI with the arrest. Evidence revealed that the hacker was an ex-employee of Amazon Web Service, which was the cloud vendor for Capital One. Although investigators later stated that the consumer data was not misused, this incident was still a huge breach for the organization and a general reminder of insider threat programs and third-party security for the users and the vendors of cloud services (Lu, 2019).

ANALYSIS

Capital One data breach incident took place on March 22 and 23, 2019, impacting 106 million users, the result of unauthorized access to its AWS-hosted cloud servers (Novaes Neto et al., 2020). The technical and regulation details about this case and its compliance report are available publicly. The organization valued the evolving technology and invested in its infrastructure and was one of the first banks to move its on-premises data centers to the AWS-hosted cloud servers.

The data breach was done using a Server-Side Request Forgery or SSRF attack, where a remote user misleads the server to run specific commands which allowed the hacker to use the server for their own requests and access the confidential data. This was possible due to the misconfiguration of the open-source WAF which was hosted in AWS cloud (Krebs, 2019). The attackers usually compromise the security of web-based applications using WAF which is deployed along with an open-source Apache web server to protect against the vulnerabilities.

Capital One data breach was discovered when a user reported about the data being published over social media. This shows that Capital One did not monitor their exposure on the internet even by using keyword search which if they had, the breach would have been discovered by them sooner. Any application that can provide access to millions of sensitive customer data and is hosted on cloud servers with just one firewall, if misconfigured will be at a potential risk, compromising the application and the underlying data (Altman, 2019).

The hacker was able to trick the firewall into relaying requests to a metadata service which is responsible for transferring temporary information on a cloud server. It was comprised of the security service's recent authorization which was used to access the resources on the cloud server. These temporary credentials were based on the permission policy, stated under the identity and access management or IAM roles. In this incident, the presence of an asterisk gave full permissions allowing the intruder to access millions of customer records (CloudSploit, 2019). SSRF is becoming an alarming vulnerability for organizations that use public clouds.

An SSRF attack is when the attacker sends a request to the vulnerable or misconfigured web server which then makes a request to the victim server providing access to the hacker with an organization's internal network. SSRF allows attackers to scan the ports and IP addresses and also interact with certain protocols, exposing the IP running on proxy servers and remote execution of any commands (Team, 2019). Organizations need to be aware and alert toward such issues and take necessary actions such as restricting the access to specific machines and raising an alarm when any new machine tries to connect. Additional actions are to limit the ports that are allowed to connect to the sensitive data and to scan any external network.

AWS has a shared responsibility model in which AWS is responsible for the cloud infrastructure and the customers need to manage the security in the cloud which includes patching and software updates, management of data, and configuration of tasks (Altman, 2019). Capital One configured its firewall incorrectly, a grave error which was its security responsibility, resulting in the loss of data. Unauthorized employees from inside and hackers from outside can access data in the cloud model of IaaS (Infrastructure as a Service), SaaS (Software as a Service), and PaaS (Platform as a Service). Misconfiguration can be a result of a lack of awareness about cloud security, appropriate controls, and ignorant inside behavior. The manual approaches to fixing the issues further increase the possibility of human error during categorizing the critical fields or while remediating an existing issue.

Data breach is becoming frequent as more companies are migrating to cloud storage services because of the financial and logistic benefits. Cloud service provides a huge amount of space and effective processing capacity, making them a popular choice among enterprises. Any financial institution or company that holds customers' sensitive data and is migrating to cloud should invest more toward their security to prevent the loss of confidential information (Novaes Neto et al., 2020). Data breaches can be a result of various reasons such as employee misuse, human error, system bugs, and malicious attacks. In the case of Capital One, the data breach was due to the misconfiguration of the firewall which occurred because of human error during the setup, providing an opportunity for the intruder to infiltrate the system and steal the data. Human error occurs because a lot of users fail to follow the instructions and guidelines (Barona & Anita, 2017).

The increasing volume of data provides an opportunity for cyber criminals to expose a vast amount of data in a single breach. These data leaks not only cause monetary problems but also raise issues for social trust and personal safety of an organization and its consumers. The cloud infrastructure facilitates the attackers to start a cyber-attack remotely on multiple machines simultaneously. A service level agreement (SLA) for organizations needs to have different specifications depending on the choice of the public or private cloud offering which includes, SaaS, IaaS, or PaaS. These SLAs need to be further evaluated and customized according to the regular feedback from customers and shareholders (Mondal, Paul, Goswami, & Nath, 2020).

During a data breach in a cloud computing environment, users do not have the choice of using a physical acquisition tool kit as the compromised server needs to be shut down completely until the backup data is retrieved. The presence of vulnerabilities in virtualization software provides an opportunity for malicious users to bypass the security restrictions and gain privileges (Sangroya, Kumar, Dhok, & Varma, 2010). Allowing a guest operating system user to run code on the host or other systems and granting guest users full access to the host's restricted files and folders are two examples of such vulnerabilities which act as a platform for data loss and hacks.

LESSONS LEARNED

The incident of the Capital One data breach was a reminder that data security is of utmost importance and an intruder can exploit banks, businesses, and enterprises if there is any misconfiguration in an organization. The lessons learned from this incident can help organizations to improve their security measures and secure customer data more efficiently. Credit monitoring and cyber-hygiene processes are important for any organization that handles customers' confidential data.

Misconfigurations are a result of human error and can be reduced by providing security teams and the system administrator with relevant and accurate training. This will help them develop a better understanding of the implementation of any specific configuration in the system (Poston, 2019). Organizations should develop strategies to prioritize training, manage attrition, and address any existing skill gaps by providing sufficient time for knowledge transfer. Mistakes are inevitable and unfortunately, errors from its security team can cause a huge data breach for any organization. To avoid these situations, internal reviews should be conducted within an organization to test the configurations and system patch status, and to analyze any vulnerabilities that could cause potential security threats. Implementing the required follow-up actions for a timely remediation of an issue is vital (Altman, 2019).

Active monitoring and efficient audit logging of the events that occurr during a security breach help in a faster discovery of the anomalies and provide a better resilience toward cyber-attacks. During the incident of Capital One, there were comprehensive action logs available which led to the accurate reconstruction of the events and helped in a better understanding of the intrusion. Organizations should have effective measures in place to maintain the data of their robust logging and monitoring system (Altman, 2019).

Organizations should provide any user with a limited level of permissions which allows them to perform only their job roles. The failure to implement the limited privilege principle can provide an opportunity for a cyber-attack (Poston, 2019). If a user requires full permission for any folder, then an organization should have a policy in place to closely monitor those activities and grant requests for access with approval from top management for only a limited amount of time. Having several layers between the sensitive data and the surface where a hacker could attack is beneficial. Companies should conduct penetration tests regularly to improve their security.

Identity access management (IAM) that handles permissions and privileges should be closely monitored, and regular checks should be in place to validate the permissions provided to users for any files or folders. If any user is found to have excessive permissions, they should be revoked immediately. Also, organizations should have dedicated teams to manage the servers by regular patching of the systems so that any potential risks or vulnerabilities can be eliminated.

Organizations need ways to continuously monitor the security performance of the data stored in physical and cloud servers. The security performance can reveal a single point of failure by highlighting the risky service providers associated with a company's vendor. The automated monitoring will help security and risk leaders to improve their understanding of the key areas of security breaches such as a compromised system or a server, outdated software, and open ports. Organizations handling confidential and sensitive user data should have strict compliance policies and the data should be protected under multiple firewall layers.

CONCLUSION

With the advancement in technology, there are new threats and vulnerabilities which an enterprise can assess more accurately when they prioritize any possible security issues and implement the relevant remediation needed. Organizations are always at risk of a breach and so it is important to have an emergency response plan or a disaster recovery policy in place. Customers should be more alert and monitor their cyber activities closely and regularly update their passwords to the accounts handling their sensitive data. With more enterprises moving on to the cloud, there is a need for a clear understanding of the security threats that could occur which will help vendors develop improved tools to provide security toward such risks and vulnerabilities.

DISCUSSION QUESTIONS

1. What is the role of security in the financial organizations that handle sensitive customer data?
2. What is the responsibility of an organization that plans to migrate to cloud?
3. What checks should the organization follow when configuring the system services on their end?
4. What are the roles of the company and the cloud vendor in preventing any data breach?
5. Who maintains the servers and needs to monitor the server and system logs to prevent an intrusion?
6. How can a well-informed employee help in noticing data breaches?

REFERENCES

Altman, H. (2019). Lessons from the Capital One data breach. Retrieved from https://www.bai.org/banking-strategies/article-detail/lessons-from-the-capital-one-data-breach/

Barona, R., & Anita, E. A. M. (2017). A survey on data breach challenges in cloud computing security: Issues and threats. In *2017 International conference on circuit, power and computing technologies (ICCPCT)*, pp. 1–8. IEEE.

Capital One. (2019, September 23). Capital One. Retrieved from Frequently Asked Questions: https://www.capitalone.com/facts2019/2/

CloudSploit. (2019, August 02). A Technical Analysis of the Capital One Hack. Retrieved from CloudSploit Blog: https://blog.cloudsploit.com/a-technical-analysis-of-thecapital-one-hack-a9b43d7c8aea

Krebs, B. (2019, August 02). What We Can Learn from the Capital One Hack. Retrieved from KrebsOnSecurity: https://krebsonsecurity.com/2019/08/what-we-canlearn-from-the-capital-one-hack/

Lu, J. J. L. F. O. C. O. D. B. (2019). Assessing The Cost, Legal Fallout Of Capital One Data Breach.

Mondal, A., Paul, S., Goswami, R. T., & Nath, S. (2020). *Cloud computing security issues & challenges: A Review*. In *2020 International Conference on Computer Communication and Informatics (ICCCI)* (pp. 1–5). IEEE.

Novaes Neto, N., Madnick, S., de Paula, M. G., Malara Borges, N. J. S. E., Moraes G. de Paula, A., & Malara Borges, N. (2020). A Case Study of the Capital One Data Breach. Available at SSRN: https://ssrn.com/abstract=3542567 or http://dx.doi.org/10.2139/ssrn.3542567

Poston, H. (2019). Lessons learned: The Capital One breach. Retrieved from https://resources.infosecinstitute.com/lessons-learned-the-capital-one-breach/#gref

Sangroya, A., Kumar, S., Dhok, J., & Varma, V. (2010). Towards analyzing data security risks in cloud computing environments. Paper presented at the *International Conference on Information Systems*, Technology and Management.

Team, N. S. (2019). What is the Server Side Request Forgery Vulnerability & How to Prevent It? Retrieved from https://www.netsparker.com/blog/web-security/server-side-request-forgery-vulnerability-ssrf/

1.5 LinkedIn Data Hack

Nishanth Chakkere Ramesh
Kapiti Coast District Council, Paraparaumu, New Zealand

Ying-Pin Lan
Hato Hone St. John, Auckland, New Zealand

CASE STORYLINE

LinkedIn is the largest professional networking site with over 600 million users as of December 2019. As such, the organization prioritizes the privacy of its users and the security of their data, and it is generally trusted by employers, jobseekers, and other users. However, in June 2012, 6.5 million passwords of users were stolen after LinkedIn was hacked by cybercriminals who were traced to Russia. These account details were posted by the hacker on a password forum, which allowed access to the stolen passwords. The stolen information comprised of email addresses and passwords of the stolen accounts. Moreover, the hacker, named 'Peace' sought to sell access to the data for a mere five bitcoins or 2,200 USD.

The motive behind the hacking remains rather unclear, especially because the hacker decided to sell the entire set of data acquired for a meager amount of 2,200 USD. One calculation shows that the amount paid per account would be 0.001 cents. However, the stolen data could be sold to multiple buyers repeatedly, and the hacker potentially earned multiples of 2,200 USD from the sale of the data. However, the value of the data being sold may diminish as and when multiple copies of the data exist with multiple parties (Hunt, 2016).

The reaction of LinkedIn was swift and an apology to the LinkedIn user community was issued. The organization also urged all users to change their passwords, and, in particular, those users whose account details were compromised were unable to access their accounts. These members received special email instructions about how to reset their account passwords. A second email from LinkedIn was sent to the members with compromised accounts, and it outlined the situation and context surrounding the data hack. LinkedIn advised to turn on the two-step verification process which includes a password as well as a mobile phone notification required to access the account (Silveira, 2012).

LinkedIn stated that it adopted additional measures, salting and hashing, to ensure the security of its account holders. Salting refers to the addition of a unique value at the end of the original password to prevent large-scale attacks from hackers. Hashing refers to the indexing of the string of data into a shorter format for easier search and retrieval. These measures add an extra layer of security (Krebs, 2016).

DOI: 10.1201/9781003264415-7

A new revelation in 2016 came as a surprise for LinkedIn when it was revealed that the 2012 data hack compromised the information of 117 million (and not just 6.5 million) accounts, their email addresses, and passwords. LinkedIn stated that the hack of 117 million accounts could be traced back to the 2012 hack, and no new security breach occurred in 2016. LinkedIn invalidated the passwords of the accounts and users were sent instructions about changing passwords. General instructions were issued to all the users of LinkedIn to change their passwords and account access details regularly (Scott, 2016).

Moreover, after the 2016 revelation about the additional data breach, LinkedIn posted certain precautionary suggestions that users can take to secure their accounts. Instructions relating to changing passwords, creating stronger passwords, and privacy tips were posted on the LinkedIn blog (Scott, 2016). LinkedIn also threatened to take legal action against the hackers and other parties who posted the hacked data online and provided access to this data for a fee.

Some LinkedIn users, however, were not satisfied with the way the organization handled the aftermath of the data breach. Members have complained about the inconsistency in receiving clarification emails despite being members since before the 2012 data hack. But some observers theorize that such emails were only sent to those users who had not changed their emails since 2012. With the findings in 2016, it is increasingly difficult for the stakeholders to monitor which accounts had their data compromised due to the sheer number of accounts. Also, there is an estimate that only about 25 percent of all LinkedIn members use their accounts on a monthly basis. Therefore, the rest of the users may be unaware or may not be bothered about the privacy recommendations because they are not invested in their LinkedIn account as much as they may be invested in their Facebook or Instagram accounts. Security experts also hold LinkedIn liable because it did not have the salt and hash measure before the data breach, especially considering that it is the largest professional networking website in the world (Hunt, 2016).

A Russian hacker named Yevgeniy Nikulin was prosecuted by the U.S. courts for executing one of the largest data breaches in history. He was also accused of hacking the data of DropBox in 2013. He was found competent to stand trial, and the case is currently under trial in the court. The LinkedIn data hack showed to the world that even a large corporation with a global user presence is not immune to data hacks and breaches. In order to maintain the trust of the users, security protocols must be created and updated at regular intervals to prevent the compromise of the security structure (Eckel, 2019).

CRITICAL ANALYSIS

The information for this case study was gathered from the official communications from LinkedIn in the aftermath of the data hack and after the 2016 revelation about the actual extent of the data hack. In addition, information was also obtained from news articles and independent commentators who covered the data hack and its aftermath. Since this was one of the largest data hacks in history, many sources provided varying opinions about the data hack itself. Some of these sources were also from LinkedIn users who had been victims of this data hack, while others were from industry commentators who provided information about how the data hack had occurred and how it could have been prevented.

The social context of the LinkedIn data hack involves transnational cybersecurity-related crimes gaining prevalence in the twenty-first century. In July 2020, Yevgeniy Nikulin was found guilty by a jury for orchestrating the data hack and is scheduled to be sentenced in September 2020 (Rosenblatt, 2020). Transnational data hacks are becoming increasingly common since the jurisdictions of the crimes are different, and it is difficult to hold the perpetrators accountable. Since the internet itself is transnational, the security of the internet, or the lack thereof, is also transnational. Measures adopted by the public and private sector to tackle such crimes have been insufficient, due to the increasing number of crimes. Therefore, an effective program to deal with such transnational crimes would be to have legal cooperation between different states all operating in a cooperative manner. The disparities among states in the legal and technical expertise show the weakness of the systems to deal with transnational cybercrimes (Sofaer & Goodman, n.d.).

The hacking of the passwords on LinkedIn was important for the hackers because most people reuse their passwords on other platforms, and therefore, the hackers may acquire access to the banking and email addresses of LinkedIn users. Such data can also be sold to third parties who may exploit the passwords for their gain, as was done in this case study. LinkedIn had also failed to encrypt the passwords of users, which made it an easy target of attack for hackers (Pagliery, 2016).

Research on LinkedIn passwords showed that there were a few interesting things about the passwords being used on the professional networking site. The top 15 base words were identified which were used in LinkedIn passwords. Words such as 'linkedin', 'love', 'blue', 'pass', and 'Anna' (individual names of users) featured on this list. Similarly, pass phrases were also used such as phrases from the Bible or famous quotes or songs. For example, 'you'll never walk alone', 'the light shines in the darkness', and 'truth sets you free' were some of the common pass phrases used. This shows that despite the passwords being long (sometimes more than twenty characters), their common usage made them easy to predict. Similarly, colors were also used in LinkedIn passwords. Unsurprisingly, blue, the color of LinkedIn's logo was the most common followed by green, maroon/red, brown, etc. (Thorsheim, 2012b).

It is also important to distinguish between online and offline data hacking. Online password hacking is more difficult because the hacker will be unable to test multiple accounts simultaneously, as measures such as CAPTCHA and account lockout policies will prevent the hacker from repeating the same tests for longer periods of time. It is unclear if or how offline hacking was conducted in the case of the LinkedIn data hack. Accessing stored password hashes is the main way in which hackers conduct offline hacking. Since LinkedIn has not provided how the data hack occurred, it may be difficult to understand the actual nature of the attack. Nonetheless, the analysis by Professor Kirsi Helkala and the team found that a significant percentage of people used the color of the company's logo in their passwords, as was discussed in the preceding paragraph (Thorsheim, 2012b).

During the initial stages after the data hack had become public, there were criticisms about how LinkedIn handled the data hack. There was a lack of response from the organization in different channels. When users logged in to change their passwords, they did not notice anything different, and no statement such as 'For security reasons we encourage you to change your password now' appeared on the screens. The delay in LinkedIn sending out emails for the reset of passwords took several days. However, critics opined that such emails should have been sent out within hours, especially had there been any subsequent hacks on the same day or within a few days. This showed that LinkedIn was not prepared to handle such a large crisis from an organizational point of view. The number of accounts for which LinkedIn recommended password resets was not made public by the company. Therefore, it is difficult to conclude whether the number of compromised accounts were equal to the number of compromised users. Additionally, there were opinions that the entire user database was breached (Thorsheim, 2016).

The 2016 revelation that the number of data hacks was even larger – more than 115 million, was another blow to the public relations of LinkedIn. Experts opine that a forced password reset of the 6.5 million initial accounts must have been conducted first, followed by the resetting of passwords of the remaining 115 million or so accounts. If one reads between the lines, the LinkedIn statement 'We have begun to invalidate passwords for all accounts created prior to the 2012 breach that haven't updated their password since that breach' may show that the details of all accounts may have been compromised. Experts also suggest that, contrary to the opinion of LinkedIn, the passwords must not be changed frequently because users may choose simpler and shorter passwords to prevent forgetting new passwords every time they are created (Thorsheim, 2016).

Another criticism faced by LinkedIn was their manner of storage of users' passwords. Irrespective of the length of the password, hackers were able to hack and obtain passwords of long and short characters. Hashing passwords is a technique to prevent the passwords from being discovered. If the hash algorithm is easy to guess, many passwords may be guessed and obtained. If the algorithms are slow to be broken down, it is very difficult to obtain the passwords. LinkedIn, prior to the data breach, stored passwords using one of the fastest-computing hash algorithms available: SHA-1, which was never designed to be used for password storage. This went against the advice of authorities such as the National Institute of Standards and Technology (NIST) and the Open Web Application Security Project (OWASP). LinkedIn also did not employ 'salting', which is adding a random string of characters after a password which makes a password unique, even if two users shared the exact same password. It was also discovered that the passwords were cracked using a program called oclHashcat, which cracks passwords using graphic cards (Thorsheim, 2012a).

The sale of the data for five bitcoins or 2,200 USD (at the time of the data breach) is also a cause for concern. Since the hacker can sell the same data repeatedly to multiple buyers and multiply the revenue received from the sale of the passwords for a mere 2,200 USD, it is difficult to keep track of who has bought the data. The hack and sale of password data were traced to an entity named 'peace', which is thought to have been a group of Russian cybercriminals (Hunt, 2016). Usually, cybercriminals seek to hack the data for various reasons such as to assert superiority in the face of political rivalry between countries (for example, between the U.S. and Russia) or for monetary reward, as was the case in this data breach.

The primary question that arose here is whether the developers of LinkedIn made the decision to use the password storage scheme, which was weak, or whether this matter did not become a priority until the data hack was revealed. Instead of speculating about the answer to this question, LinkedIn's adoption of a strong risk management policy is important. The risk management policy must be aligned with the Systems Development Life Cycle so that the system security can be reviewed from time to time (Thorsheim, 2012a).

The risk management framework provides insights into how the risk management system must function. The following steps may be followed by LinkedIn, or other organizations dealing with information security, to prevent data breaches:

1. Executive Governance and Support: The top management must acknowledge that risk management is essential for the sustainability of the business processes in the long term. The senior management must also commission the risk management framework and establish it in a systematic manner.
2. Framework Design: This step involves formally assessing the current levels of risk in the organization and what steps can be taken to reduce the current risk levels. The risk appetite of the organization must also be decided so that it acts as a benchmark for risk management.
3. Framework Implementation: The implementation of the risk management framework can be divided between the various levels of management. To reduce the difficulty in implementation, the organization may follow the Software Development Life Cycle or agile methods used in traditional IT implementation in the organization.
4. Framework Monitoring and Review: The risk management team reviews the risk management framework by monitoring its performance and whether the performance is in alignment with the organizational risk management goals.
5. Continuous Improvement: The processes in risk management can be tweaked to ensure that the overall goals are being achieved. The gaps between the actual outcomes versus expected outcomes need to be assessed and corrective action must be taken. This is called Gap Analysis. The performance measures used in the risk management process provide the information required to assess the risk management performance (Whitman & Mattord, 2018).

This case study showed that organizations, big and small, must be constantly updating their data security policies to prevent large-scale data hacks. Preventing such attacks by using expensive computational storage of passwords is better than facing reputational challenges and losses after the hack has taken place. LinkedIn could have taken a more proactive step in ensuring that the users were kept informed about the data hack after the hack had occurred. While LinkedIn did not shy away from its responsibility, it could have informed users about the data hack sooner to reduce users' anxieties. This case shows that information security is an iterative process, which must include pre-crisis and post-crisis management plans, in addition to general risk management plans (Whitman & Mattord, 2018).

LEARNINGS FROM THE CASE

The LinkedIn Data Hack case highlighted that even for a large corporation with millions of users, the ability to ensure information security is an ever-present challenge. Information security policy is, therefore, crucial to not only ensure the security of data but to also contribute to the overall success of the organization. The Bull's Eye Model provides insights as to how organizations like LinkedIn can effectively formulate and manage information security policy. Policies are guidelines that determine the behaviors internal to the organization. As the first layer of defense, policies ensure that the responsibility for information security is divided across all stakeholders in the organization. In this case, LinkedIn's security policy can be broadened to include users who are responsible for setting strong passwords and enabling two-step verification, and the professional networking site itself in terms of salting and hashing the passwords. Policies also serve as reference documents should there be any legal disputes. The network of LinkedIn must also be secured by ensuring that login partners like Google also ensure information security from their end. Systems refer to computer and manufacturing systems, and their physical security is usually not a threat. Applications, based on desktops or mobile devices, should have adequate security. This is relevant to LinkedIn as the site can be accessed as a mobile application, in addition to the standard desktop view (Boyd, 2017).

Beyond policy, the LinkedIn data breach highlighted the prompt response from the organization due to communication procedures that were implemented before the data breach. This ensured that the users were kept informed about the gravity of the situation, as well as corrective measures that needed to be taken to resolve issues. Guidelines, which are optional recommendations for users, were also released by LinkedIn after the discovery of the data breach of 117 million accounts in 2016. These guidelines involved changing passwords, management of profile privacy, and creating strong passwords. However, communication through email directed at victims of the hack could have been more comprehensive, streamlined, and uniform. Standardizing this communication to users would ensure that all victims, and not just a few, are informed about the data breach and the remedial measures being taken. Such measures will improve user trust on LinkedIn (Silveira, 2012).

Another dimension of learning from this case is the legal and ethical aspects of the data breach. LinkedIn employees must be trained in the ethical and legal aspects of using the system so that the employees are well-prepared to tackle future disruptions to data privacy and security. The expected behavior of employees when disruptions do occur must be covered in the legal and ethical practices training so that the employees can proactively engage with the users to mitigate their doubts and concerns.

Collaboration between the business (LinkedIn) and the government also serves as a synergetic method to implement stricter laws for the prosecution of data breaches. In the case of LinkedIn, the accused was extradited to the U.S. to stand trial. The courts found him to be competent to stand trial and the case is currently being deliberated. This is an example of how the supportive relationship between the government and the business can improve the responses to data breaches and other cybersecurity-related crimes. Laws only have an effect against crime if the penalties are duly administered. However, such relationships may lead to corrupt practices if the businesses request favors from the government. Therefore, such relationships must be clearly codified to prevent misunderstandings and manipulations (Eckel, 2019).

When massive data breaches such as LinkedIn are publicized, phishing attacks may occur. Scammers pretending to be LinkedIn may contact LinkedIn users to retrieve sensitive personal information under the guise of investigating the data breach. Although such attacks have not been explicitly mentioned in this case, precautions must be taken to prevent scammers from taking advantage of an already fragile situation. LinkedIn could introduce procedures for informing users on how to distinguish between authentic emails from LinkedIn versus phony emails sent as part of phishing attacks (Hunt, 2016).

The lessons from the LinkedIn data breach can be applied to other data breaches as well, especially those involving large companies and millions of users. The data breach also serves as a learning case within the organization, so that more robust information security measures may be adopted. LinkedIn also may devise plans to contain the spread of the leaked data from one party to another. Dealing with inheriting a data breach is also important. The senior management of LinkedIn had undergone changes between when the 2012 data breach was discovered and the 2016 amplification of numbers. Although the senior management did not preside over the data breach, they are still responsible for the aftermath of the data breach. This shows that successive top-level management of a company must be willing to accept previous mistakes and data breaches, and find feasible solutions for the same (Hunt, 2016).

These major learnings from the case of the LinkedIn data breach in 2012 and the subsequent discoveries made in 2016 are key to addressing the ever-present challenges of information security.

DISCUSSION QUESTIONS

1. How effective is two-step verification during login for access to LinkedIn accounts?
2. Why is the sale of LinkedIn user data to third parties a cause for concern?
3. What is the importance of quick responses and updates from organizations in the aftermath of a data breach?
4. How can LinkedIn password policies be strengthened to act as the first defense against data breaches?

REFERENCES

Boyd, K. (2017). The Bull's-eye Model. Retrieved from https://kalaup.wordpress.com/2017/02/09/the-bulls-eye-model/

Eckel, M. (2019). U.S. Judge Finds Alleged Russian Hacker Nikulin Competent To Stand Trial. Retrieved from https://www.rferl.org/a/nikulin-hacker-trial-competency/29975018.html

Hunt, T. (2016). Observations and thoughts on the Linked in Data Breach. Retrieved from https://www.troyhunt.com/observations-and-thoughts-on-the-linkedin-data-breach/

Krebs, B. (2016). As Scope of 2012 Breach Expands, LinkedIn to Again Reset Passwords for Some Users. Retrieved from https://krebsonsecurity.com/2016/05/as-scope-of-2012-breach-expands-linkedin-to-again-reset-passwords-for-some-users/

Pagliery, J. (2016). Hackers Selling 117 Million LinkedIn Passwords. Retrieved from https://money.cnn.com/2016/05/19/technology/linkedin-hack/

Rosenblatt, J. (2020). Russian Convicted of LinkedIn Hack Spent Years in Custody. Retrieved from https://www.bloomberg.com/news/articles/2020-07-11/russian-convicted-of-massive-linkedin-hack-by-u-s-jury

Scott, C. (2016). Protecting Our Members. Retrieved from https://blog.linkedin.com/2016/05/18/protecting-our-members?version=meter+at+1&module=meter-Links&pgtype=article&contentId=&mediaId=&referrer=http%3A%2F%2Fnews.google.com%2F&priority=true&action=click&contentCollection=meter-links-click

Silveira, V. (2012). An Update on LinkedIn Member Passwords Compromised. Retrieved from https://blog.linkedin.com/2012/06/06/linkedin-member-passwords-compromised

Sofaer, A. D., & Goodman, S. E. (n.d.). Cyber Crime and Security The Transnational Dimension. Retrieved from https://www.hoover.org/sites/default/files/uploads/documents/0817999825_1.pdf

Thorsheim, P. (2012a). The Final Word on the LinkedIn Leak. Retrieved from http://securitynirvana.blogspot.com/2012a/06/final-word-on-linkedin-leak.html

Thorsheim, P. (2012b). Linkedin Password Infographic. Retrieved from http://securitynirvana.blogspot.com/2012b/06/linkedin-password-infographic.html

Thorsheim, P. (2016). LinkedIn's Poor Handling of 2012 Data Breach Comes Back to Haunt Them. Retrieved from https://www.linkedin.com/pulse/linkedins-poor-handling-2012-data-breach-comes-back-haunt-thorsheim

Whitman, M. E., & Mattord, H. J. (2018). *Management of Information Security* 6th Edition. Cengage.

1.6 Zoombombing
A Technical Perspective

Joe Mani Kuttikatt
Z Energy NZ, Wellington, New Zealand

Rion Jovial Luis
TussockLine, New Plymouth, New Zealand

SUMMARY OF STORYLINE

Zoom Video Communications, Inc. (Zoom) is a communications technology company headquartered in San Jose, California. It offers a cloud-based peer-to-peer web platform for video and online chat services and is used for teleconferencing, telecommuting, distance education, and social relations. In contrast to their competitors, Zoom works to reduce the infrastructure cost and the business strategy of the company focuses on offering an easily usable medium with a high degree of employee efficiency. Eric Yuan, former Vice President of Cisco WebEx Corporate founded Zoom in 2011. The company has more than 2500 employees, working across various geographies. The net revenue of the company is more than $600 million. Due to the global pandemic, Covid-19 Zoom services were actively used in all sectors for Business meetings, daily update calls, educational institutions for lecture purposes, and by people in same or different geographies or for social interactions. Zoom was considered a secure platform to share sensitive information.

According to cyber experts, around 530,000 Zoom accounts were hacked and sold on the dark web for less than a dollar and some for free (Mathews, 2020). The hacked accounts included several colleges, banks, universities, and business organizations like health providers and software services. The purchased accounts included a lot of sensitive pieces of information like email addresses, passwords, personal meeting URLs, and host keys. A host key is a 6-digit number that is used to control the meeting. The details provided or captured were stated as valid information by top cybersecurity companies like Cyble and Insights.

According to Lapin (2020), the FBI has also reported several cases of porn, vulgar images, and threatful language disrupting Zoom conference calls. After the continuous occurrence of similar issues, the FBI has also warned against using Zoom for sharing sensitive information. According to Maor (2020b), the kind of attack done by the hackers is called a "credential stuffing attack" where they attempt to reuse the combination of passwords used before by the users and try their luck on users' accounts. Therefore, the people who have re-used their passwords to log in to Zoom are potentially vulnerable to the attack.

DOI: 10.1201/9781003264415-8

CRITICAL ANALYSIS

As per the cyber experts, the attack was made by trying different combinations of credentials which had been previously hacked, thus most of the victims are organizations or people who have re-used their same passwords. However, this has not affected bigger firms as most of them use the Single Sign On (SSO) login feature. But soon after the disruption in services started, Zoom usage was stopped and banned in many places in America. The founder of the company also apologized for the privacy and security issues found with the application and was swift in bringing up different measures to use Zoom without any disruptions, enabling users to use the application without "zoombombing" incidents.

Upon further analysis of the situation, there are several matters which are of grave concern about when the user accounts are being hacked. Users not only experience disruptions in video conferences or calls but much more. According to the experts, more than 90% of people join Zoom using the Google sign-in option if they don't have an SSO login mechanism and very few create a separate login for Zoom and with the Google ID, the hacker can monitor the user's activities. The first thing the intruder will do is to go to the "passwords.google.com" where all the saved passwords for different applications like social media and bank accounts are visible. Here within a single click on enabling the view, the hacker will be able to see all of a user's passwords. If using an Android device, when one goes to the site "google.com/android/device manager", the hacker will be able to perform a couple of actions to the Android device like resetting the device with a different password or even erasing all the data in it. But the main part is that the malicious actor will be able to track the places you visited, the time of the visit, and all tasks performed. Using "photos.google.com", the pictures uploaded to Google Photos can be accessed. In addition to this, most of the people using social networking sites like Facebook and Instagram use their Google accounts to sign in. While logging into social networking sites with password protection, if these are linked to Google, hackers can get the password in seconds.

One could just click on the password field and type F12 to view the console and change the field type from "password" to "text" so that the password will be visible. Another important feature by Google is "activity.google.com" wherein all the activities performed by the user have been tracked. Although all these are very good features provided by Google to revisit the things performed, if the same is obtained by a hacker, then any benign or malicious intrusion can be implemented. The different ways to secure an account is by going to "accounts.google.com" and changing the password. Google provides an add-on security feature, the two-step verification, where the user can protect his Google account using a one-time password feature to his registered mobile number.

If the user has used Google ID to log in to Zoom, the first step is to change one's password in Google ID. Users should never use the primary Google account in any of the third-party applications. Many suggest creating a dummy Google account for third-party applications and making sure the same ID is not used in one's Android device.

Looking at how the accounts were hacked the first and foremost thing to do, according to Graham (2020) is to change one's Zoom password to log in to Zoom.

There are a couple of ways to use Zoom without getting affected by "zoombombing". Any stranger will be able to join the Zoom call with the meeting ID if the meeting is not password protected hence the best way is to require meeting passwords and let in people as decided by the meeting's host.

According to Maor (2020a), attackers made encroachments on the old database with numerous usernames and passwords. First, the attackers collected different usernames and passwords available in various venues and black markets even on the dark web. During the analysis done on breach experiences, researchers stated that the databases were used back in 2013. After infiltrating the databases, attackers prepared a configuration file like Open bullet or SNIPR to validate the databases having the same credentials which are mapping to the hacker's application of choice. Unfortunately, this time, Zoom was the victim.

Once the intruders locate the configured tool in the application, they initiate the credential stuffing method. To avoid detection, the attackers use multiple bots. Multiple attempts to the same IP would be suspicious activity; hence, the bots or malicious actors always use a lag in between different attempts and still look for multiple valid login attempts. If the target application is not protected with multiple authentication methods like a captcha or employing bot detection, the intruder is notified when the login is successful. Once the login is done, they can collect multiple data. In the context of Zoom, the hacker can collect multiple data like email ids, passwords, meeting URLs, and the host keys. Finally, when all the data are collected, the hacker dumps all the data into a new database and shares or sells them on the dark web. As per experts, credential stuffing is not the only way of stealing sensitive data; the most common way is by phishing attacks.

According to recent studies, the core issue lies in reusing the same password in multiple places. Most people tend to produce a strong password having a capital letter, small letter, special characters, numbers, etc., with more than eight characters. But the likelihood of forgetting one's password is high hence people tend to use the same password in multiple platforms. Thus, if the account is hacked in any one forum, the intruders try to reuse the password in multiple places which results in users being victims of a cyber-attack.

The zoombombing attack was effectively handled by the team. The founder also stated that due to mandatory social distancing measures put forward by the regulatory bodies, the application was widely used by both businesses and consumers where the usage increased exponentially. The following is a diagram (Figure 1.6.1) which supports the founder's claim of the exponential growth of Zoom by illustrating the increased number of daily meeting participants from December 2019 to April 2021.

There are various measures provided by the company officials to get rid of "zoombombing". The first is to use authenticated URLs to enter the meeting or else there are high chances of becoming a victim of phishing attacks. Second, create private Zoom meetings with the password, wherein the host can decide the attendees. Additionally, manage the screen sharing feature where only the host will be able to share the screen. Fourth, use the latest version of Zoom where they have included the security feature to create default passwords for meetings. Also, use strong passwords to get rid of Brute force attacks. Zoom will also create virtual waiting rooms by

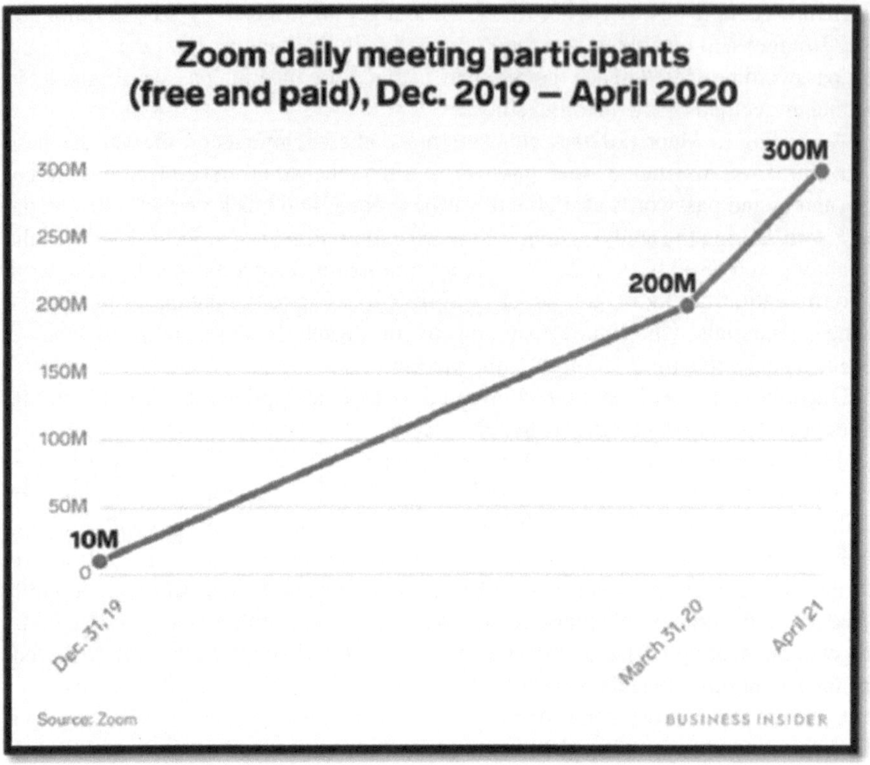

FIGURE 1.6.1 Exponential increase in users during the pandemic.

default to stop zoombombing because here the host will get to know the people joining. The organizations should also educate the employees on the importance of creating strong passwords and accessing the URLs, otherwise, there are high chances of getting attacked by phishing emails. Parmar (2012) gives an example of how understanding the behavior of an employee in social media a phishing email was sent out by a hacker which got clicked and the hackers were able to access the mainframe of tech giant Google.

LESSONS LEARNED

Based on the critical analysis done there are best practices to do when using Zoom and what to keep in mind while setting passwords for different applications. The importance of being informed on the applications we deal with, and its harmful effects, is also understood. The following are the ideal practices to follow:

1. Use strong passwords to avoid getting attacked by Brute force attacks and make sure the same password is not used anywhere else.
2. Make sure you click only on authorized URLs or else, you might be a victim of a phishing email attack.

3. Organizations should educate people to use the applications and make sure they are aware of all the features available.
4. Always try to use the application with the latest updates so that you don't miss any of the added security features.
5. Don't use Google ID for third-party applications – try to create a new one.
6. Secure your Google account with available secure options like the two-step verification technique.
7. Create not public but private Zoom meetings with passwords and share the meeting link directly with people (Zaveri, 2020).
8. The screen share facility should be only available for the host.
9. The meeting host should let in people who should be there in the meeting group.
10. Change your passwords frequently to increase your security.

DISCUSSION QUESTIONS

1. Why shouldn't a user have the same passwords used in multiple accounts?
2. Why it is important to know the application well before sharing sensitive information?
3. What are the different steps the management team should take to avoid accidental mistakes?
4. What is the importance of using the original software with the latest patches?
5. What is the importance of having two-step verification methods when the user tries to log in to a system?

REFERENCES

Graham, J. (2020). Use Zoom? These 5 Safety Tips Can Keep the 'Zoombombing' Hackers Away. Retrieved from https://www.usatoday.com/story/tech/2020/04/02/how-to-keep-zoombombing-hackers-away-zoom-safety-tips/5113080002/

Lapin, T. (2020). FBI Warns of Hackers Hijacking Online Zoom Meetings, Classes. Retrieved from https://nypost.com/2020/03/31/fbi-warns-of-hackers-hijacking-online-zoom-meetings-classes/

Maor, E. (2020a). Recycling Credentials in Four Easy Steps. Retrieved from https://intsights.com/blog/recycling-credentials-in-four-easy-steps

Maor, E. (2020b). Zooming in on the Target: Cybercriminals Automate Attacks Against Remote Workers. Retrieved from https://intsights.com/blog/zooming-in-on-the-target-cybercriminals-automate-attacks-against-remote-workers

Mathews, L. (2020). 500,000 Hacked Zoom Accounts Given Away For Free On The Dark Web. Retrieved from https://www.forbes.com/sites/leemathews/2020/04/13/500000-hacked-zoom-accounts-given-away-for-free-on-the-dark-web/#65daf2d858c5

Parmar, B. (2012). Protecting against Spear-phishing. *Computer Fraud & Security*, *2012*(1), 8–11. doi:10.1016/S1361-3723(12)70007-6

Zaveri, P. (2020). Zoom Will Enable Meeting Passwords and Virtual Waiting Rooms by Default to Help Stop 'Zoombombing' Incidents. Retrieved from https://www.businessinsider.com.au/zoom-security-passwords-waiting-rooms-stop-zoombombing-2020-4

1.7 Case Study: Ransomware Attack on IT Firm Collabera

Joe Mani Kuttikatt
Z Energy NZ, Wellington, New Zealand

Rion Jovial Luis
TussockLine, New Plymouth, New Zealand

SUMMARY OF CASE STORYLINE

This case is about a recent cyber-attack on the Information Technology Giant named Collabera. The company Collabera Inc. is an American firm founded in 1991 and is headquartered in Basking Ridge, New Jersey. The company delivers services such as professional Information Technology recruiting, staffing, consulting, and business services to companies globally. The estimated revenue generated by the company is at an enormous 856 million USD. The company has 16,000 employees ensuring the smooth flow of its operations across different geographies.

The company was initially called Global Consultants Inc. After 6 years, it was taken over by entrepreneur Hiten Patel. For more than 25 years, Collabera has been at the top of the list for IT and professional staffing and is also known for its flexibility and providing high customer satisfaction.

According to Jay (2020), the company operates in different domains like banking, financial, technological, healthcare, and communication Industries. Due to the global pandemic COVID-19, IT businesses were impacted in many ways, and so the main aim of the company was to solve business functions effectively. For this, Collabera introduced the Virtual Workforce Model which focused on different key components like the working environment of the employees, the collaboration of teams, the process of delegation, the nature of work, demand management, and different improvements to address the success of the workforce model.

According to Nichols (2020), Collabera was attacked by a malware known as the 'Maze Ransomware' in June 2020. As per the memo issued by Collabera to its employees, the data stolen by hackers included employees' names, personal addresses, contact details, SSN (Social Security Number), DOB, employee payroll benefits, passport details, and Visa Immigration details. This was a huge blow to the

DOI: 10.1201/9781003264415-9

top, middle, and low-level management since they have an enormous customer base and a big list of potential clients in different sectors namely Intel, Amazon, IBM, Bank of America, and so on.

Soon after the identification of the unauthorized access to their system through malware, the company decided to approach law enforcement authorities to confine the situation. They have also taken support from cyber defense companies to identify the different ways to handle the situation and to know the adverse effects caused.

As a defensive approach, the company has also provided its clients with an Indicator of Compromise (IOC) which is a set of forensic information from the log file to identify the potential threats on the system and the network to protect their database.

As part of company policy, they provide Experian Credit cards to their employees. After the cyber-attack took place and as an extra precautionary measure, the company advised its employees to monitor their financial statements and credit card transactions and to report if any unauthorized or fraudulent transactions have taken place.

According to Correspondent (2020), after the breach was identified Collabera immediately retrieved access to their backup files and instantly sent an inquiry to cyber defense companies and other sources to understand the nature and scope of the cyber-attack.

CRITICAL ANALYSIS

The case describes the recent cyber-attack which took place on Collabera Inc. in July 2020. This information was revealed and spotted by a cyber-security firm named Cyfirma which is dedicated to finding cybersecurity threats mainly for government and businesses. Collabera became a victim of a malware attack popularly known as 'Maze Ransomware'. As per the analysis done by experts, the kind of attack was like the recent attack on IT giant Cognizant.

According to recent reports from cybersecurity firms, there has been a huge number of ransomware attacks in different sectors causing huge losses to governments and businesses. Morgan (2017) states that ransomware has grown into different and new varieties of families since the year 2013, and its expected growth is more than 350% every year, targeting mainly Windows operating systems. The attack that happened to Collabera is not different from other ransomware attacks; the intruders used similar techniques like Credential stuffing/stealing through phishing emails and collecting data from different stages.

This is the kind of pattern used by attackers to steal the information. Initially, the attackers or hackers deploy malicious programs to the infected systems to take control over the files, steal then lock them from retrieval. Hackers then take a copy of the existing data files and demand a huge amount of money to recover the files, or else they threaten to release the sensitive data on the dark web or in public domains where anyone can access it.

As per the investigating cyber firm, the actors behind the attack were a Russian-based hacking group named ATP28 or Fancy Bear. The group works in an organized manner, and they use very basic methodologies to initiate their attack. First, they try to find vulnerable files or loopholes in a dedicated system or domain using different common attacks like faking domain names, using spear-phishing emails, and finding a waterhole in websites and other different vulnerabilities in servers. Second, they use other configuration management frameworks like Core-shell or PowerShell to deploy the attack, download malicious modules, and find other ways to intrude and continue the operation. After that, the group shows its persistence inside the environment, the key feature of an Advanced Persistent APT attack. The attacker needs to exploit the system only once to get full control over the system. After that, they try to find credentials to get access to other external computers by using active directories where all the data related to users, groups, network shares, endpoints, and servers are stored just like an organization's database. Finally, legitimate services and hash techniques are used to infect more endpoints. Once the data is obtained from databases and emails the data is exfiltrated. With the obtained sensitive information, the malicious actors threaten the victims and demand money.

In the case of Collabera, the stolen information included employees' names, personal addresses, contact details, SSN (Social Security Number), DOB, employee payroll benefits, passport details, and Visa Immigration details. Due to the recent pandemic and COVID crisis, the company was highly dependent on the Virtual Workforce Model to meet customer demands and the Work from Home facility to continue the smooth flow of operations. The company was able to predict that the demand for services would peak due to the current crisis. Hence, most of the employees continued their contribution by working from home. While employees are working from home and especially when many deal with sensitive information or projects related to clients in fortune 500 companies, the company must take extra precautions to make sure the data is safe. Most of the employees are using a Virtual Private Network (VPN) to connect from home to their servers. According to KABANDA (2020), VPN enables employees to work offshore or onshore with secure access. In addition to VPN, it should be equipped with Intrusion Detection systems (IDS) and Intrusion Prevention systems (IPS) that secure from external intruders. According to Stallings (2015), there should be firewall restrictions to protect networks from cybercriminals. Besides, this protection also completes the basic function like securing the operating system. This process involves installing the operating system with the latest patches making its security measures more effective.

Even though most of the employees were using VPN to connect to their servers or work domains, the Maze Ransomware was equipped to identify the known vulnerabilities of the virtual private network especially the pulse VPN and how to break it. If the user hasn't updated with the latest patch or using counterfeit software.

The organizations should take the responsibility to educate their employees and business partners on the security risks. There should be security audits conducted at least once every quarter by inspecting the systems and determining whether all the employees are using the latest updated software and the latest patch of operating systems. There should be a dedicated security team to inspect and monitor the logs and other potential areas of attacks that makes the system vulnerable. They should be continuously monitoring the secured gateways and firewalls to see if any loopholes are present and rectifying the issue before intruders or attackers hit upon the gateways or firewalls.

LEARNING OUTCOMES

From this case, we learned a lot of steps and precautionary measures to be taken while working with sensitive information and VPN networks. Although Collabera is one of the top firms in its domain and even after achieving numerous awards in various venues, they were still attacked by cybercriminals. This proves that every company needs to be aware of the various ways to secure themselves from the different attacks.

A company needs to have an effective and alert cyber-security team which monitors the organization's systems and networks prominently. Informing the staff on the different ways and techniques to protect themselves is also important. With VPNs being hugely popular and in use due to the pandemic, educating employees on how to deal safely with sensitive information is paramount. The right procedure to follow is to use the correct licensed software which has the firewall and security patch updated and prohibit the use of counterfeit modules to be safe from unwanted intruders. The employees must be aware of the best practices they can follow to perform safe operations by performing two-step authentication while connecting to a virtual network with desktop applications like pulse secure code where unique codes are generated every 5 seconds and refreshed instantly. Besides, to always keep backups ready during intervals when you deal or work with critical information, using unique and hard passwords which are difficult to crack is important and very good practice.

Furthermore, the company management team should inform or create awareness among employees about the different methods or approaches taken by cybercriminals to steal data and equip their system with the latest antivirus and firewalls in the market so that they can reduce their risk of getting attacked. Antivirus and firewalls will track and block any suspicious links and files.

Specific to the case, a common question for company leaders is whether to pay or not to pay once they get hit by some malware attacks like Ransomware. But, as per the information revealed by cyber experts, contacting some technical experts to understand the situation and the intensity of the damage caused and information that was breached is always better. More importantly, the questions to be answered are – What went wrong and why, how the company and employees can fix problematic areas, and what are the different situations to consider so that an attack doesn't happen again in the future?

The company should also implement some automated measures to monitor their logs continuously so that any unauthorized access or entries can be easily spotted, and personnel can take the necessary measures before the damage occurs. Even though companies have security expert teams who monitor manually, it's a tedious and hectic job to do because the size of data is in terabytes and there are high chances of misses.

DISCUSSION QUESTIONS

1. What are the best practices that employees should follow while dealing with sensitive data?
2. What should the management team do to create awareness among the employees on cyber-attacks?

3. What are the different steps to be taken by the employees while working on Virtual Private Networks? And how can the security team make sure employees follow the guidelines accurately?
4. Should an organization pay or not pay for Ransomware attacks?
5. What are the different ways in which companies protect themselves from being a victim of Ransomware attacks?

REFERENCES

Correspondent, T. (2020). IT Service firm Collabera Suffers the Maze Ransomware Attack. Retrieved from https://content.techgig.com/it-service-firm-collabera-suffers-the-maze-ransomware-attack/articleshow/76977348.cms

Jay, J. (2020). IT Services Major Collabera Suffers Maze Ransomware Attack. Retrieved from https://www.teiss.co.uk/collabera-maze-ransomware-attack/

Kabanda, G. (2020). Performance of Machine Learning and other Artificial Intelligence Paradigms In Cybersecurity. https://www.computerscijournal.org/vol13no1/performance-of-machine-learning-and-other-artificial-intelligence-paradigms-in-cybersecurity/

Morgan, S. (2017). Ransomware Damage Report. Retrieved from https://www.herjavecgroup.com/ransomware-damage-report-2017-edition/

Nichols, S. (2020). Collabera Hacked: IT Staffing'n'services Giant Hit by Ransomware, Employee Personal Data Stolen. Retrieved from https://www.theregister.com/2020/07/14/collabera_ransomware/

Stallings, W. (2015). Operating System Stability. Accessed on 27th March, 2019.

1.8 Neuralink
A Neural Technology Company

Anjali Krishnan
University of Canterbury, Christchurch, New Zealand

Shameer Thottoli
Acclaim Group Limited, Auckland, New Zealand

CASE SUMMARY

Neuralink was founded in 2016 and is a venture supported by Tesla CEO Elon Musk and SpaceX. Neuralink researches the brain–machine interface (BMI) to create devices that can be implanted inside human brains. Neuralink aims to resolve issues concerning brain disorders and bridge the gap between artificial intelligence (AI) and humankind. The headquarters of the company is in San Francisco and its chief executive officer is Jared Birchall. In the launch video [1], Elon Musk has given three reasons why he chose to support this new venture. The reasons were to "understand and treat disorders of the human brain, conserve and improve human brains, and develop a well-aligned future". The initial aim of the company was to create an interface that will be directly implanted inside a human brain. This brain interface is intended to treat injuries or trauma. However, the company's ultimate hope is to upgrade humans in such a way that humans can interact with computers or AI at the speed of thought. Elon Musk has stated that he founded this company for the sole purpose of aiding human beings to keep up with future AI advancements.

In 2019, Neuralink introduced an approach that uses surgical robots to implant gossamer-thin wires into a human's brain that will be connected to an exterior computer processing unit. The company also shared its hope that in the future these connections will be made wireless to achieve maximum mobility. In addition, they have revealed that the company has run tests on apes and mice with successful results. Neuralink also confirmed that human trials would be conducted that year. Elon Musk has shared his belief about how lossy it is to convert thought to input using an old-fashioned keyboard and mouse. Musk has faith that this technology will help humans better interact with computers and eventually mitigate the risks of AI overpowering humans. He believes that uncontrolled AI with very less to no regulations poses a

DOI: 10.1201/9781003264415-10

great threat to the human race as a whole. Hence, he claims Neuralink is the antidote to the risks posed by unmanaged AI in the future.

A tweet made by Musk on Twitter regarding Neuralink was published by The Daily Mail [2]. The article highlights the tweet made by Musk as a reply to a Twitter user's question, about whether the Neuralink brain chip will help to eliminate depression or addiction. Musk tweeted in response, "For sure. This is both great and terrifying". Musk has explained that the process of implanting electrodes will be done by surgical robots and removing a chunk of the skull with only an unnoticeable scar. When asked about how the human body will react to this device, Musk replied that there is a very small probability of rejection. According to Musk, humans are already essentially cyborgs, as missing mobile phones or laptops make humans feel like they've lost a limb. He also states that a full brain interface will be developed within 25 years and that it will result in "symbiosis between humans and AI".

Symbiosis of humans and AI means that a person's neurons will be linked to an AI extension of themself. Recently, an article published by Futurism discusses how Musk replied affirmatively when a Twitter user asked whether Neuralink will stimulate the release of good hormones like oxytocin, serotonin, etc. [3] Musk has also confirmed that Neuralink can stream music into a person's brain without the use of headphones. The official site of Neuralink has a link to a white paper [4] published on the technology and its clinical applications.

CASE ANALYSIS

Musk believes that Neuralink is both great and terrifying. This statement testifies that even the founder of Neuralink is awestruck by evaluating the scope of numerous possibilities and advancements in this field. He also does not shy away from openly admitting the terrifying aspects of Neuralink. Many others share the same apprehensiveness and have raised concerns about introducing a foreign body into the brain.

Most devices that are capable of interpreting brain waves have widely used practical applications in the medical field. These can be classified as unidirectional devices because the communication between the brain and the device is in one direction; no feedback from the machines goes back to the brain. When Musk stated that Neuralink can stimulate the production of chemicals in human brains, this was seen as a clear indication that Neuralink is capable of bidirectional communication between human brains and external computing devices. Neuralink's founder also indicated that their device is capable of training human brains to eliminate addiction and depression. This confirmed that Neuralink is not only just interpreting brainwaves coming from the human brain but also providing input. This makes Neuralink terrifying because it allows an external device to have control over brain functions. While external factors have always shaped brain chemistry and development, these external factors have all been organic.

There is a common belief that advancements in the field of AI have the potential to cause human extinction. In his Neuralink launch video [1], Musk explained why he started this venture. One of the reasons was to have a well-aligned future between AI and humans. Musk fears a future where AI overtakes humans and can cause harm to the human race as a whole. In an article published by Oxford Student [5], Neuralink is a solution that doesn't limit itself to "what humans do" but goes beyond "what humans are". Neuralink is creating a way for humans to achieve symbiosis with AI, thereby ensuring a well-aligned future between humans and AI. However, security concerns remain over how advantageous this alignment will be for humans.

The primary goal of Neuralink is to treat patients with brain disorders and cure mental health issues. As per the article published by 'Vox' [6] until now Neuralink has only done trials on animals and is awaiting approval from the US Food and Drug Administration (US FDA) to perform human trials. To implant electrodes – threads that are as thin as hair – a hole needs to be drilled inside the skull. The products from Neuralink purportedly have the potential to positively impact the quality of life for millions of people, such as the 5.4 million people who are paralyzed in the US alone. The chief surgeon of Neuralink, Matthew MacDougall stated that the company's main focus is on patient safety and gradually they aim to achieve a non-invasive procedure – likened to LASIK eye surgery – instead of the current invasive surgery. Their target customer base is individuals who are completely paralyzed because of spinal cord injuries. The article claims that Neuralink is among the foremost ventures that are dedicated to biohacking. The term biohacking is defined as research on improving human beings' performance – physically and cognitively – with the help of radical experiments conducted on human bodies. Like other biohackers, Neuralink also complains about the lack of support from the medical community. We posit that the concerns shared by the medical community – whose focus is on patient safety, including healthcare data protection – are reasonable and that the resulting restrictions imposed must be strictly adhered to by Neuralink if they wish to further expand their research in this area.

In another article titled, "Security Watch: Elon Musk's NeuraLink Links Brains to iPhones via Bluetooth" published by 'Threatpost' [7], the security implications of implanting a Bluetooth-enabled electronic chip into the human brain are discussed. Security experts unanimously agree that Neuralink needs extensive cybersecurity scrutiny – separate from, and additional to, FDA approval as a medical device – to continue marketing its products. Connecting an iPhone or any other mobile device via Neuralink to a human brain through Bluetooth introduces security threats such as tracking, spamming, and spoofing. At the very least, Neuralink must create a mechanism to detect and recover in the event of an attack. Strong encryption techniques must be used, along with regular security updates for the device. Even though Neuralink claims that its products will have a positive impact, they can be misused. Neuralink products could be used to maliciously manipulate the human brain. Human error, while unintentional, could also result in undesirable outcomes. Neuralink is a novel system and security experts need to identify security issues and provide safeguards.

LESSONS LEARNED

Neuralink is applauded, as well as criticized, by many experts in information security. Some experts view Neuralink as a protection against possible security threats posed by AI advancements. Other experts are concerned about the security lapses that can occur with Neuralink and may have devastating consequences. The possible information security threats that Neuralink might be subjected to include intellectual property compromises, spying or intrusion, error or failure committed by humans, information blackmailing, sabotage, software threats, technical failure in software and hardware, and theft [8].

Since Neuralink's intended customers are vulnerable people with brain disorders, mental health issues, or disabilities, more discussions must happen to highlight ethical issues that can arise. Users, employees, and stakeholders must undergo ethical and legal education [8]. By charting proper compliance requirements, policies, and frameworks, most cases of potential misuse can be addressed. The three main unethical behavior classifications that Neuralink needs to take care of include ignorance, accident and intent [8].

DISCUSSION QUESTIONS

1. Is Neuralink a blessing in disguise or is it going to make humans vulnerable to greater cybersecurity threats?
2. Is the threat of advanced AI a risk that society needs to address?
3. What is the best solution to protect against attacks from AI?
4. Will we inevitably become one with AI as cyborgs?

REFERENCES

1. https://www.youtube.com/watch?v=r-vbh3t7WVI
2. https://www.dailymail.co.uk/sciencetech/article-8530581/Elon-Musk-says-brain-chip-Neuralink-train-area-linked-depression-addiction.html
3. https://futurism.com/the-byte/elon-musk-neuralink-stream-music-brians
4. Musk, E. (2019). An integrated brain-machine interface platform with thousands of channels. *Journal of medical Internet research*, *21*(10), e16194.
5. https://www.oxfordstudent.com/2020/05/19/will-elon-musks-neuralink-shape-the-future-of-humanity/
6. https://www.vox.com/future-perfect/2019/7/17/20697812/elon-musk-neuralink-ai-brain-implant-thread-robot
7. https://threatpost.com/elon-musk-neuralink-brains-iphones-bluetooth/146552/
8. Whitman, M. E., & Mattord, H. J. (2013). *Management of information security*. Nelson Education.

1.9 Securing the Internet of Things

Amelia Samandari
University of Canterbury, Christchurch, New Zealand

CASE SUMMARY

Smart cities require the integration of different technologies that will become part of the critical infrastructure of our society. Integration of these technologies, such as artificial intelligence (AI), data analytics, and sensors, introduces further avenues for generating revenue. For example, businesses can use the information generated by smart cities to provide targeted advertising to consumers. A smart city can track an individual's movements and actions, cross-referencing them to generate a buyer profile. Using this buyer profile, consumers can receive very personalized ads for products or services they are likely to purchase. The integration of technologies that enable smart cities is referred to as the Internet of Things (IoT).

The IoT encompasses any device that is capable of network connectivity, and of gathering and dispersing data. As such, the IoT contains objects such as mobile phones, laptops, home appliances, and vehicles. These communicate with other IoT objects as well as additional entities, for instance, human beings. The goal of the IoT is to provide a platform between diverse technologies. This connection of physical objects opens doors for new applications and supports intelligent decision-making (Al-Fuqaha et al., 2015).

The IoT is designed to improve our quality of life through the "automation of everything around us", connecting physical things through their virtual IoT counterparts (Al-Fuqaha et al., 2015). In pursuit of this improved quality of life, IoT objects and networks have seen a swift rise in recent years. The number of IoT objects is projected to reach 212 billion deployed globally by the end of 2020 (Gantz & Reinsel, 2012).

The IoT has long been identified as one of the emerging technologies in the field of Information Technology (IT). As noted in Gartner's 2013 IT Hype Cycle, a graph that "represents the emergence, adoption, maturity, and impact of applications on specific technologies", the IoT would reach market adoption in 5–10 years (Gubbi et al., 2013). Indeed, the IoT is a technological revolution encompassing the convergence of information, communication, and technology. When physical objects have locatable, addressable, and readable virtual counterparts, ubiquitous computing can be realized through our use of these smart devices (Mahalle & Railkar, 2015).

In 2013, the Internet was cited to consume approximately 5% of the total energy generated globally (Gubbi et al., 2013). This has only increased, and will continue to do so, with the demand IoT applications can place on the Internet. The number of

DOI: 10.1201/9781003264415-11

connected machines has grown by 300% over the last 5 years (Manyika et al., 2013). In 2011, there was a 250% increase in the volume of machine-to-machine traffic (Shafiq et al., 2012) and by 2022, machine-to-machine traffic flows were expected to constitute up to 45% of global Internet traffic (Taylor, 2013). The annual financial impact of the IoT is estimated to reach the range of $2.7 trillion to $6.2 trillion by 2025 (Manyika et al., 2013).

As the IoT grows in scale and scope, users' options for interacting with their physical and virtual environment will increase. Through novel and efficient use of enabling technologies like RFID, Bluetooth, and ZigBee, the IoT can be used for value-adding applications (Mahalle & Railkar, 2015).

Security vulnerabilities in the IoT can occur as a result of the dynamic network topology (Mahalle & Railkar, 2015). This changing topology is due to node mobility and can exacerbate the effects of vulnerabilities commonly found in low-power IoT devices that have weak physical security. Mobility is an integral factor in the value proposition of the IoT. Therefore, since the dynamic changes of the IoT are reflected in the changes to its attack surface, corporate security policies and strategies must be adaptable to cater to the evolving conditions of systems using IoT devices.

CASE ANALYSIS

Various definitions have been created to describe the nature and functionality of the IoT.

The Radio Frequency Identification (RFID) group defines the IoT as "the world-wide network of interconnected objects uniquely addressable based on standard communication protocols". In the IoT context, entities represent devices, services, and users (Mahalle & Railkar, 2015).

The Cluster of European Research Projects on IoT says that:

> 'things' are active participants in business, information and social processes where they are enabled to interact and communicate among themselves and with the environment by exchanging data and information sensed about the environment, while reacting autonomously to the real or physical world events and influencing it by running processes that trigger actions and create services with or without direct human intervention.
>
> (Gubbi et al., 2013)

IOT APPLICATION SCENARIOS

The IoT's main appeal is its widespread application scenarios. The possibilities for gathering and connecting so much data are endless. As such, the goal of the IoT is to have all one's devices connected. This includes the creation of smart cities, health monitoring devices, earthquake warning systems, and smart homes. Cities such as Amsterdam and Madrid are already implementing IoT solutions to manage much of their infrastructure, and work is currently being undertaken by LINZ's Smart Cities Coalition to create a technologically advanced infrastructure in Christchurch, New Zealand.

Applications of the IoT can be classified into four domains (Gubbi et al., 2013):

1. Personal and Home
2. Enterprise
3. Utilities
4. Mobile

Industry 4.0 calls for increased automation, interconnectedness, and data dissemination, and the IoT can meet these goals. Two technologies that enable the interconnectedness required to fully realize the IoT are RFID and Wireless Sensor Networks (WSN). RFID creates a system of automatic identification. This is done through the design of microchips that have wireless data communication capabilities and can act as electronic barcodes. By using RFID, physical devices can have a digital presence. This technology can be applied in many domains, especially in retail and supply chain management (Welbourne et al., 2009). WSNs provide efficient remote sensing using miniature, low-cost, low-power devices. The use of a large WSN enables the collection, processing, analysis, and dissemination of information. This is accomplished by data being shared between sensor nodes for collection and processing. This data can then be sent to centralized or distributed systems for analysis and eventual dissemination (Akyildiz et al., 2002).

IoT scenarios describe interconnected systems that use smart technologies to provide services in a variety of contexts (Al-Fuqaha et al., 2015). Common IoT scenarios are those involving smart homes, health monitoring, environmental or agricultural sensing, and personal devices such as laptops and cell phones. The IoT scenarios outlined in this section are categorized as Lifestyle or Environmental based on their area of societal impact. Lifestyle scenarios describe IoT networks that are contained in a physical structure, such as smart homes and buildings. This also includes scenarios that affect the quality of life for multiple parties, such as smart city and smart hospital scenarios. Environmental scenarios affect the external surroundings and encompass traffic monitoring and agricultural sensing.

In the context of these scenarios, the interconnected nature of data between applications and domains can be seen. For instance, a home IoT network can monitor data for energy usage in a house. This data can be made available to the electricity company, providing the company with data that can be used to optimize electricity supply and demand. To optimize energy management, home equipment such as refrigerators, washing machines, and heat pumps can be managed by another IoT network that oversees utilities (Alkar & Buhur, 2005). In this way, data about energy usage is transferred between different applications and domains. This facilitates more effective decision-making and management.

Lifestyle

First and foremost, as with all IoT networks, the goal of smart homes is to improve our lives (Cook et al., 2013). Smart homes will have monitoring and operating capabilities for home appliances and systems, allowing residents to remotely open their garage door, turn on the house lights, prepare a cup of coffee, and turn on the heaters before walking into their homes (Al-Fuqaha et al., 2015). To perform to their fullest extent, smart home systems need to interact not only with the internal environment of home devices but also with external environments, such as smart grids, that are not under the control of the smart home system (Komninos et al., 2014). Another home IoT network has been proposed for elderly care, providing monitoring with the goal of "reducing hospitalization costs through early intervention and treatment" (Hu, 2016).

Smart buildings can control and manage automated building systems, such as lighting, security, entertainment, and Heating, Ventilation, and Cooling (HVAC), by connecting these systems to the Internet. Therefore, system management becomes independent of human intervention. This results in the ability to optimize maintenance, energy consumption, and reaction time in the event of failures (Finch, 2001).

Smart grids support smart homes and buildings by monitoring energy consumption. Using smart grids, suppliers can manage power by providing power proportionate to the population consumption. This increases the quality of service because suppliers can meet their consumers' needs. In addition, greater efficiency is achieved, and the risk of power failures is reduced (Komninos et al., 2014).

Smart cities improve the quality of life in cities by making it easy and convenient to find information of interest (Gea et al., 2013). Smart healthcare can be seen in patients and medicine that have embedded sensors and actuators. The physiological status of patients can be monitored through the collection and analysis of data. This data can then be sent to processing centers that can suggest the further steps that may need to be taken. Concerning hospital hygiene, a hospital in Ohio used IBM's RFID implementation to track handwashing, avoiding infections that caused annual losses of approximately $30 billion and about 90,000 deaths (Michaelsen et al., 2013).

Environmental

Environmental monitoring can refer to enterprise management as well as natural environment management. In the context of enterprise management, building occupants, lighting, the water network, and HVAC can be monitored using a network of sensors. A similar network can be used for monitoring items of interest in agricultural lands, such as soil parameters or irrigation. By using these monitoring IoT networks to collect data, informed decisions can be made (Liqiang et al., 2011).

Another valuable environmental factor to monitor is the effect of urban traffic. Greenhouse gas emissions, deteriorating air quality, and noise pollution are all unwanted side effects of urban traffic. By adding the data collected by WSN to the data from existing traffic control systems, specialized information can be provided to commuters. In this way, commuters will have accurate, up-to-date information that comprehensively outlines high-traffic areas. This will encourage commuters to take routes that are less congested, reducing noise pollution and the concentration of greenhouse gas emissions, as well as preventing rapid deterioration of air quality (Kumar et al., 2005).

To control the transportation network, intelligent transportation systems have been proposed. These systems aim to make driving more enjoyable (Gerla et al., 2014). This is done by increasing the availability, reliability, efficiency, and safety of the transport infrastructure (Talcott, 2008). To support an intelligent transportation system, the United States Department of Transportation is establishing requirements that will ensure all new automobiles have systems in place for vehicle-to-vehicle communication (Al-Fuqaha et al., 2015). An extension of traffic monitoring is the support it can provide for logistics management. Using these traffic IoT networks, a business can make their product transportation systems more efficient (Lin et al., 2005).

Security Challenges of the IoT

Security governance ensures that business needs and priorities are at the forefront of security decision-making. Business needs, communication structures, attack surfaces, and attack vectors will continue to change throughout the lifecycle of an IS, but the focus should always be placed on securing the most critical information assets and systems.

The IoT also poses a challenge for security management, because the dynamic nature of the IoT can more rapidly change the attack surface. This has implications for the organization's security policy, such as enforcement methods, device use regulations, and the operation and design of IS systems. Security risks involving data and system integrity can be effectively mitigated with blockchain.

Blockchain is an ever-growing list of linked records. These links are established and made secure through a cryptographic hash of the previous block. The data collected by IoT devices can be shared using blockchain. There is accountability for all entities interacting with the data because blockchain records are immutable and persistent. As a result, all entities that interact with the data can be granted the same level of access and visibility. Blockchain's distributed nature supports real-time action verification. All actions must be approved by a majority of entities before they can be undertaken. Communal verification of actions can limit the likelihood of improper actions being realized. Through the integration of blockchain, IS can benefit from the value provided by IoT devices while mitigating many of their security concerns.

IS creates value in organizations by providing an organized strategy for the creation and dissemination of information. As such, information systems require a level of openness to operate and provide value. The most secure systems are closed systems, as these systems operate in isolation and are therefore not vulnerable to threats. However, an IS that is designed as a closed system cannot provide value because such a system would be unable to collect the required data or send the information generated to relevant parties.

Therefore, security governance must consider an appropriate balance between openness and security. This balance is measured by the risk we accept to take. Generally, making an information system more secure introduces a risk of inefficiency due to the lack of openness. In cases where security measures are enacted, it is believed that the threat posed by an attack vector far outweighs concerns about efficiency.

Lessons Learned

The heterogeneous nature of the IoT introduces a myriad of security issues that need to be addressed. Adoption of this technology should not include the caveat that the people and systems using it are at a higher risk of attacks due to inadequate security management.

In certain application-specific sensing networks for the IoT, such as air quality monitoring, sensor modules can be attached to phones. This elevates smartphones from participatory devices to critical IoT nodes in these networks (Honicky et al., 2008) and increases the importance of ensuring their security. Developing participatory sensing platforms creates opportunities for low-cost sensing of the user's immediate environment. As a result, participatory sensing platforms provide the most accurate and timely indication of tested parameters as experienced by the user when compared to fixed infrastructure sensor networks. Participatory sensing is also more valuable than its fixed counterpart as it generates context-specific information (Campbell et al., 2006).

Heterogeneous networks cannot easily guarantee the security and privacy of their users. The IoT is inherently heterogeneous and as such, security and privacy are important considerations to make when deploying IoT devices. The Internet-connected objects that make up the IoT include vehicles, phones, appliances, and living things. These varied IoT objects can monitor and control physical objects. However, there is currently no common standard or architecture for IoT security (Al-Fuqaha et al., 2015).

IoT devices communicate wirelessly, have resource and computational constraints, and may be deployed in open environments. This makes the virtual and physical objects represented by these devices vulnerable to attacks from malicious entities (Roman et al., 2011).

Security management is "fundamentally important" in IoT environments (Bao & Chen, 2012). Secure things, devices, and network entities are the core components of IoT security (Mahalle & Railkar, 2015), and security and privacy models used for security management need to represent these components. In this way, it is possible to determine the security measures and attack mitigations that must be implemented. Appropriately addressing the security concerns surrounding IoT networks plays a significant role in the overall success of the IoT.

DISCUSSION QUESTIONS

1. What effect does security governance have on IS that use the IoT?
2. What effect does security management have on IS that use the IoT?
3. Why does the IoT pose unique security concerns for security management?

REFERENCES

Akyildiz, I. F., Su, W., Sankarasubramaniam, Y., & Cayirci, E. (2002). Wireless sensor networks: a survey. *Computer Networks*, *38*(4), 393–422.

Al-Fuqaha, A., Guizani, M., Mohammadi, M., Aledhari, M., & Ayyash, M. (2015). Internet of things: A survey on enabling technologies, protocols, and applications. *IEEE Communications Surveys & Tutorials*, *17*(4), 2347–2376.

Alkar, A. Z., & Buhur, U. (2005). An internet based wireless home automation system for multifunctional devices. *IEEE Transactions on Consumer Electronics*, *51*(4), 1169–1174.

Bao, F., & Chen, I. R. (2012, September). Dynamic trust management for internet of things applications. In *Proceedings of the 2012 international workshop on Self-aware internet of things* (pp. 1–6).

Campbell, A. T., Eisenman, S. B., Lane, N. D., Miluzzo, E., & Peterson, R. A. (2006, August). People-centric urban sensing. In *Proceedings of the 2nd annual international workshop on Wireless internet* (pp. 18-es).

Cook, D. J., Crandall, A. S., Thomas, B. L., & Krishnan, N. C. (2013). CASAS: A smart home in a box. *Computer*, *46*(7), 62–69.

Finch, E. (2001). Is IP everywhere the way ahead for building automation? *Facilities*, *19*(11/12), 396–403.

Gantz, J., & Reinsel, D. (2012). The digital universe in 2020: Big data, bigger digital shadows, and biggest growth in the far east. *IDC iView: IDC Analyze the Future*, *2007*(2012), 1–16.

Gea, T., Paradells, J., Lamarca, M., & Roldan, D. (2013, July). Smart cities as an application of internet of things: Experiences and lessons learnt in barcelona. In *2013 seventh international conference on innovative mobile and internet services in ubiquitous computing* (pp. 552–557). IEEE.

Gerla, M., Lee, E. K., Pau, G., & Lee, U. (2014, March). Internet of vehicles: From intelligent grid to autonomous cars and vehicular clouds. In *2014 IEEE world forum on internet of things (WF-IoT)* (pp. 241–246). IEEE.

Gubbi, J., Buyya, R., Marusic, S., & Palaniswami, M. (2013). Internet of Things (IoT): A vision, architectural elements, and future directions. *Future Generation Computer Systems*, *29*(7), 1645–1660.

Honicky, R., Brewer, E. A., Paulos, E., & White, R. (2008, August). N-smarts: networked suite of mobile atmospheric real-time sensors. In *Proceedings of the second ACM SIGCOMM workshop on Networked systems for developing regions* (pp. 25–30).

Hu, F. (2016). *Security and privacy in Internet of things (IoTs): Models, Algorithms, and Implementations*. CRC Press.

Komninos, N., Philippou, E., & Pitsillides, A. (2014). Survey in smart grid and smart home security: Issues, challenges and countermeasures. *IEEE Communications Surveys & Tutorials*, *16*(4), 1933–1954.

Kumar, P., Ranganath, S., Weimin, H., & Sengupta, K. (2005). Framework for real-time behavior interpretation from traffic video. *IEEE Transactions on Intelligent Transportation Systems*, *6*(1), 43–53.

Lin, H. E., Zito, R., & Taylor, M. (2005, September). A review of travel-time prediction in transport and logistics. In *Proceedings of the Eastern Asia Society for transportation studies* (Vol. 5, pp. 1433–1448).

Liqiang, Z., Shouyi, Y., Leibo, L., Zhen, Z., & Shaojun, W. (2011). A crop monitoring system based on wireless sensor network. *Procedia Environmental Sciences*, *11*, 558–565.

Mahalle, P. N., & Railkar, P. N. (2015). *Identity management for internet of things* (Vol. 39). River Publishers.

Manyika, J., Chui, M., Bughin, J., Dobbs, R., Bisson, P., & Marrs, A. (2013). *Disruptive technologies: Advances that will transform life, business, and the global economy* (Vol. 180, pp. 17–21). McKinsey Global Institute.

Michaelsen, K., Sanders, J. L., Zimmer, S. M., & Bump, G. M. (2013). Overcoming patient barriers to discussing physician hand hygiene: do patients prefer electronic reminders to other methods? *Infection Control & Hospital Epidemiology*, *34*(9), 929–934.

Roman, R., Najera, P., & Lopez, J. (2011). Securing the internet of things. *Computer*, *44*(9), 51–58.

Shafiq, M. Z., Ji, L., Liu, A. X., Pang, J., & Wang, J. (2012). A first look at cellular machine-to-machine traffic: large scale measurement and characterization. *ACM SIGMETRICS Performance Evaluation Review*, *40*(1), 65–76.

Talcott, C. (2008). Cyber-physical systems and events. In *Software-intensive systems and new computing paradigms* (pp. 101–115). Springer, Berlin, Heidelberg.
Taylor, S. (2013). The next generation of the Internet revolutionizing the way we work, live, play, and learn. *CISCO, San Francisco, CA, USA, CISCO Point of View, 12*(6).
Welbourne, E., Battle, L., Cole, G., Gould, K., Rector, K., Raymer, S., … & Borriello, G. (2009). Building the internet of things using RFID: The RFID ecosystem experience. *IEEE Internet Computing, 13*(3), 48–55.

Research Cases

1.10 Connected Vehicles
An Era of Communications Technologies, Cybersecurity, and Innovation

Sasha Pugh
Ryman Healthcare, Christchurch, New Zealand

Juliet Samandari and Hannah Yeo
University of Canterbury, Christchurch, New Zealand

INTRODUCTION

Market demand is driven by the value perception of customers; a value perception that often changes in accordance with the introduction of new technologies that improve a customer's life. Businesses must operate in a constant state of evolution if they wish to fulfill the value perception of their customers and retain a competitive advantage in these dynamic markets. The car manufacturing industry is a prime example of how businesses utilize new technologies, in accordance with their customers' value perception, to generate a competitive advantage by providing valuable products that are tailored to the needs of their customers.

Connected vehicles are the latest innovation in the automotive market. These connected vehicles have in-built wireless communication systems, such as Wi-Fi, Bluetooth, or visible light communication, with many connected vehicles offering multiple wireless communication options. Wireless communication systems, and indeed technology in general, are subject to risk. This risk stems primarily from the systems' vulnerability to cyberattacks. In the context of vehicle manufacturing, this increased risk is illustrated by the varied cyberattacks that have been launched against connected vehicles: cyberattacks that include tracking a vehicle's location, accessing information from the vehicle's infotainment system, and changing the limit on a vehicle's pressure sensor (Hayes, 2020). Therefore, having contingencies in place to mitigate the risk that these technologies pose is critical for businesses utilizing communication technologies in their products. By mitigating this risk, businesses can capitalize on the value of communication technologies for their customers.

DOI: 10.1201/9781003264415-13

The competitive strategy that automobile manufacturers use is differentiation, and integrating new technologies into their vehicles contributes to this differentiation. Through the integration of new technologies into their vehicles, automobile manufacturers hope to gain a competitive advantage, market share, and larger profit margins. General Motors, Ford, and Toyota have pledged to only sell connected vehicles by 2020, and other automobile manufacturers have pledged to achieve similar goals by 2022 (Hayes, 2020). Tesla, which is often considered the standard for connected vehicles, has integrated communications technologies into all their cars, and from mid-2018 has offered a premium level of this technology in their cars. This premium level is included with the purchase of the vehicle and provides constant connectivity over both the cellular and the Wi-Fi network (Tesla, 2020).

The process of integrating new technologies into a vehicle is rife with challenges, as each new feature added to a vehicle provides an opportunity for a hacker to try and access the vehicle's system. In addition, re-engineering vulnerable parts is likely to be a prolonged process due to the existing industry standards around the development of products. This is because there are no existing industry standards defining best practices for maintaining cybersecurity in automobiles. Furthermore, finding employees skilled in both cybersecurity and the automotive field is difficult. A 2019 survey by Synopsys and SAE International found that 63% of organizations test less than half of their technology for security. In addition, the survey also found that 30% of surveyed organizations did not have a group or program in place that focused on cybersecurity (Hayes, 2020).

The importance of testing security was demonstrated when Fiat Chrysler automobiles had to recall approximately 1.4 million vehicles to add remedial technology in 2015 (Hayes, 2020). This recall occurred upon the revelation that both nonessential functions, such as radio and climate control, as well as essential functions, could be controlled remotely. In fact, a test conducted on these vehicles demonstrated a hacker's ability to deactivate the brakes. On a larger scale, in 2018, a white hat hacker gained access to thousands of vehicles through a GPS tracker application (Hayes, 2020). This gave the hacker control of tens of thousands of connected vehicles. The hacker was then able to access the locations of these vehicles, cause automatic doors to open, and even turn off the engines of moving vehicles (Hayes, 2020).

Such an attack would not only affect individual cars but also the city in which these cars are driving. If only 10% of cars during peak hours are hacked, this percentage would be enough to prevent emergency vehicles from being able to reach their destination quickly through the resulting traffic as well as affect regular traffic flow at peak hours (Hayes, 2020). This highlights the importance of vehicle manufacturers providing high-quality security measures for connected and autonomous vehicles because the vehicles' security can greatly affect customer safety and customer perception of the vehicle manufacturer. This will have a significant impact on the company's brand image and reputation. Therefore, any competitive advantage gained by the addition of the technology will be lost if connected vehicles are not secure.

The technology used to connect cars is an area full of possibilities for future growth and for car manufacturers to differentiate themselves from their competition. Connected cars are becoming more prominent and this will change the way that we view vehicles as connectivity gives cars more capabilities. Connectivity also raises concerns about the privacy of connected vehicles, as GPS-enabled cars can provide a relatively easy way to track an individual's movements.

Connected vehicles are going to affect our lives in the future and this makes it an important and relevant topic to research. This is especially so since large vehicle companies such as Ford and Toyota aim to sell only connected vehicles by 2020, with other companies following suit in 2022 (Hayes, 2020). This means that connected vehicles may soon be the only automobiles available for people to use and drive every day. When the time comes that people will have to use connected vehicles, being able to trust and rely on the technology in connected vehicles to maintain safety for all individuals is critical. Customers trust those vehicle systems will keep their loved ones safe, making the safety of vehicles very valuable personally for consumers. If issues within this emerging industry are not addressed, there could possibly be costly impacts for society, companies, manufacturers, governments, and consumers.

CASE STORY

This case analysis discusses the growing phenomenon of connected cars in the vehicle market. Maeirhofer predicts that this market will grow by 270% until 2022 (2019). This is due, in part, to new legislation in the European Union (EU) requiring there to be an automatic emergency calling function installed in each vehicle, therefore requiring the cars to be connected. The market increase is also linked to the growing Chinese vehicle market that has been wanting new innovations, such as connectivity, in their products. The widespread use of 5G is also imperative for the growth of connected cars as this network allows for faster communication of more information (Maeirhofer, 2019).

Consumers and governments can see how connected vehicles add many benefits for their users as compared to a standard vehicle. As mentioned, with the new legislation in the EU, cars will be able to send out a distress signal if there is an emergency, such as an accident or a flat tire. This distress signal can be used to contact emergency services or the closest service center, depending on the situation. Another benefit is the advanced infotainment system which will allow users to access streaming services such as Pandora and Spotify, as well as the ability to access other applications and content through the car. Along with these applications, there is also the opportunity to have a digital wallet in an individual's car, so that the user no longer has to worry about forgetting their credit cards when leaving the house. This digital wallet can also be used to automate road tolls. Virtual assistants such as Siri or Alexa could also be added into the cars, allowing all infotainment functions in the car to be used hands-free. Another benefit is the advancement of the navigation system. A car could be linked to real-time traffic updates, the weather, and the owner's calendar. A connected vehicle is, therefore, able to consider current road conditions to inform the user when they should leave for an appointment and also what is the best way to reach their destination (Janakiram, 2016).

Another opportunity presented by connected vehicles is the ability to communicate with other vehicles, authorities, service stations, and roadside infrastructure. This communication makes possible improved safety and greater efficiency of vehicles on the road. A connected car sends periodic updates to a service center to ensure that the car is functioning as it should. A connected car's internal tracking system monitors the car's speed and drivers can use sensors to prevent speeding and/or collisions. There is also the possibility of monitoring the driver using an internal camera, thereby ensuring that the driver is not fatigued or distracted. Many new vehicles have an automatic shut-off mechanism that stops the vehicle's engine when the vehicle is at a standstill for a given period. This mechanism has been introduced to reduce the fuel consumption of the vehicle. However, the limited information that a vehicle uses to make this decision (i.e., if the given time has elapsed, then the engine is turned off) results in the shutoff mechanism often being suboptimal. If vehicles waiting at a traffic light are connected, and can, therefore, communicate with the traffic lights, these vehicles will know when they are stopping because a traffic light is red as well as when the lights are scheduled to change. As a result, connected vehicles can make more informed choices regarding when to invoke the shutoff mechanism and when to restart the engine, thus improving the overall fuel efficiency of the vehicle (Janakiram, 2016).

Many automobile companies have implemented different levels of vehicle connectivity, often with noticeable and appreciated improvements in usability and performance. However, there are also concerns regarding the security of a connected vehicle. The protections established for connected vehicles must consider the sensitive nature of the information that these vehicles store. Because the nature of vehicle connectivity necessitates the establishment of points of communication to and from the vehicle, these points of communication absolutely must not become points of entry that hackers can exploit to gain unauthorized access to a vehicle's systems and settings. In addition, legal and political groups must be significantly involved in ensuring that there are guidelines and regulations surrounding vehicle connectivity (Ismail, 2018).

This research case study will analyze the case of implementing connectivity in vehicles from five different perspectives: ethics, security, legislation, governance, and the use of emerging technology. Customer expectations for vehicle design and functionality have evolved to include vehicle connectivity and influence the value-adding activities that vehicle manufacturers must now support to meet customer needs. Also, vehicle manufacturers must maintain a competitive advantage with the higher threshold for differentiation that vehicle connectivity has introduced. In addition, the analysis includes consideration of the method by which vehicle connectivity is being governed.

ANALYSIS

ETHICS

Cars generate and store many types of data that can be turned into information such as safety statements, the use of vehicle functions, and personal information from the infotainment system. Unbeknownst to the average user, this information has been collected for years whenever the vehicle has been serviced via the vehicle's diagnostic port. With the introduction of connected vehicles, this information is more easily accessible because many car manufacturers have established online data transfers for their connected vehicles. In addition, dealers may retain legal permission to access the data generated by vehicles they sell by including a clause to this effect in the sales contract signed by the buyer. The European Commission has also made it mandatory for cars to have a SIM card installed that can be used to send information to emergency services, thereby increasing the connectivity of cars in the EU (Maeirhofer, 2019).

Authors of a 2017 study by KPMG of vehicle manufacturers found that 67% of the participating organizations either did not perform analytics on the vehicle data they collected, due to a lack of infrastructure to support analysis, or they simply did not have access to the data generated by their vehicles. Implementing specific tools for gathering data remains quite expensive. To then turn this data into information car manufacturers are able to sell is also difficult (KPMG, 2017). This begs the question of where car manufacturers will sell this information, and to whom. The processing and eventual sale of data may not be taking place now, but this seems likely to occur in the near future; especially because increased connectivity of vehicles will see increases in the amount of data generated and the accessibility of this data. Currently, the data that is collected could be the car's location, black box data, vehicle status, and service status. This data is used in contexts such as: identifying when the car next needs to visit a service station, discovering the location of a car requiring roadside assistance, or helping law enforcement ascertain what happened after a car accident takes place.

The consequentialist system of ethics places the utmost importance on the outcome of actions, thereby determining that actions resulting in favorable outcomes are ethical (Sinnott-Armstrong, 2019). When one considers this scenario from a consequentialist ethical viewpoint, the availability of the personal and vehicle data mentioned above is ethical because this availability results in a favorable outcome for both the user and the manufacturer. The owner of the car can resolve issues or receive assistance more efficiently. This is also favorable for the manufacturer, as this added feature is likely to increase satisfaction with the product since vehicle owners see the collection and use of this information to be of value. This does, however, assume that the information that is collected is used solely for the purpose of improving and adding value to the user's experience when using connected vehicles.

The deontological ethical viewpoint is where a set of ethical rules and frameworks is established and all actions must uphold these rules to be deemed ethical (The Ethics Centre, 2016). From a deontological viewpoint, data collection from vehicles is unethical because most users are unaware that this data is being collected and that vehicle manufacturers can access this information. Using the deontological ethical viewpoint, vehicle manufacturers are obligated to inform users when they collect data, including the types of data collected and how they are being used. Vehicle manufacturers should also provide customers with the ability to opt out of sharing vehicle data if they so wish, thereby preserving every individual's right to privacy.

If the driver or passenger's cell phone is connected to the vehicle, information such as call logs and contact lists can be stored in the car and therefore accessed. This information is likely not something the cell phone's owner would want to be accessible to the vehicle manufacturer. As such, this situation brings up the issue of whether car manufacturers having access to the driver or passenger's personal information is ethical if it is stored in their car. From a stockholder perspective of ethics, where gaining profit is the most important element for stockholders, actions are considered ethical if they meet the desired outcome of generating profit (Pearlson & Saunders, 2013). Therefore, collecting information from connected cell phones is ethical if vehicle manufacturers remain within existing legal frameworks. This is ethical because this information, used effectively, could increase profit for the company in many ways. For instance, cell phone information could be used to customize the experience of the user when they are in the vehicle. This information could also tailor a vehicle manufacturer's advertising by identifying connectivity features which individuals like and use in their cars. The cell phone information could even be sold to third parties that value this information because they can use it to create more comprehensive advertising profiles for individuals.

In 2011, a hacker was able to gain access to a police car's video monitor in real time. The hacker could also view, download, and delete stored video recordings, as well as upload other recordings. This is quite worrying, as there would likely be many confidential topics discussed in police vehicles. In the context of social contract theory, where the outcomes of action must be in the best interests of society for that action to be ethical, for individuals outside law enforcement to have access to these recordings is not ethical (Pearlson & Saunders, 2013). The recordings should remain confidential because providing access to sensitive information does not benefit society and its value of privacy. From the viewpoint of stockholder theory, this scenario and the disclosure of this vulnerability will damage the vehicle manufacturer's reputation, causing company profit to drop. Therefore, stockholder theorists would consider the accessing of data from the police vehicle as unethical.

Security Risks

Connected vehicles have many associated security risks. These risks stem from accessing the vehicles physically, remotely, or through applications that interact with these connected vehicles. For example, cars that have keyless entry can be unlocked by mimicking the signal transmitted by the car's key. If the car is parked outside a house and the key is located near the entrance, as it often is, one individual can stand near the car while the other stands near the front door. The individual near the car sends the car's signal to the second individual who then transmits this into the house. The car key then responds with its own signal, which is then passed back to the first person who sends this key signal to the car. This allows them to gain access to the vehicle by fooling it into believing the key is nearby. This is referred to as a relay attack and is relatively easy to implement provided the individual is moderately knowledgeable. A relay attack is also quite inexpensive: this attack has been launched successfully, using two devices created at under $11 each by a research group, against two different cars and is also likely to work on many more models as many cars use the same chips for their keyless entry system (Greenberg, 2017). Devices to carry out this relay attack can also be bought online for just over $100 from sites such as eBay and Amazon, with some of these devices continuing to be effective from over 100 meters away (Evans, 2020).

Another security risk is the ease of gaining access to a vehicle's engine control units (ECUs). Most hackers aim to access ECUs in order to gather information from the vehicle or to obtain control of it. The vehicle's ECUs is a computer component that takes data from vehicle sensors and uses it to adjust the components controlling the engine's physical mechanisms. All vehicle data goes through the controller area network (CAN) and this network is used for communication between the different components in the car. This means that vehicle elements, such as the buttons for windows and airbags, have a path of communication. Therefore, if a hacker gains access to the CAN, they can access and control almost every aspect of the car, including tampering with locking systems and disabling airbags (Osborne, 2018). Once the CAN is accessed, all the hacker requires to launch a successful attack is the ability to alter a CAN data frame and its fields (Hayes, 2020). This was carried out on a Tesla Model S in 2016. A paper published by Keen Security Lab of Tencent outlines the methodology and code used to gain access to the Model S's CAN through the vehicle's Wi-Fi connection. Once Keen Security accessed the CAN, the lab used it as the gateway to other systems within the vehicle (Nie, Liu, & Du, 2017).

Security risks in connected vehicles can even arise from the simplest connections. Two outsourced GPS services, installed in connected vehicles under the names of ITrack and Pro Track, made vehicle accounts with default passwords that owners of these connected vehicles did not change. This allowed a hacker to access over 27,000 accounts between the two GPS services; access that could have allowed the hacker to turn off the vehicle's engine if it was going below a set speed of 12mph (Lemos, 2020). As well as mobile applications, cloud-based automotive services can also provide an entry point to the vehicle. These cloud-based systems often do not have the same level of security as the car itself, making them the easiest point of access into the car's control system.

When evaluating security, the three main elements to consider are confidentiality, integrity, and availability. In the case of connected vehicles, the most prominent of these elements would be concern over confidentiality. The security breaches show that there have not been sufficient processes put in place to prevent access to the car.

One strategy to try and minimize what is accessible through this vulnerability is the use of a hypervisor. A hypervisor is a software that allows multiple processes to be run within virtual machines, a reproduction of a computer system. This reduces the points of entry, and therefore vulnerability, and improves the overall security of the system since the processes running the elements in the car (i.e., the braking system, infotainment system, airbags) are now running separately in different virtual machines. Because there is no direct connection between the virtual machines, using a hypervisor separates vehicle elements and prevents hackers from using connections from a compromised element to access other elements in the vehicle. Most importantly, using a hypervisor provides a better separation of critical elements from the infotainment system. A hypervisor also allows different operating systems to be run simultaneously on a vehicle's central hardware. Therefore, multiple processes that rely on varying operating systems can now all be run on a single physical machine. This could allow older, more reliable operating systems to be used for critical functions while a newer operating system can be used for noncritical operations (Hayes, 2020).

Governance

Governance, in terms of technology, is the study and allocation of decision rights that define expectations, grant authorities, and ensure performance. As such, decision rights contribute significantly to the governance of technology in connected vehicles. An analysis of governance and decision rights would not be complete without the inclusion of a discussion regarding organizational structure. This is because the organizational strategy of a firm can determine the entities that are governing, monitoring, and making decisions about technology.

One way to consider the governance of connected vehicles is by using agency theory (Pearlson, et al., 2015). Agency theory refers to the most efficient distribution of risk in the firm. In the ideal case, decision-makers should be held responsible for all decisions: those that result in favorable outcomes, as well as the ones that suffer poor consequences. A problematic method of governance is when If decision-makers are not accountable for their actions. Agency theory proposes that conflicts of interest between the owners of the firm and its employees result in an "agency problem". Within vehicle manufacturing firms, the agency problem could occur if the owner of the firm finds value in creating a reputation of their vehicles being safe and secure, while the firm's employees may simply value their paycheck, thus creating a conflict of interest. Alternatively, a firm could experience a case where the employees value the security and safety of the connected vehicles they are producing, while the owner values cost reduction and speed of manufacturing. The owner will therefore cut costs and may eliminate developing security features in favor of quickly getting the vehicle to market.

Companies, such as vehicle manufacturers, also establish governance of their products through their organizational structure. Consider the IT department in a company with a centralized organizational structure that brings all staff, hardware, software, data, and processing into a single location. This centralized structure is desirable because it allows companies to achieve higher levels of governance maturity and control (Pearlson, et al., 2015). For a vehicle manufacturer, this centralized structure would, therefore, be the best structure to implement, as opposed to a decentralized structure, as the IT department would play a considerable role in configuring the technology and security required to support connected vehicles.

The organizational structure of an IT department greatly affects the allocation of decision rights in a company. This is important because the IT department is most likely in charge of cybersecurity for connected vehicles. Having a centralized organization structure creates clear, definitive roles within departments and companies, outlining the allocation of decision rights that are related to security in connected vehicles (i.e., the level of security that must be upheld, how to maintain security, etc.). Using a centralized structure also makes it easier to establish a system of accountability and responsibility that can be used to manage the consequences of decisions that are made.

Companies in the vehicular industry have started to increase their investments in governing the security risks associated with connected vehicles (Lohrmann, 2016). One method that companies in the vehicular industry are employing to address these security risks is utilizing bug bounty programs (Lohrmann, 2016). A bug bounty program is a company-hosted event, or program, where hackers and researchers are invited to purposefully launch cyberattacks against the company's latest connected vehicle models (Lohrmann, 2016). The purpose of the event is to find any bugs, loopholes, or security risks that are unknown to the company and malicious actors could exploit (Lohrmann, 2016). If an event participant is successful in their cyberattack, they are usually paid a generous bounty reward because they have done the company a favor by discovering a weakness in their vehicles. Companies that use bug bounty programs have experienced positive outcomes. The programs identify flaws in connected vehicle systems and demonstrate an organization's commitment to maintaining a high level of security in their vehicles, an element of connected vehicles that is especially valued by customers. Publicizing bug bounty programs also allows customers to gain trust in participating connected vehicle companies. These programs make customers feel safer because they can see the value the company places on safety, as demonstrated by the company openly taking preventative measures to maintain the safety of their vehicles. This strengthens the company's image. However, the success of bug bounty programs relies on the participation of skilled hackers and researchers. Companies must promote an appealing bounty reward if they wish to attract highly skilled individuals. If a company does not post an appealing reward and the bounty program only attracts more novice individuals, the program could be unsuccessful in cases where more novice individuals are unable to find weaknesses that a highly skilled attacker could exploit.

Another aspect of these bug bounty programs is clarifying what outcome, in terms of finding security issues, is deemed a success. From the company's point of view, a program could be considered successful if no security issues were found. This is because such a condition implies that there are no security issues with the company's connected vehicles. However, this condition could also imply that none of the hackers involved in the program were skilled enough to find the security issues in the vehicles. Consider a company that uses bug bounty programs as their only method of governing the security of their connected vehicles. There is the possibility that the vehicles are later found to have bugs that were previously unknown. This illustrates the need to use multiple methods of governance, rather than relying solely on one method, to ensure that a company's approach to governance is comprehensive.

Alternatively, if bugs and issues were found in the company's connected vehicles this raises questions whether their vehicle's current security measures will be noted as unsuccessful or successful. Previous bug bounty programs, where bugs and security issues have been found, have been deemed a success because the programs allowed companies to eliminate flaws and improve security measures in their vehicles.

Usually, bug bounty programs are an effective method of self-governance that companies can use. However, companies should not rely solely on the outcome of bug bounty programs. Companies must also have a second governance method to double-check the security of their connected vehicles before manufacturing.

External parties, such as the UK government, are also developing many projects around connected and autonomous vehicles (Centre for Connected and Autonomous Vehicles, 2018). One of these projects, i-Motors, focuses on standardizing methods of communication with connected vehicles, storing of collected data, and interaction between connected vehicles and other systems (Centre for Connected and Autonomous

Vehicles, 2018). Another important development is the 5Star: Automotive Cybersecurity Through Assurance project (5Star, n.d.). This funded project is researching an assurance framework for assessing the cybersecurity of vehicles. Other projects include UK CITE, a real-world lab that supports testing the interaction between connected vehicles and communications infrastructures, improving Internet of Things (IoT) and Automated Driving Systems (ADS), researching emissions produced by connected vehicles, and developing technology to support connected vehicles. For vehicle manufacturers to have clear and defined methods of governance is important. Equally important is the contribution of external parties, such as government agencies, in defining methods of governance for an automobile industry that is integrating new communication capabilities that are subject to many challenges.

Some companies do not even test the security in their vehicles before sending them to the market (Hayes, 2020). This is due to an organizational structure that does not account for security testing of vehicles, a lack of defined decision roles, and the limited or nonexistent cybersecurity and technology knowledge of employees and decision makers. The connected vehicle industry has developed quickly, with established automobile companies rapidly modifying the functions offered by their vehicles to keep up with changing trends and evolving customer demands. Within this culture of rapid change in the market, automobile companies have often rushed to bring vehicles to the market with the aim of capitalizing on current market trends. Owners, manufacturers, and individuals responsible for decision rights need to understand the technologies that are being integrated into vehicles. Automobile companies can then establish best practices for governance that prevent vehicle recalls, reduce security flaws, eliminate bugs, and minimize the potential for connected vehicles having a negative impact on society.

Emerging Technology

Consumers have come to expect innovation in the products they purchase. Companies must constantly update the products they offer to cater to this consumer expectation. Therefore, a way to generate value for products is by investing in and implementing emerging technologies. However, companies must carefully and strategically identify when integrating emerging technology into their products is appropriate. A company must use new and innovative technology to stay on the leading edge. However, a company should never find itself on the "bleeding edge" of technology, where the use of a technology that is novel, and therefore unreliable, incurs a significant cost to the company in order to implement the technology correctly. If a company implements an emerging technology before it is properly tested, then the technology does not add value to the product but rather can devalue the it.

Many consider connected vehicles to be the steppingstone to autonomous vehicles. Autonomous vehicles are the first phase in the introduction of greater automation in our everyday spaces (Koul, 2020).

The vehicular industry and wise consumers expect companies that can implement secure, connected vehicles to have the ability to produce cars with more automation that are also secure because these companies have established the infrastructure and skills that will support the successful design and implementation of a secure autonomous vehicle. Security is always a major concern regarding emerging technology; autonomous vehicles need to be reliable and secure (Spencer, 2019).

Vehicle manufacturers wanting to improve their autonomous systems through machine learning need a wide variety of samples of cars interacting with the road in distinctive situations, such as if an individual jaywalks in front of the car or how to parallel park. With connectivity between vehicles and their manufacturers, data from all situations the car is in can be shared. As connected vehicles become more prevalent in all parts of the world, the likelihood of unique data being collected increases. This would result in more valuable data being collected. Tesla is an example of a car manufacturing company that has been able to collect unique data due to the widespread use of its vehicles.

There is also the possibility to integrate the consumer's personal driving style into the vehicle's automation. This allows the vehicle to be customized to give an agreeable experience to the vehicle owner according to their own personal preferences. This will be valuable to consumers; for consumers to find autonomous vehicles appealing, they must be safe, reliable, and have a comfortable user experience (Kuderer et al., 2015).

Connected cars will be able to share information with each other, such as how the conditions are for driving. This will be an important element for implementing autonomous vehicles as they will require a safe and reliable form of communication with other vehicles in order to drive safely. There is also the possibility that connected vehicles could communicate with roadside infrastructure, such as traffic lights. In this case, the traffic light will know when a car is approaching, as well as how many cars may be waiting at each side of an intersection. Vehicles could also be informed of when a light is about to turn green in order for the vehicle to conserve fuel. This would be implemented by having sensors attached to roadside infrastructure that can then send information to the vehicles (Ismail, 2018).

There are also opportunities for vehicles to share the information they may gather regarding the condition of the road. Vehicles could be connected to a central facility that is responsible for the maintenance of roads. In this way, the vehicles will inform the road maintenance facility if there are areas that need repairs, such as lane markers that need to be redrawn or signs that have been damaged. This will be especially important as autonomous vehicles will rely on road markers such as these. Communication between vehicles and maintenance facilities has the possibility to reduce the number of accidents that occur, as connected vehicles could now inform other vehicles if there are dangerous road conditions (Ismail, 2018).

However, companies need to be aware that consumers may not find all technologies that are added to connected vehicles appealing. Depending on the nature of the technology that has been added, such as its susceptibility to attack or its ability to introduce efficiencies to the user experience, consumers will either be disillusioned by the technology or will find the technology to be a value-adding element of the vehicle. If consumers do not see the value in the technology that has been added, this addition becomes a waste of resources. This is where ethical handling of user information can ensure the success of emerging technologies developed for connected vehicles by adding technologies that manufacturers know consumers will want to purchase.

Connectivity in cars heralds the launch of major changes in the functionalities that cars provide. Connectivity introduces many valuable additions that make the use of vehicles safer and easier for customers, with many opportunities for these additions to be developed further. However, simply implementing new technology will not necessarily add value to the vehicle for consumers. Vehicle manufacturers need to be aware of what consumers find valuable to implement these value-adding functionalities in their vehicles. Put simply, connected vehicles must maintain the value-adding tenants of safety, cybersecurity, and a user-friendly experience if they wish to satisfy consumer needs.

LEGALITY

With the increasing number of connected vehicles and the emergence of autonomous vehicles, the risks related to vehicle cybersecurity and driver data privacy are yet to be sufficiently covered by the legislation. In 2015, the Security and Privacy in Your Car Act (SPY Car Act) was introduced by the US Senate with the aim of protecting US consumers from security and privacy threats to their vehicles. This bill required any vehicle manufactured for sale to be protected against the misuse of the vehicle's ECUs or its driving data, including information about the vehicle's location, owner, driver, or passengers (S. 1806, 2015). Practically, this means that vehicles with accessible data or ECUs should have the capabilities to detect, report, and stop any attempt to steal driving data or interfere with the control of the vehicle.

In terms of preserving the privacy of driving data, the SPY Car Act of 2015 proposed limits on what manufacturers can and cannot do with the driving data they collect from consumers. The bill also outlined consumer rights to driving data. Translating this into vehicle requirements means that vehicles must be capable of notifying consumers of the collection and use of their driving data. In addition, vehicles must provide the option of terminating this data collection.

There were two readings of this bill. However, the Senate did not pass this introduced legislation on vehicle cybersecurity and privacy. Another attempt was made two years later by the same US senator to pass the SPY Car Act of 2017, which essentially was a slightly revised version of the SPY Car Act of 2015. Unfortunately, this bill also did not make it past the second reading.

In like manner, other countries have attempted to introduce legislation that ensures manufacturers are incorporating cybersecurity and data privacy practices into their vehicles. For example, the UK government issued guidance principles in 2017 that compel manufacturers to develop connected vehicles that are well protected from cyberattacks (GOV.UK, 2017a). These guidance principles are not only intended for manufacturers but for all parties in the automotive industry, to make sure that cybersecurity is properly considered at every stage of the vehicle's life cycle. Of the eight principles listed, the most notable ones required for vehicle systems to be designed resilient to cyberattacks and able to appropriately respond when its defenses fail, as well as the storage and transmission of driving data to be secure and able to be controlled by the consumer (GOV.UK, 2017b).

Since cyberattacks on connected vehicles have become an issue in recent years, legislation is still in its beginning stages. However, taken together, the introduced bills of the SPY Car Act in the US and issued guidance principles in the UK are effective first steps in establishing much-needed industry standards. An important addition to these first steps is to not only require manufacturers to abide by but to ensure compliance with the attachment of civil penalties to violations of not meeting those requirements for connected and autonomous vehicle cybersecurity and driving data privacy (Bollinger, 2017).

On the other hand, regulators in the automotive industry are in fact preparing minimum standards for vehicle cybersecurity. In 2020, the expectation is that the World Forum for Harmonization of Vehicle Regulations under the United Nations Economic Commission for Europe (UNECE) will finalize its vehicle cybersecurity regulations (Deichmann et al., 2019). This means that cybersecurity will be made a definitive requirement for the sales of future connected vehicles and manufacturers will have to consider cybersecurity as an integral part of their core business processes. These regulations are believed by industry experts to mark the start of a new era for significant security standards regarding connected car cybersecurity; standards that are to be officially regulated (Katz, 2020).

A possible explanation for why there are bills not being passed and the lack of regulation by laws in many countries could be due to these governments and lawmakers not knowing of or understanding the technology within connected vehicles. Due to the lack of understanding, there is a lack of urgency regarding how significantly this technology could affect society and the lives of connected vehicle users.

LESSONS LEARNED

ETHICS

If vehicle manufacturing companies want to carry out ethical practices, they must be more transparent about the data they are collecting as well as what they are planning on doing with this information. They should also be clear regarding the options that are available for vehicle owners if they wish to opt out of certain aspects of data collection.

Vehicle manufacturers need to understand that having the added convenience that comes with the data being gathered and then shared using connected vehicles only gives an advantage in the current market if individuals trust vehicle manufacturers to have ethical practices for how they collect data, as well as how they use it. In general, consumers adhere to a more consequentialist school of thought. Consumers are, therefore, generally of the opinion that the collection and use of their information are ethical, so long as there are positive consequences for them. This makes it important for car manufacturers to be able to make use of this information in a way that provides a better experience for the consumer than they would have had without this information being collected. However, there do need to be options that can cater to privacy-concerned consumers, so they have a way to still use a connected vehicle while limiting the data that can be collected about them.

A stockholder perspective would consider this collection of data to be ethical as it will result in an increase in profit. However, social contract theory would be less clear as to the ethics of data collection from connected vehicles. The ability to customize a vehicle, especially in the context of an individual who is willing to give their information to have a better experience, can be seen as a benefit to society as it helps these users. However, other users who are unhappy about the collection of their data will view this data collection as a detriment to society and therefore an unethical practice.

Currently, there is no governance surrounding the collection of information and for what this information is used. The introduction of this governance would likely increase ethical practice if vehicle manufacturers were to try and uphold the given standards of governance; especially because businesses that work within a stockholder framework mainly focus on the gain of profit and the only other element that is considered, when ascertaining the ethical nature of their business practices, is the legality of their actions.

If car manufacturers continue to keep the data collected proprietary to their organization, this proprietary data is likely to result in a noticeable competitive advantage once the manufacturers have learned how to use the data to generate valuable information. This information will allow vehicle manufacturers to tailor their vehicles according to customers' preferences. Using the information, manufacturers could also track the applications and functionalities that customers interact with frequently. Vehicle manufacturers could, therefore, increase the value of their vehicles by making these functions easily available.

This case highlights the importance of providing clear benefits from data collection and information generation to customers. For vehicle manufacturers, keeping their customers apprised of the information being collected, and to perhaps allow for opt-out options if customers do not wish to have their information collected is important. Data collection is generally considered ethical, and therefore acceptable, by users when they feel that value is gained from the sharing of their data. If there is trust that the vehicle manufacturing company uses shared data to improve the user's experience, users will feel more comfortable having their data collected. Vehicle manufacturing companies being aware of their own ethical framework is important, as well as the ethical framework their customers generally subscribe to, to align business processes with these frameworks with the goal of providing value to the business and its customers.

Security Risks

Connected vehicles have not been implemented in a way that mitigates the many security risks inherent to the use of a wireless connection. For connected vehicles to have a lasting appeal, customers will need assurance that their car is secure. When keyless entry was first introduced, it was a valuable addition that increased the vehicles' ease of use. As a result, customers were interested in purchasing new car models with this feature. However, keyless entry has become an undesirable feature as the security risks, and the ease with which individuals can hack into cars with keyless entry, have been made known to users. Therefore, keyless entry is no longer seen as being a valuable addition to a vehicle.

One of several steps taken to try to mitigate risks is the use of a hypervisor solution. While this is a promising development, as using a hypervisor reduces vulnerabilities, it does not address methods to protect the vulnerabilities that still exist in connected vehicles. Implementing a hypervisor simply reduces the points of entry to a vehicle's system. It does not implement protections for the vulnerable points of entry that remain.

Being the first vehicle manufacturer to have a connected car available for consumers or to have a proprietary connectivity implementation in its vehicle, will only provide a competitive advantage if viewed to be an addition that is of value to customers. If connected vehicles are not secure, the value of any added functionality will not outweigh the associated privacy concerns.

Vehicle manufacturers have not displayed an understanding of the technology that they want to implement and have subsequently left the related security risks unaddressed. To start solving security issues that may arise, vehicle manufacturers need to be aware of the factors that can cause security issues. This can be done through the integration of cybersecurity into the manufacturing process of vehicles, either by creating an in-house cybersecurity department or by outsourcing this aspect of production. Making cybersecurity a part of the manufacturing process will give cybersecurity the same importance as the other elements of the vehicle. This increased focus on the area of cybersecurity will likely improve the security of connected vehicles because the main cause of security risks is the lack of testing and cybersecurity knowledge in vehicle manufacturing companies.

Increasing the importance of cybersecurity in vehicles would cause shifts in a vehicle manufacturing company's processes. For instance, a vehicle would not only be tested for the function of the car itself but also for the robustness of all its technological components and their security. This would generate value as the integration of technology testing into production would increase the reliability of vehicle connectivity and would also create assurance for customers that the vehicle manufacturer is upholding a satisfactory level of security for all elements of the vehicle. This process would add value to the vehicles and would likely give an advantage to vehicle manufacturers that are implementing the process, over those that have not.

Governance

Bug bounty programs have proven to be a successful method for connected vehicle companies to govern themselves and the products they manufacture. Companies such as Tesla, Chrysler, and General Motors have been actively participating in bug bounty programs since 2016, and these are a proven method for vehicle manufacturers to self-regulate and improve the security of their vehicles.

The creation of these programs allows connected vehicle companies to monitor and find the security issues in their products while reducing the resources (i.e., money, time, employees) that are spent on this. Although these programs do offer a bounty to those individuals who find bugs and security issues, the monetary reward given is less than the costs involved should these companies support a specialized in-house team focused on finding security issues. Publicizing the bounty programs is not only a method to find security risks; it also promotes the company's brand image and vehicles. The one cost of the bounty reward not only pays for finding security issues in the company's vehicles but also for marketing these vehicles.

Additionally, within vehicle manufacturing companies, there need to be defined roles assigning the responsibility of decision rights associated with the cybersecurity of connected vehicles. Decision rights need to be addressed so responsibility can be taken for both positive and negative consequences of decisions.

The organizational structure within vehicle manufacturing companies can also influence how to internally govern the technologies used by connected vehicles. Having centralized structures for IT departments has proven to be the most common and successful organizational strategy for supporting governance. Centralized structures allow ease of access to resources while also defining roles and decision rights within an IT department. Vehicle manufacturers should consider changing their organizational structure to a centralized one, or at least changing the structure of their IT department to be centralized because connected vehicles incorporate a lot of technology. Their success relies heavily on the effectiveness of the IT department. Addressing the organizational structure of a business and its departments, as well as defining decision roles, resolves the agency problem in the company and supports successful outcomes for the business's products.

If vehicle manufacturers develop a deeper understanding of the technology used in their connected vehicles, it would lead to more informed decisions being made within the organization and would enhance the performance and success their vehicles would have in the market. An excellent grasp of the technology would allow decision-makers to understand why staff within the IT department have made decisions while providing these decision-makers with the opportunity to make sure the IT department is working according to company policies and standards. Ultimately, this increase in knowledge simultaneously helps the business and its products long term because decision-makers can make more informed decisions that benefit their company and generate value for their connected vehicles.

Emerging Technologies

The lessons learned for vehicle manufacturers regarding emerging technologies are significant. In the race to gain a competitive advantage by integrating new technologies into vehicles and supporting the development of connected vehicles, some vehicle manufacturers have paid the price for their hastiness with product recalls. The vehicular industry is rapidly developing with vehicle manufacturers identifying opportunities to gain a competitive advantage, expand market share, and increase profits by integrating new technologies. However, this is often done without taking the time to ensure that these technologies meet customer needs, are not subject to cybersecurity risks, and maintain the safety of their vehicles. Vehicles must achieve these criteria to be successful in the market.

There are a lot of opportunities in the connected vehicle market for success. Vehicle manufacturers need to slow down their development processes to ensure the success of these new technologies in the market. To recall, fix, and revise technology that is developed too rapidly will be more costly for vehicle manufacturers than to implement processes that take longer and result in the correct development of technology. If a vehicle manufacturer is to introduce a new technology that has been carefully considered and developed, the gain in market share will be significantly higher than if the technology is developed quickly and with flaws.

Legal

From the analysis of the legal development and regulations of connected vehicles, the UK is a leading example of putting legislation in place that will cause manufacturers to design out cybersecurity threats in their development of connected vehicles. The successful issue of the UK government's guidance principles puts pressure on both developers and manufacturers regarding the approach of their business processes for vehicle systems. These guidance principles also keep them accountable in the case of cyberattacks. In other notable attempts, the US has made positive steps to pass the SPY Car Act addressing the business, cybersecurity, and privacy issues associated with connected vehicles. Unfortunately, even after the bill was revised and reintroduced a second time, neither has successfully made it past the draft stage.

A lesson learned, through their response in introducing legislative drafts and issuing principles, is that governments have been made aware of the cybersecurity concerns surrounding connected vehicles. However, the government's understanding of the technologies used by connected vehicles remains underdeveloped. Because governments are unaware of the full scope of requirements necessary to maintain the cybersecurity of these technologies, these authorities do not heavily mandate the introduced legislation. The current legislation introduced in the UK for one, although successfully issued, only focuses on guidance for developers and manufacturers to incorporate cybersecurity protections. These principles do not consider other aspects of the cybersecurity of connected vehicle technologies. For instance, they make no mention of the need to monitor appropriate data collection, enforce the ethical use of information, protect data privacy, or ensure legal consequences for cyberattacks.

Legislation is required to govern how vehicle manufacturers implement connectivity in their vehicles. Legislation also provides accountability if vehicle manufacturers neglect to do security analysis when integrating cybersecurity protections into their connected vehicles, as this negligence on the part of the vehicle manufacturer could lead to cybersecurity breaches. Legislation is also required to impose restrictions on the accessibility of vehicular communications technologies and to discipline individuals or companies that misuse their access to vehicle information or controls that are available through the vehicle's connected systems.

The rollout of UNECE's vehicle cybersecurity regulations in 2020 sees the introduction of cybersecurity regulations that will pave the way for further legislation to be introduced by industrial regulators and governments for the purpose of maintaining adequate security standards (Katz, 2020). If governments recognize the many risks related to the implementation of connectivity in vehicles, they will likely be more active in introducing new legislation to oversee the management of these risks. This active management would involve stringent regulations around the cybersecurity and privacy of connected vehicles. With the use of regulations such as UNECE, governments can implement informed legislation that is better suited to addressing the individual and societal concerns and effects of connected vehicles.

This research case study shows the value to society of the future development of connected vehicles as long as it is an ethical, highly secure, legally regulated, and regularly governed industry. Most countries that provide vehicle manufacturers with high standards should introduce additional laws and globally recognized standards to which manufacturers should comply. For the continued success of connected vehicles, a collaborative effort involving governments, vehicle manufacturers, and consumers is needed to create regulations that guarantee consumer safety and security when using connected vehicles while ensuring that vehicle manufacturers have the best intentions when producing these vehicles for the market and that they are held responsible for their decisions.

CONCLUSION

This research case has analyzed the security risks that have occurred in vehicles due to the addition of vehicle connectivity. The research case considered this issue from several perspectives: the ethics of privacy, security concerns, governance, the novel technology being used, as well as legislation to cover the issue.

Clearly, connected vehicles will rapidly become more prevalent in society. For integrating new technologies to be a successful business strategy, ensuring that the added features afforded by connected vehicle systems are secure is imperative. This case study identified areas where vehicle manufacturers can be successful in implementing secure connected vehicles, as well as highlighted areas where improvements can be made to maintain the security of connected vehicles.

There are several lessons to be learned from our analysis that one should take into consideration by connected vehicle manufacturers, governments, and lawmakers. These lessons consist of the need for vehicle manufacturers to follow practices that support the ethical use of consumer data, establish an increased cybersecurity focus, provide clearly defined decision roles, and cultivate an in-depth understanding of the technology being implemented to identify the testing methods that are most appropriate. There are merits of adopting a centralized organizational structure to support the lessons listed above and increasing lawmakers' understanding of the technologies used in connected vehicles and their associated concerns, for the purpose of implementing appropriate legislation to address these concerns. If these considerations are dismissed, vehicle manufacturers will face a loss of profit, a reduction in market share, and a damaged reputation with instances of vehicle cybersecurity breaches likely to rise. If the vehicle industry and its various companies adhere to these recommendations and lessons, the benefits of establishing a long-term competitive advantage, increasing market share, expanding the profit margin, and developing a trustworthy and reliable reputation will follow. In addition, because connected vehicles are an in-demand product for consumers, the connected vehicle industry has a significant potential for growth.

In the future, vehicle manufacturers, as well as governments globally, will have to create and maintain up-to-date laws and regulations for connected vehicles to ensure that vehicle manufacturers are upholding high standards of cybersecurity and privacy. This will help prevent security issues as well as enhance safety for customers and society.

Connected vehicles are going to become more prevalent in the future, and for consumers, vehicle manufacturers, and legislative bodies to fully understand the risks involved is important. Only then will manufacturers be able to implement the necessary safeguards in the vehicles themselves and for there to be suitable legislation in place, including options for users to customize their privacy and security settings. This will allow connected vehicles to be used while the associated security risks are minimized, resulting in the value that is added by maintaining connectivity in vehicles.

DISCUSSION QUESTIONS

1. What ethical challenges do connected vehicles pose for society at-large?
2. How would you address the ethical challenges in #1 above?
3. In which direction(s) do you believe connected vehicle technology will evolve over the next 5 years? 10 years?
4. What will have the most influence on vehicle manufacturers' attention to security? Explain.

REFERENCES

5StarS. (n.d.). What Is 5StarS? https://5starsproject.com/

Bollinger, B. (2017, April 1). The security and privacy in your car act: Will it actually protect you? *North Carolina Journal of Law & Technology*, 18(5), 242. https://scholarship.law.unc.edu/cgi/viewcontent.cgi?article=1337&context=ncjolt

Centre for Connected and Autonomous Vehicles. (2018). UK connected and autonomous vehicle research and development projects 2018. https://assets.publishing.service.gov.uk/government/uploads/system/uploads/attachment_data/file/737778/ccav-research-and-development-projects.pdf

Deichmann, J., Klein, B., Scherf, G. & Stutzle, R. (2019, October). *The Race for Cybersecurity: Protecting the Connected car in the Era of New Regulation*. McKinsey & Company. https://www.mckinsey.com/~/media/McKinsey/Industries/Automotive%20and%20Assemby/Our%20Insights/The%20race%20for%20cybersecurity%20Protecting%20the%20connected%20car%20in%20the%20era%20of%20new%20regulation/The-race-for-cybersecurity-Protecting-the-connected-car-in-the-era-of-new-regulation.ashx

Evans, C. (2020, April 22). Keyless car theft: What is a relay attack, how can you prevent it, and will your car insurance cover it? leasing.com. https://leasing.com/car-leasing-news/relay-car-theft-what-is-it-and-how-can-you-avoid-it/

GOV.UK. (2017a, August 6). New generation of smart cars will now be better protected from hackers. GOV.UK. https://www.gov.uk/government/news/new-generation-of-smart-cars-will-now-be-better-protected-from-hackers

GOV.UK. (2017b, August 6). The key principles of vehicle cyber security for connected and automated vehicles. GOV.UK. https://www.gov.uk/government/publications/principles-of-cyber-security-for-connected-and-automated-vehicles/the-key-principles-of-vehicle-cyber-security-for-connected-and-automated-vehicles

Greenberg, A. (2017, April 24). Just a pair of these $11 radio gadgets can steal a car. Wired. https://www.wired.com/2017/04/just-pair-11-radio-gadgets-can-steal-car/

Hayes, J. (2020, March 17). Hackers under the hood. *Engineering and Technology*. https://eandt.theiet.org/content/articles/2020/03/hackers-under-the-hood/

Ismail, K. (2018, November 14). Connected car experiences in 2019: Exploring the possibilities. CMSWire. https://www.cmswire.com/digital-experience/connected-car-experiences-in-2019-exploring-the-possibilities/

Janakiram, M. S. V. (2016, June 6). How will consumers benefit from connected cars? Forbes. https://www.forbes.com/sites/janakirammsv/2016/06/06/how-will-consumers-benefit-from-connected-cars/#453492a24a05

Katz, A. (2020, January 15). Our 2020 prediction: Automotive cybersecurity will finally be regulated. VDOO. https://www.vdoo.com/ja/blog/automotive-cybersecurity-standards-and-regulations

Koul, R. (2020, May 28). Internet-connected cars are first step towards autonomous driving: 5G to be backbone of connectivity. Financial Express. https://www.financialexpress.com/auto/industry/internet-connected-cars-are-first-step-towards-autonomous-driving-5g-to-be-backbone-of-connectivity/1973473/

KPMG. (2017). The connected car is here to stay: Driving digital transformation in the automotive industry. https://assets.kpmg/content/dam/kpmg/nl/pdf/2017/sector/automotive/the-connected-car-is-here-to-stay.pdf

Kuderer, M., Gulati, S., & Burgard, W. (2015, May). Learning driving styles for autonomous vehicles from demonstration. In 2015 IEEE International Conference on Robotics and Automation (ICRA) (pp. 2641–2646). IEEE.

Lemos, R. (2020, January 7). Car hacking hits the streets. Dark Reading. https://www.darkreading.com/edge/theedge/car-hacking-hits-the-streets/b/d-id/1336730

Lohrmann, D. (2016). Auto industry bug bounty programs point to our security future. Government Technology. https://www.cnet.com/roadshow/news/2019-automotive-cyber-hack-security-study-upstream/, https://www.usatoday.com/story/tech/2019/07/05/tesla-gm-hire-hackers-make-your-connected-car-safer/1594717001/

Maeirhofer, M. (2019, April 2). An overview of connected cars and the automotive industry. Mavoco. https://www.mavoco.com/overview-connected-cars/

Nie, S., Liu, L., & Du, Y. (2017). Free-fall: Hacking Tesla from Wireless to Can Bus. Briefing, Black Hat USA, 1–16. https://www.blackhat.com/docs/us-17/thursday/us-17-Nie-Free-Fall-Hacking-Tesla-From-Wireless-To-CAN-Bus-wp.pdf

Osborne, C. (2018). The most interesting internet connected vehicle hacks on record. ZDNet. https://www.zdnet.com/article/these-are-the-most-interesting-ways-to-hack-internet-connected-vehicles/

Pearlson, E. K., Galetta, F. D., & Saunders, S. C. (2015). Governance of the Information Systems Organisation. *Managing and Using Information Systems: A Strategic Approach*, 1(6), 5–16.

Pearlson, K. E., & Saunders, C. S., (2013). *Strategic Management of Information Systems*, 5th Edition International Student Version. Wiley.

Sinnott-Armstrong, W. (2019). Consequentialism. In *The Stanford Encyclopedia of Philosophy* (Summer 2019 Edition), Edward N. Zalta (ed.), https://plato.stanford.edu/archives/sum2019/entries/consequentialism/

Spencer, J. (2019, May 6). *Why is Cyber Security Important in 2019?* Security First Corp.

SPY Car Act of 2015, S. 1806, 114th Cong. (2015).

Tesla. (2020). Premium connectivity. https://www.tesla.com/support/connectivity

1.11 SilverPush
A Case of Convenience versus Privacy

Ying-Pin Lan
Hato Hone St. John, Auckland, New Zealand

Nishanth Chakkere Ramesh
Kapiti Coast District Council, Paraparaumu, New Zealand

CASE BACKGROUND

Along with the development of technology, people gradually got used to having smart devices such as smartphones, smartwatches, and smart TVs in their houses. These devices make people's lives easier by providing real-time service and a wide selection of personalized content. These smart devices analyze user habits and suggest suitable products and content to users. While collecting the users' data, the smart devices also create a comprehensive profile of users that enables the companies to analyze and predict what a user is looking for by using Artificial Intelligence software. With these users' data at hand, an advertising company can separate the customer segments precisely and target potential customer groups to achieve higher sales, while users can save the time of browsing the internet to find their dream products. Technology companies consider this feature a win-win situation; customers enjoy the convenience of smart devices, and the advertising companies can provide personalized content to each segment, effortlessly.

While users enjoy the benefits of smart devices, some people feel uncomfortable about this technology. If a smart device can precisely think for a user, does it mean the machines are collecting personal data in the background? Or are technology companies selling an individual's privacy to an advertising company? As a result, some people claim that these devices are recording the owner's conversations and personal information, and transferring these data to a third party without permission or consent from users. SilverPush, a software that is developed by an Indian company based near Delhi, has been deployed and installed in several smart TV brands, including Samsung and Sony, to collect users' information through the audio beacon and microphone (Nanalyze, 2017). These items as installed cause a wide range of concerns about information privacy worldwide.

DOI: 10.1201/9781003264415-14

In March 2016, the Federal Trade Commission (FTC) in the United States sent warning letters to SilverPush. It claimed that SilverPush's software could monitor a smart device's microphone and listen to the audio beacon that is connected with smart TV advertisements ("FTC issues warning letters to app developers using 'Silverpush' code," 2016). An audio beacon, which is emitted by televisions, cannot be heard by human ears, but smart devices can detect these signals and transfer them to data. SilverPush applied this technology to collect audio beacons and recordings from the receiver on a Smart TV and then send them to a third-party server for analysis. After receiving the analysis report, SilverPush could generate a detailed datasheet of the users' viewing patterns, such as channels watched, preferred content, and total time spent on a program to determine the user behaviors. If a Smart TV has this software installed, the TV will pick up the signals silently and identify whether the signals belong to the same person. Once the devices confirm that the signals are from the same user account, the software will transfer the data to a remote server. This software is particularly useful when a user uses the same account for his or her smart TV and smartphone. With the collected personal data, SilverPush can deliver personalized advertisements to a user's phone and other devices, including other applications under this specific user account. The primary purpose is to allow advertisers, marketers, and technology companies to harvest the raw data and predict each consumer's behavior. With the comprehensive database, the marketing companies can precisely deliver advertisements and products to their target customers. At the same time, these data help companies to not only gather the content users are consuming, but also to figure out consumers' reactions when they are watching, and which channels they check most often. This function enables the companies to record the patterns through multiple devices and to store them in the database.

After this tracking procedure was reported, SilverPush was criticized for violating the privacy policy of recording users' conversations and behaviors without their consent. Some people were furious when they found out that the software could continue monitoring their viewing habits even when the app was off. FTC states in their warning letter that SilverPush should have informed its users about this data collection procedure and the function of monitoring the viewing habits. Besides, FTC requested SilverPush to ask for permission from users before applying the audio beacon function to collect the consumers' data. Finally, SilverPush should employ a self-regulatory program and submit a quarterly report to the FTC explaining the usage of audio beacon and consumer consent-related issues. If SilverPush does not comply with the requests, the developers will be considered as violating the FTC Act 2013, which is used to protect privacy disclosures in mobile applications.

However, SilverPush announced on their official website that they have never broken the law and that they adhered to all regulations. The audio beacon is simply a signal to secure the smart device's output. SilverPush has never used this technology to collect user's data and transfer it to any third party. Also, SilverPush claimed that they did not record users' voices since the software listens to only a specific audio pattern to activate and function. In addition, SilverPush argued that users agree to the terms and conditions automatically once they purchase the Smart TV. Therefore, SilverPush denies the charges and liabilities from the FTC and insists that the company is doing the "right thing."

This case is worth investigating because, first, smart devices are gradually becoming a necessity in people's lives, but few users are aware of the potential threat they cause to user privacy. Second, research shows there are some cases in other countries where companies that were collecting personal data without the consumer's consent were charged and punished with considerable penalties. Although SilverPush has denied the charges, FTC regarded the behavior of receiving the audio beacon without user consent as a violation of its policies. Finally, this case enables readers to use this case to reflect on other points of view, including user agreements, smart homes, and implications to policymakers. Therefore, SilverPush is the ideal case study to represent risks in security management.

LITERATURE REVIEW

SilverPush is not the first company that applies eavesdropping software to smart devices. There are other claims indicating that technical giants such as Facebook, Amazon, and Twitter are trying to harvest consumers' data by collecting the users' activity patterns in diverse ways. For example, does a virtual assistant collect users' information while performing daily tasks? Once a user has finished talking with friends about something, the user will see the related advertisement in the Google search results, Facebook feeds, or Twitter notifications. This user is certain that he/she did not use the exact words as the search term on any search engine but, for some reason, the related products appeared on his/her smartphone screen. Does the smartphone know what a user is looking for and does it give what he/she wants without command? Does a smart device really think for humans?

The answer to these questions is "maybe." Hachman (2014) sought out an answer by conducting an experiment with several smartphones, an iPhone, a Microsoft phone, and a Google Android. First, he placed the phones on a desk and repeated a specific word, such as Burger King, a few times per day. A week later, he checked the pre-set Facebook account, Google account, and Twitter account on the three separate phones. Interestingly, he received some notifications about fast food, delivery service, and, of course, burgers. Next, he tested the term "laptop" and followed the same procedure as the above experiment. The result was similar in that all three phones delivered advertisements such as a notebook brand, 3C stores, and online shopping websites. He did not use these three phones to perform any search or post any comments related to the testing term. Although this is not an academic experiment, it indicates that this concern might be real; the smart devices are collecting users' personal information and recording the conversation in the background.

Consumers like convenience, and that is why people use smart devices to connect to the internet and perform daily tasks. Users can search for answers and suggestions on the internet, and all the information is at their fingertips. Also, the Internet of Things (IoT) helps to accelerate the penetration rate of smart devices and brings a new era to people's lives. This technology (see Figure 1.11.1) refers to connecting sensors, machines, and people together by using internet communication and cloud storage to build an intelligent network. However, IoT also has its challenges such as information security and privacy issues (Carritech Telecommunications). Once a user connects all the devices to the internet, the personal data may be exploited by scammers to obtain

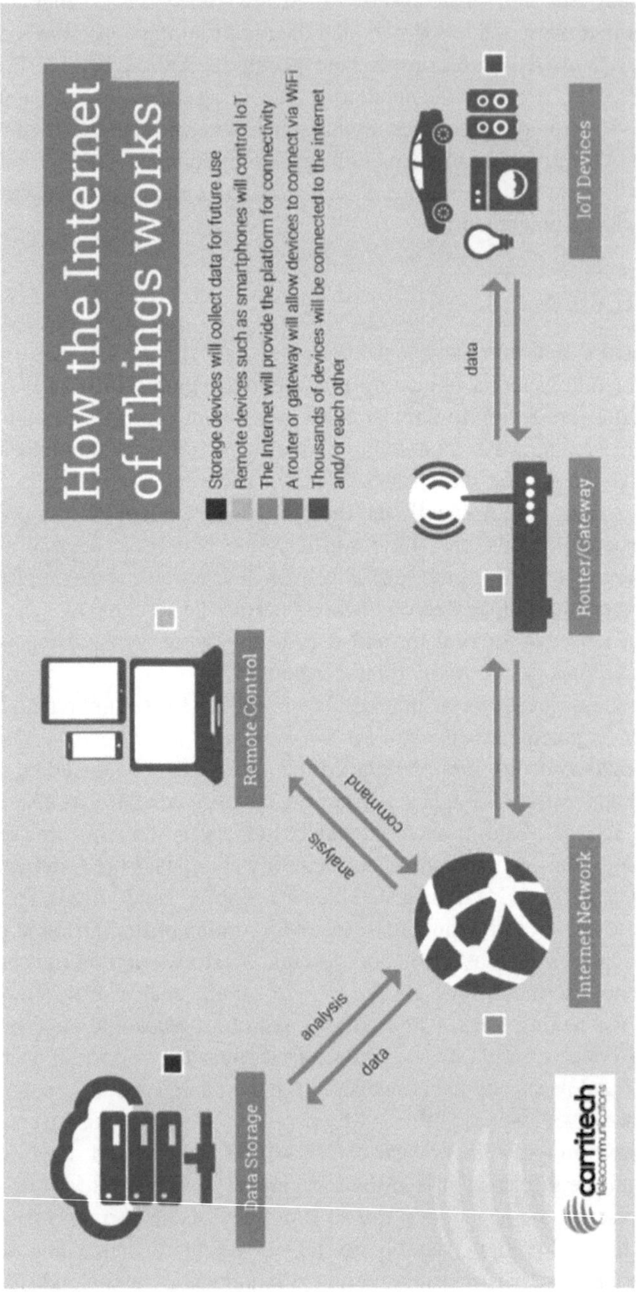

FIGURE 1.11.1 Simplified explanation of how the Internet of Things works (Carritech Telecommunications, Internet of Things: Explained).

personal information (Andersson, You, & Palmieri, 2018). One of the solutions for IoT data leak concerns is to provide IoT nodes with data authorization. This system can examine the data with authorization, confidentiality, integrity, and certification to protect sensitive information. In addition, it can help in creating an anonymous ID to protect from potential attackers so that their real personal identity will be safe. If the balance between availability and information security can be achieved, then the threats of IoT can be dramatically reduced (Carritech Telecommunications).

Some people argue that technology companies should state clearly in the terms and conditions whether they will use the personal data collected for other commercial purposes to reduce the concerns of privacy violation. For example, if a company is planning to provide personal information for a better-targeted advertisement, the company must inform the user of the agreements and explain the possible consequences. However, due to the increasing applications and capabilities of smart devices, consumers are less likely to read user agreements (Terms of Service; Didn't read, n.d.). They will, instead, choose to accept all the terms and conditions when registering to use the services and products (McMahon, 2018). Bashir, Hayes, Lambert, and Kesan (2015) also indicate that privacy agreements are usually long and wordy, hence the users are unlikely to read it through before using the apps. Their research shows that some readers did try to read the agreements but found out that the terminology was difficult to understand. That is, there are several knowledge gaps where an information asymmetry has taken place between users and service providers, and this leads to dissatisfaction with user experience. As a result, users face a choice between agreeing to all the terms and conditions regardless of the possible consequences or refusing to use the app completely. This unfair situation will increase in the coming years with the development of IoT, and the informational asymmetry between consumers and service providers is likely to worsen.

In addition, while IoT and smart devices provide great convenience to users, the technology has the potential risk of data leaks due to security vulnerabilities. Since the applications adopt user permission-based agreements, malicious apps might have a chance to access personal information, including contacts, phone numbers, and addresses when users install it and agree to the terms and conditions. In some cases, the app will ask the users for more permissions than the app needs. For example, a camera app might request access to contacts, and a PDF app needs the right to connect to audio files. However, the difficult part is that the user cannot identify malicious apps from the regular apps by analyzing permission requests. Therefore, Cha and Pak (2018) introduce a system that can hide user-related data and replace it with virtual data. When an app asks to access the user data, the system will direct the request to the virtual data location, which protects the users' real data. This system can also detect apps that leak personal data or transmit data to third parties in the background. Once the system finds malicious apps, it will send notifications to users and suggest users remove them for information security reasons.

Besides the virtual data introduced above, Perera et al. (2020) bring up the Privacy-by-Design (PbD) framework guidelines for IoT applications. The instructions consist of 45 privacy protection features, such as hidden data routing and encrypted data processing, that monitor software engineers to follow specific principles while developing the functions of the smart devices. The result shows that despite the different understandings and awareness of information security, the PbD framework helps software engineers perform significantly better in complying with privacy policies while designing IoT applications.

Besides IoT, the concept of a smart city is popular among tech-savvy citizens. A smart city, comprised of Smart Buildings, Smart Mobility, Smart Public Services, Smart Integration, Smart Water, Smart Energy, Smart Logistics, and similar facilities, is widely expected to be the next generation of modern society. This concept of a smart city claims to be able to dramatically improve the convenience, efficiency, and security of people's lives. By connecting everything and sharing the data on the cloud, people can access the information at any time, anywhere, and with any device. However, some people realize that information security and privacy will be hard to protect if people live in a city which uses smart devices instead of traditional mechanical devices. If a company wants to make the concept of a smart city become a reality, all the devices need to be able to recognize their owner by their identity, voice pattern, or fingerprint. The identity smart devices can record data and store it in the cloud platform, which puts the owner's personal information at risk (Li, Dai, Ming, & Qiu, 2016). This privacy leakage concern is caused by data over-collection, which indicates that smart devices keep collecting users' data more than what is necessary. This problem is different from malware and data over-collection which is hard to define and solve. Since users have agreed to the terms and conditions before using the devices, the smart devices have the permission and authorization to do so. Therefore, before people step into a smart city-like society, privacy concerns will be the main obstacle and point of contention. Li et al. (2016) suggest a framework that can upload all users' data to the cloud and have it encrypted. Whenever an app needs the information, the cloud will send permission requests to the user for approval. This method will not only save computing resources and storage but will also protect the data from leaking to external parties.

In addition to a smart city, a wearable device is another device that is discussed when it comes to conflicts between convenience and privacy (Safavi & Shukur, 2014). Wearable devices, such as wristbands and smartwatches, help to record health information and supervise users' activities. This technology is used to remotely monitor patients' wellness and exercise habits, which allows doctors to take prior actions to ensure their patient's wellness before serious health complications develop. Also, some people use wearable devices for recreational purposes. These devices usually come with built-in apps that can set goals and provide guidelines for training and exercising purposes. A recreational or fitness device provides real-time images and feedback that users can use to track their activities and record their improvements. Wearable devices are equipped with motion sensors that can monitor users' heart rate, walking steps, and movements to calculate their health conditions. This device has changed the way medical care is given and has improved the effectiveness of healthcare. The tech giants, including Apple, Google, and Samsung, have launched their own devices such as HealthKit, Google Fit, and S Health, respectively. However, if a person starts using a wearable device, it means the person has agreed automatically to the terms and conditions and that this device might transfer the data collected to a third party. This agreement leaves patients no choice but to allow these devices to continue receiving the data while knowing that there is potential for data leakage.

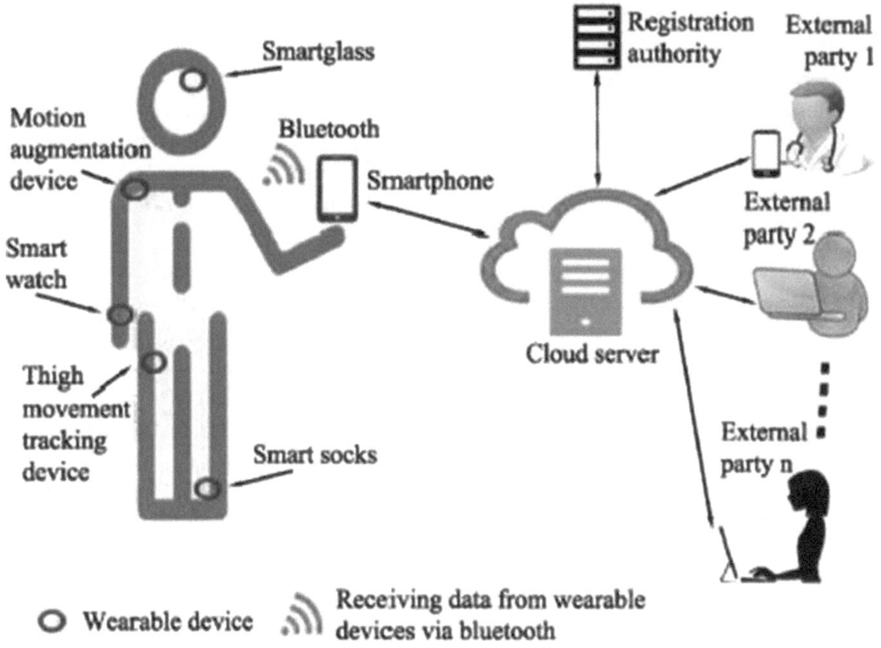

FIGURE 1.11.2 Data acquisition, transfer, and analysis from wearable devices (Das, Zeadally, & Wazid, 2017).

Although some countries have passed laws to regulate these activities and to ask companies not to use the data collected for commercial purposes, there is no guarantee that the companies will follow the privacy safeguards and guidelines in their products (Figure 1.11.2).

CRITICAL ANALYSIS

From the literature review, people may wonder why these technology companies want to gather personal information even though this action might break the law. Falcioni (2016) states that one of the purposes is to collect this personal information and construct a database that allows organizations to analyze and provide better advertisements. With individual preferences and recordings such as web browsing, viewed products, and watched videos, the technology companies can then use the algorithmic model to perform the analysis and predict users' behaviors.

The most famous case is the Target prediction of a teen girl's pregnancy (Hill, 2012). Target's system was able to know that this teen girl was pregnant before her father. This prediction comes from the data collection and data analysis. If a shopper registers for a membership card, all the purchasing records will be stored in the database and Target will analyze this data and provide an individual marketing strategy to each customer. This database includes all sorts of personal information, including email address, credit card, and date of birth, which allows Target to perform data mining on each customer. In this case, Target's analyst tracked a teen girl who had purchased calcium and zinc tablets in the past few weeks. In addition, she also bought a large quantity of scent-free soap and extra-big bags of cotton balls. Based on the data analysis result, Target's system suggests that there is an 87% chance that she is pregnant, and the delivery date will be around the end of August. As a result, Target started sending pregnancy-related advertisements to this teen girl. However, her father did not know of this information, and he thought that Target had sent the wrong marketing brochures.

This case shows the power of data mining and prediction model that with a comprehensive set of personal data of users, a company can then send the correct marketing messages and thereby increase their sales. If a marketing company can deliver a personal marketing message that thinks for customers, it can increase sales and customer satisfaction by providing precise solutions. However, Hill argues that if Target can predict the pregnancy date, which is a sensitive topic, who knows what else Target can predict? Also, Target is not the only company that profiles consumers by collecting their data, other companies are also eager to obtain this data to increase their marketing precision and generate higher profits.

Besides Target, YouTube also utilizes personal data to deliver personalized content (Deng, Yan, Sang, & Xu, 2015). With a Google account, each viewer on YouTube has a record of the videos watched and content preferences. Google then uses the algorithmic model to figure out a suitable video list with advertisements tailored to each viewer. Before, Google used profile and history behaviors to predict a user's interests. This method focused on a static strategy that performed better in the long term. However, since consumers' attention spans are becoming shorter with dynamic messages from the internet, Google decided to adopt a cross-platform data analysis system, such as a social website and multimedia sharing software, to grab the short-term interest changes of users. The implication is that if a user logs into their Google account, they will be continuously monitored by Google's algorithmic model. In a simplified summary, YouTube manages its users' recommendations by monitoring what videos a user clicks on, those which they watch, and whether the person signals that they enjoyed the videos (Nate, 2018).

Nevertheless, Target and YouTube collect personal data under the consent of customers while other companies developed software and applications to collect personal data without notice. In the case of SilverPush, they developed cross-device tracking software to collect personal data from multiple devices. The SilverPush system integrates devices that SilverPush wants to monitor, finds the mobile users' mobile audience by tracking their choices and actions, and retargets the users using multiple channels. With the data set, SilverPush can construct an accurate prediction model for an individual user. This system helps the advertising company to deliver the products and services better. On SilverPush's official website, they claim that their software can synchronize real-time advertisement across different devices and achieve 95% accuracy in advertisement detection. However, all these activities were done without consumers' agreement. SilverPush states that once a consumer uses the device which has SilverPush's software, this utilization will be considered their agreement, the giving of consent, automatically. Some people suggest that governments should establish data security-related laws to protect consumers from threats. However, most of the current regulations are from the previous generation and the legal department cannot keep up with the speed of technological development and update the laws in time. In the case of a smart home, the house is installed with technological devices that perform tasks automatically. The devices provide convenience, safety, and security using monitoring apps, and entertainment to the users. However, Basarudin, Yeon, and Yusoff (2019) state that, in Malaysia, the current Consumer Protection Act 1999 (CPA 1999) does not protect smart home buyers. Despite the benefits mentioned above, the owners might suffer from nonconformity, incompatibility, and information security laws that may fail to address consumer-related issues. Zheng, Apthorpe, Chetty, and Feamster (2018) claim that smart homeowners trust that the manufacturers will protect their privacy but would not be sure whether these procedures are in place.

The report also shows that most of the users do not understand appropriately the privacy risk of the smart devices surrounding them. Therefore, Basarudin et al. (2019) suggest that the device developers, service providers, and manufacturers need to be aware of their responsibilities to ensure that the smart homes provide an information safety and data-security-affirming environment. This sense of awareness cannot be obtained without the imposed laws being regularly updated along with the regulations. Since there are more technological devices and IoT products available in the market, the current CPA 1999 must consider including the software and IoT concepts in its scope. Besides, CPA 1999 should introduce a certification system to regulate smart devices and make it compulsory to pass all the checks before releasing them to the market. With such comprehensive systems, the laws can ensure that the manufacturers and software developers adhere to the information security requirements.

Besides Malaysia, the UK Government introduced three new regulations on IoT devices to ensure privacy ("'Bad news for hackers': UK govt announces new IoT cybersecurity law," 2020). First, all IoT and smart devices must be equipped with a unique password and cannot be reset to the factory settings. Second, the companies must establish a public point of contact (customer service) to tackle reported privacy issues quickly and easily. Third, companies must inform the minimum length of time for a smart device's security to update, both in-store and online. The National Cyber Security Centre (NCSC) has passed the laws to make IoT devices more secure and reliable ("UK plans IoT law," 2020). Also, the practice of providing customer service to solve the defects and vulnerabilities in time is a crucial process to align with the NCSC's regulations. This procedure follows the UK's Design Code of Practice, which was launched in 2018 for the IoT industry. The primary purpose is to build cybersecurity and protect the users' privacy, which is a great relief to consumers when they purchase smart home devices. Several international companies entered this agreement, including HP, Panasonic, and Centrica Hive.

Australia dealt with the potential privacy violations caused by smart devices in the health sector by updating regulations (Parker, Bero, Gillies, Raven, & Grundy, 2019). Since the laws of health-related applications were vague, the Australian government applied the concepts from the US Food and Drug Administration (FDA). This framework covers the areas of medical devices, privacy, and digital content. The primary purpose is to protect consumers' privacy when they use digital health smart devices. Since the devices will record all the health-related data such as heartbeats, pulse, and blood pressure, which are sensitive data, the Australian government decided to apply stronger regulations on the app's developers and companies to improve their privacy and safety.

The United States also had announced in 2017 its intention to modify the FTC to take a series of actions against companies in the IoT industry for collecting personal data without permission (Murphy, 2017). Since there will be around 50 billion smart devices which connect to the internet, unauthorized access to send and receive personal data should be stopped.

With the new regulations in place, the FTC issued orders and fines to companies breaking the laws. In February 2017, VIZIO, a smart TV manufacturer, was sued by the FTC because VIZIO was collecting consumers' data about programs watched, IP addresses, and Wi-Fi information without their consent. VIZIO then sold the data, including what the consumers were watching, how they watched it, user behaviors, and website visit history to third parties, and obtained revenue. Consumers' privacy was breached by those third parties who used their data to perform targeted marketing. In this case, VIZIO was ordered to pay $1.5 million to the FTC due to the violation of privacy laws. Also, VIZIO had to disclose their behavior and have the consumers' affirmative consent before collecting and sharing the data with third parties. VIZIO also needs to present a privacy program to monitor and audit its products every two years and report to the FTC.

Another case was InMobi Private Limited, an advertising company, which was accused of collecting personal information through released applications and directing the advertisements to children without their parents' consent. The FTC claimed that InMobi misrepresented the users' location and privacy to target children in their advertisement campaign. InMobi paid $950,000 for violating the Children's Online Privacy Protection Act (COPPA) and had to implement a privacy program that protects children's personal information.

According to Singh, Millard, Reed, Cobbe, and Crowcroft (2018), there are more than 120 countries which have established data-protection laws and regulations across different businesses. The most common laws followed are the OECD Privacy Guidelines, the General Data Protection Regulation (GDPR) in the European Union, and the Data Protection Directive (95/46/EC). However, some of the European laws were from the 1970s and have not been updated much since they were released. Therefore, the new regulations must review data protection regarding fairness, integrity, and confidentiality. In the case of an IoT system, updating laws in time can be a challenging task. First, any type of data can be personal data thereby restricting the governing rules that cover information and devices. Second, having detailed information and compliance obligations for data control and processing is difficult due to the unique and dynamic characteristics of IoT. Third, since most of the data is not stored on the devices but is stored on the cloud servers, restricting the data transfers across physical devices will not be a realistic practice. Finally, some existing regulations may not be applicable to the IoT-based environment. For example, a system that can generate automated decisions may find it challenging to prove that its functions are for "meaningful information under logic."

Besides text data and personal information, virtual assistants activated by voice are also part of the privacy scope. Pfeifle (2018) claims that governments often focus on text data and ignore voice data, which also contains personal information. In addition, in the United States, Federal laws responded slowly compared to the changing technologies. Although state legislatures have tried to move faster, they often are not proactive. As a result, companies should apply privacy procedures to their products and ensure the communications stored are appropriately protected.

Pfeifle suggests six approaches to achieve this goal. First, companies should only store the recordings for a brief time. For example, Amazon keeps the records of Alexa for one year or until users request to delete them; Apple keeps the voice clips from Siri for no more than two years. Second, companies must inform users how the recordings are used. By notifying users of their recording data, this system can maintain transparency and comply with the laws. Third, the software needs to display a visual indicator when the data is transmitted and collected. For example, the device should show a red light when the recording starts and a green light when the recording ends. This signal can give users more control over their privacy. If the device records the conversation without light indicators, the government should act quickly to correct this behavior. Fourth, companies should not have control of the smart device remotely. In the case of SilverPush, users worry that their Smart TV could be switched on and it may start recording without giving any indication or obtaining users' consent. Technology companies should block the "back door" or terminate the always-on function to give the control back to the users. Fifth, most of the data processing should be handled on the device instead of in the cloud servers. Since technology companies control cloud storage, there is a concern that the companies may use the data for their own purposes. In addition, the data on the devices should be encrypted to prevent unexpected attacks and hacking activities. In the case of Apple and the FBI, if the information was stored on the cloud, the FBI can access the information, which may violate privacy regulations. Since the laws cannot catch up with the changing pace of technology, companies must take proactive measures to protect users' privacy. Finally, companies should make the terms and conditions clear and easy to understand for users. Most of the users do not bother to read and cannot understand the articles in the agreement. Technology companies have the liability to explain and notify users in advance if users' data is used for other purposes.

Although most smartphone users or smart device owners understand the risks of data leakage, they have limited knowledge about actions to be taken, according to Alsaleh, Alomar, and Alarifi (2017). They investigate how users protect their personal information against security threats based on users' attitudes and perceptions. In their research, Alsaleh et al. examined several features, including encrypted instant messages, malware applications, passwords, and location signals. The instant messages test results show that 33% of people did not know that the messengers encrypted their content, while 57% thought there was no such measure applied in the software. For the suspicious applications test, 67% of the participants engaged in downloading free applications, which are supposed to be paid apps, from third-party platforms. Additionally, 63% have no anti-virus app installed on their smartphone, and 53% of them said that they have never paid attention to the requests, such as the right to access contacts, audio, and camera while installing applications. For the password setting, 30% of the users use important dates or phone numbers to create their passwords, which poses a significant risk of being hacked. Finally, for the location signal, 30% of the participants said they shared the location with their friends only while 57% of them thought that there is no security- and privacy-related risk in sharing their current location.

Interestingly, if these people turn off the location service, it is because of the computing resource concerns regarding saving their battery life rather than due to privacy concerns. From the observations stated above, we know that most of the participants were not aware of the risks they were exposed to when it came to privacy matters. To tackle these problems, Alsaleh et al. suggest that, first, make security warnings shorter, clearer, and more understandable, rather than showing them as an extensive list of technical text. The message must briefly point out the potential risks and consequences for users if they execute the required action being shown on the screen. Second, developers may set up security indicators in the software to determine whether users are protected from potential threats. The software may then analyze each application installed and automatically block suspicious applications that show strange behaviors. If users want to use a third-party application instead of the official applications from authorized app stores, this software must display a warning sign and suggest to users not to install the application to avoid data leakage and scams.

To tackle the privacy issue, Parker et al. (2017) suggested in the report that governments should establish or update the legislation and guidelines to include smart devices. The critical areas that the regulations should consider are consumer privacy, data security, promotion & advertising, and professional ethics.

Consumer Privacy

Privacy legislation should protect users' personal data from unauthorized collection, usage, and disclosure of sensitive information like name, address, and ID. Unauthorized data collection and behaviors such as the sharing of personal information should be warned and penalized by authorities. As a result, the developers will follow the practices of transparency and will obtain consent regarding proper protection procedures from the users. Also, companies should respect everyone's interests and must obey the code of ethics. Parker et al. (2017) suggest that governments can raise public awareness regarding privacy issues and the potential implications of user behavior. This action can put pressure on the developers and companies to endorse a higher standard of user privacy. If a third party wants to obtain the users' data, the company must follow the guidelines to disclose the collection methods, obtain consent from the users, and explain how the data will be used.

Data Security

According to Parker, 93% of applications do not have encrypted data storage and 29% of the data transmission was not encrypted. This finding shows the threats to data security. Therefore, developers should consider adopting adequate security protections in their applications, within the established legislative guidelines. The government should also encourage the public to openly discuss the implications of data security in the digital age and the practical solutions for data breaches.

Promotion and Advertising

An advertising company might need the users' data to provide personalized content and advertisements. Understandably, the companies will try to obtain this data, both from legal channels and through illegal means. A targeted marketing strategy will indeed be efficient and save costs, not only for the advertisers but also for consumers. If a company wants to use personal data for commercial purposes, they must disclose that in their terms and conditions page clearly, with an easy-to-understand explanation. If a user disagrees with sharing his/her data, the company must not try to collect the information nor share it with third parties. The government must establish related laws to stop advertising companies from stealing personal data for their own benefit.

Professional Ethics

Companies should develop applications based on the ethics of accountability, truthfulness, and transparency (Parker et al., 2017). Therefore, agreements such as privacy policies, practice rules, and consent procedures should be appropriately enforced instead of just an "I agree" tick box. This measure aims to not only limit the developers' legal liabilities but also to protect consumers while they are using these applications.

Pirzada (2020) provides a sample simplified privacy template which should include the following three elements:

1. **What information do you collect?** In your policy, you should disclose all the information your Action collects. This includes information that you may collect automatically, such as server and HTTP logs, data transmitted by the Actions on Google API to you, and usage information. This also includes information that you get from the user, either directly or via the permissions API. You should also disclose whether you collect any persistent identifiers (like the Google ID).
2. **How do you use the information?** In your policy, you should disclose how you use the information you collect. For example, you may use all the information to provide certain services to users, to recognize them the next time they use your Action, or to send them promotional emails.
3. **What information do you share?** In your policy, you should disclose the circumstances when you share information. For example, you may share information with third parties as part of the service (like restaurant reservation Action), with other users (like a social network forum), with marketing partners, or with service providers that assist with your service (like hosting companies of technology platforms).

LESSONS LEARNED

The lessons learned in this case may be analyzed according to the solutions to different issues that emerged from this case. First, users must be made aware of what they are signing up for when they begin to use a product or service. Often, users will give their assent to a terms and conditions page without reading the details mentioned on these pages. Similarly, disclaimers provided by service providers are disregarded by the users. This may be because the users are eager to set up and use the product or service and think that this information is not consequential to their use. They may also not understand the terms and conditions because of the technicalities of the words used and the legal language of the text. More importantly, some of the terms and conditions are very lengthy, and most service users may not have the time to read this in its entirety before accepting and signing in.

Therefore, it is important to simplify the terms and conditions so that they are readily understandable to the users. For example, there is an online add-on service called "Terms of Service; Didn't Read," which provides ratings and labeing (based on quality and understandability) for website terms and private policy. Class A ratings are particularly good, while Class E ratings are unbelievably bad. The ratings for many popular websites, such as Wikipedia, Soundcloud, YouTube, and Google, are available. Moreover, this feature is developed through the contribution of internet users. Therefore, it is more relatable to other users of these services. The terms and privacy policies of these websites are ranked based on their protection of user privacy, ability to terminate the service at any time, use of personal information to target advertisements, etc. Therefore, the users have a simple and clear understanding of user terms before they continue to use a particular service (Terms of Service; Didn't Read).

SilverPush could have simplified its user terms and conditions. More importantly, it could have simplified how user consent was received and applied to target the advertisements. Not notifying the users about tracking and using their online activities is also a serious breach of trust which may cause future problems in building and maintaining organizational reputation. Users must be informed beforehand of the functionalities of SilverPush (such as data collection) and its dealings with third-party organizations in the analysis of the collected data. Also, the targeting of advertisements based on clandestine tracking is immoral and will be seen as done solely to make profit from eavesdropping. The use of "Terms of Service; Didn't Read" will provide users with the required information to understand the privacy policy of SilverPush in a simplified manner. Users must also be proactive in understanding the functionalities of applications like SilverPush which they may not use directly but are installed in the devices that they use.

The second lesson learned is balancing convenience with data risks. The IoT has made users' lives more convenient through increasing interconnectedness and without requiring human interference in the transfer of data. This has improved the quality of life for users by managing their schedules, preferences, bookings, leisure, work, and many other aspects. Also, users increasingly expect that IoT will take care of their minor conveniences so that they may focus their efforts on the larger issues in life. This creates a situation where there is a delicate balance between user convenience and privacy of user data. Often, companies are unable to balance these two aspects. To provide users with quality recommendations based on their preferences, the companies must access and analyze their data – there is no other way. However, the method by which such data is accessed and whether the user is aware of such data access is a crucial point that many companies ignore.

Many analysts believe that companies need not choose one over the other in terms of user convenience and user data security. There are important questions that companies must address with reference to personalized advertising through accessing user data:

1. Is the data acquired from the users' devices useful in improving user experience or creating relevant advertisements?
2. Is the data collected proportionate to the advertisement targets being planned?
3. Does the user know that their user data is being collected to create personalized advertisements and campaigns?

The solutions to these questions provide valuable learnings for SilverPush in the given case. The answers may be provided as follows according to the analysis of Forbes magazine:

1. The user's permission must be explicitly asked for and obtained before gathering or using data.

   This means that companies such as SilverPush must explicitly ask for user permission. This must be done in a simplified way, so users understand what aspects they are giving their permission for. Only after obtaining such information should the software start accessing user data to provide targeted advertisements to the users.
2. Customer trust must be built over the long term.

   Short-term missteps such as unauthorized data access will cause long-term issues when customers will be wary about trusting the service provider with their new products or services. This is especially true if trust is lost through the exposure of unethical data collection practices. Trust once lost is exceedingly difficult to restore.
3. Ensure that customer data is advantageous to them.

   Users are not necessarily against the use of their data for improved recommendations and advertisements. In fact, they may look forward to such messages because it reduces their efforts and burdens in looking for information by themselves elsewhere. However, it is critical to obtain their permission before their data is used. Subsequently, customer data can be used to make recommendations to them so that it is advantageous to them – this is a win-win situation for both parties (Gaul, 2019).

Another learning that can emerge from this case is how often and how quickly government laws and regulations can be updated to keep up with the changes in technology, especially with regard to user privacy. The changes in technology seem to be too rapid for laws to keep up, and therefore, government agencies may not be able to provide protection to citizens when their data is used without their permission. Moreover, if the company is based overseas, local governments will be incapable of regulating their misdemeanors and inactions. For instance, SilverPush is based in India, and its products are installed in third-party appliances which are used around the world. In such circumstances, blocking the service can be the highest penalty that may be imposed.

Therefore, an international consensus must be developed to deal with such matters since user data acquisition is not limited to the boundaries of a single country. The European Union's GDPR is a transnational law which regulates data privacy among the 27 European Union countries (European Union, n.d.). This law gives users access to permissions relating to the use and retention of their data. Users can also withdraw their permission at any time if they feel that their data is being misused or that the utility of sharing their data is not meeting its purpose for advertisements or recommendations. Simplifying the regulatory environment around such user permissions is also an objective of this law. Instead of technical and lengthy terms and conditions, simple and understandable policies may be presented to users. The law also prevents automatic data access decisions that may be taken without the users' knowledge (European Union, Complete guide to GDPR compliance).

Data privacy and leverage of user data to improve the quality of services provided to users are global matters. Therefore, creating an international organization that evaluates laws and provides guidance for such purposes would be feasible. The transnational nature of technology and data usage needs to be managed transnationally.

A framework may be developed for companies to use when balancing privacy issues with data collection. The data processing ecosystem consists of Individuals, Developers, Business Associates or Partners, Commercial Products/Services, Civil Society, Research/ Education, Manufacturers, Public Services, and Suppliers. The interests of all these parties must be considered when preparing the framework. The National Institute of Standards and Technology (NIST) has introduced many such frameworks which assist organizations. Distinguishing between three of them – cybersecurity risks, cybersecurity-related privacy risks, and privacy risks is important. The first is associated with cybersecurity incidents arising from loss of confidentiality, integrity, or availability; the third is associated with privacy events arising from data processing. Cybersecurity-related privacy events arise from factors of both. After such distinction, each of these risks must be Identified, Governed, Controlled, Communicated, and Protected (National Institute of Standards and Technology, 2020).

The final learning that this case may provide is that the scope of why companies access user data must be well defined. TH is not arbitrary because it will then provide the companies with an excuse to collect user data for obscure and unnecessary reasons. Business data usage must be limited to the following purposes:

1. To improve customer experience so that the services are customized at the individual level.
2. To improve recommendations based on remarketing principles.
3. To integrate the diverse needs of the customer such as a mobile application scheduling users' times based on previous activities in a day (Freedman, 2020).

DISCUSSION QUESTIONS

1. Can user convenience be balanced with ensuring user privacy when obtaining user data?
2. Who should be held accountable for unauthorized data access – third-party software providers whose software is installed on the devices or the device manufacturers?
3. How can user permission be simplified so that users understand for what and to whom they are giving their consent?
4. How easily can users opt out of the permissions they give when they are concerned about their privacy?
5. Should matters relating to data privacy and user convenience-based advertising be regulated at the government level or company level?

REFERENCES

Alsaleh, M., Alomar, N., & Alarifi, A. (2017). Smartphone users: Understanding how security mechanisms are perceived and new persuasive methods. *PloS One*, *12*(3), e0173284. doi:10.1371/journal.pone.0173284

Andersson, K., You, I., & Palmieri, F. (2018). Security and privacy for smart, connected, and mobile IoT devices and platforms. *Security and Communication Networks*, *2018*, 1–2. doi:10.1155/2018/5346596

"Bad news for hackers": UK govt announces new IoT cybersecurity law. (2020). *Total Telecom Online U6*. https://totaltele.com/bad-news-for-hackers-uk-govt-announces-new-iot-cybersecurity-law/

Basarudin, N. A., Yeon, A. L., & Yusoff, Z. M. (2019). Regulating smart home technology devices under Malaysian legal framework. *Commonwealth Law Bulletin*, *44*(4), 607–624. doi:10.1080/03050718.2019.1657031

Bashir, M., Hayes, C., Lambert, A. D., & Kesan, J. P. (2015). Online privacy and informed consent: The dilemma of information asymmetry. *Proceedings of the Association for Information Science and Technology*, *52*(1), 1–10. doi:10.1002/pra2.2015.145052010043

Carritech Telecommunications. (n.d.). Internet of Things: Explained. Retrieved from https://www.carritech.com/news/internet-of-things/

Cha, Y., & Pak, W. (2018). Protecting contacts against privacy leaks in smartphones. *PloS One*, *13*(7), e0191502. doi:10.1371/journal.pone.0191502

Das, A. K., Zeadally, S., & Wazid, M. (2017). Lightweight authentication protocols for wearable devices. *Computers & Electrical Engineering*, *63*. Retrieved from https://www.sciencedirect.com/science/article/abs/pii/S0045790617305347

Deng, Z., Yan, M., Sang, J., & Xu, C. (2015). Twitter is faster: Personalized time-aware video recommendation from Twitter to YouTube. *ACM Transactions on Multimedia Computing, Communications, and Applications (TOMM)*, *11*(2), 1–23. doi:10.1145/2637285

European Union. (n.d.). (Complete guide to GDPR compliance). Retrieved from https://gdpr.eu/

Falcioni, J. (2016). A machine that thinks for you. Retrieved from https://memagazineblog.org/author/mechengmag/page/2/

Freedman, M. (2020). How businesses are collecting data (and what they're doing with it). Retrieved from https://www.businessnewsdaily.com/10625-businesses-collecting-data.html

FTC issues warning letters to app developers using "Silverpush" code. (2016). *The Computer & Internet Lawyer*, *33*(6), 22.

Gaul, D. (2019). Striking the balance between convenience and privacy: Here's what consumers want. Retrieved from https://www.forbes.com/sites/forbestechcouncil/2019/09/11/striking-the-balance-between-convenience-and-privacy-heres-what-consumers-want/#344649de5134

Hachman, M. (2014). Battle of the digital assistants: Cortana, Siri, and google now. Retrieved from https://www.pcworld.com/article/2142022/the-battle-of-the-digital-assistants-windows-phone-cortana-vs-google-now-vs-siri.html

Hill, K. (2012). How target figured out a teen girl was pregnant before her father did. Retrieved from https://www.forbes.com/sites/kashmirhill/2012/02/16/how-target-figured-out-a-teen-girl-was-pregnant-before-her-father-did/#7803dfb26668

Li, Y., Dai, W., Ming, Z., & Qiu, M. (2016). Privacy protection for preventing data over-collection in smart city. *IEEE Transactions on Computers*, *65*(5), 1339–1350. doi:10.1109/tc.2015.2470247

McMahon, K. (2018). Tell the smart house to mind its own business! Maintaining privacy and security in the era of smart devices. *Fordham Law Review*, *86*(5), 2511.

Murphy, K. A. (2017). Recent FTC regulation of the Internet of Things. *Business Lawyer*, *73*(1), 289–293.

Nanalyze. (2017). How audio beacons monitor you via smartphone. Retrieved from https://www.nanalyze.com/2017/05/audio-beacons-monitor-smartphone/

Nate. (2018). YouTube algorithm. Retrieved from https://karaandnate.com/travelcon/screen-shot-2018-09-16-at-10-22-18-am/

National Institute of Standards and Technology. (2020). NIST privacy framework: A tool for improving privacy through enterprise risk management. Retrieved from https://www.nist.gov/system/files/documents/2020/01/16/NIST%20Privacy%20Framework_V1.0.pdf

Pirzada, M. (2020). Sample privacy policy template. Retrieved from https://www.privacypolicies.com/blog/privacy-policy-template/

Parker, L., Bero, L., Gillies, D., Raven, M., & Grundy, Q. (2019). The "Hot Potato" of mental health app regulation: A critical case study of the Australian policy arena. *International Journal of Health Policy and Management*, *8*(3), 168–176. doi:10.15171/ijhpm.2018.117

Parker, L., Karliychuk, T., Gillies, D., Mintzes, B., Raven, M., & Grundy, Q. (2017). A health app developer's guide to law and policy: A multi-sector policy analysis. *BMC Medical Informatics and Decision Making*, *17*(1), 141–143. doi:10.1186/s12911-017-0535-0

Perera, C., Barhamgi, M., Bandara, A. K., Ajmal, M., Price, B., & Nuseibeh, B. (2020). Designing privacy-aware Internet of Things applications. *Information Sciences*, *512*, 238–257. doi:10.1016/j.ins.2019.09.061

Pfeifle, A. (2018). Alexa, what should we do about privacy? Protecting privacy for users of voice activated devices. *Washington Law Review*, *93*(1), 421–458.

Safavi, S., & Shukur, Z. (2014). Conceptual privacy framework for health information on wearable device. *PloS One*, *9*(12), e114306. doi:10.1371/journal.pone.0114306

Singh, J., Millard, C., Reed, C., Cobbe, J., & Crowcroft, J. (2018). Accountability in the IoT: Systems, law, and ways forward. *Computer (Long Beach, Calif.)*, *51*(7), 54–65. doi:10.1109/mc.2018.3011052

Terms of Service; Didn't Read. (n.d.). Retrieved from https://tosdr.org/

UK plans IoT law. (2020). *Computer Fraud & Security*, *2020*(2), 3–3. doi:10.1016/S1361-3723(20)30014-2

Zheng, S., Apthorpe, N., Chetty, M., & Feamster, N. (2018). User perceptions of smart home IoT privacy. *Proceedings of the ACM on Human-Computer Interaction*, *2*(CSCW), 1–20. doi:10.1145/3274469

Section 2

Behavioral and Social Cases

Teaching Cases

2.1 Microsoft 365 Phishing Case

Roydon Dominic Correya
MYOB, Christchurch, New Zealand

Tushar Ashok Sagathia
ASB Bank, Auckland, New Zealand

CASE SUMMARY

Microsoft, one of the biggest technology giants, obtained a court order allowing them to seize control of six domains that were used in phishing operations against Office 365 customers, including in campaigns that leveraged COVID-19 lures. According to court documents obtained by ZDNet, Microsoft has targeted a phishing group that has been targeting the company's customers since December 2019.

The phishers functioned by sending emails to businesses that hosted email servers and enterprise infrastructure on Microsoft's Office 365 cloud service. To deceive users these emails were spoofed to look like they came from fellow employees or a trusted business partner. When users tried to open the file, they were redirected to install a malicious third-party Office 365 app created by the hackers. The application, if installed, granted the attackers complete access to the victim's Office 365 account, its settings, the user's files, the content of their emails, contact lists, notes, and others.

Microsoft said that by using a third-party Office 365 app, hackers gained all the access they needed to users' accounts without actually collecting their passwords – receiving an OAuth2 token instead. OAuth2 Token works by giving client verification to the organization that has the client account and approving third-party applications to get access to the user account. OAuth 2 gives approval streams to web and work area applications, and Smart or mobile devices.

Some of these phishing attacks succeeded for three reasons. The first was the app was made to look like it was created by Microsoft and was an official, safe-to-use application. The second was that the Office 365 environment is geared toward the modularity provided by third-party apps, either custom-created by companies or readily available on the Office 365 AppSource Store, and users are used to installing apps regularly. Third, the hackers used a clever technique where the app's installation link initially took users to the official Microsoft login page. However, attackers then redirected users to the malicious app once the authentication succeeded, giving users the impression that they were using a Microsoft-vetted application.

DOI: 10.1201/9781003264415-17

Microsoft filed a civil case on June 30 this year, targeting six domains that hackers used to host their malicious Office 365 applications. The six domains used by attackers were as follows namely: Officeinventories.com, Officehnoc.com, Officesuited.com, Officemtr.com, Officesuitesoft.com, Mailitdaemon.com.

The company noted that the group's initial attacks began using business-related themes, but they quickly changed to emails carrying coronavirus-themed documents once COVID-19 became a global pandemic.

HACKERS' END GOAL WAS A BEC ATTACK

In a statement issued on Microsoft's blog, Corporate Vice President, Customer Security & Trust, Tom Burt said the malicious third-party apps were used to gain insights into victims' inner structure. This enabled the attackers to follow up with specific attacks called Business Email Compromise (BEC), a form of cybercrime. In February, the FBI said that BEC scams accounted for half of the cybercrime losses reported to the FBI Internet Crime Complaint Center (IC3) in 2019. According to the FBI, companies lost $1.77 billion to BEC scams in 2019, with an average loss of $75,000 per report. This case also marks the fourth time in the past year that Microsoft filed a legal case to take control of malicious domains in its continuous efforts of legal war against cybercrime.

In March 2020, Microsoft's legal team seized control over domains operated by the Necurs botnet. Before this, in December 2019 Microsoft took down 50 domains operated by North Korean state-sponsored hackers. Nine months earlier in March 2019, Microsoft took control of 99 domains operated by Iranian government-backed hackers. As a security measure, Microsoft would later buy the corp.com domain in March 2020, so the domain wouldn't fall into the wrong hands.

Necurs botnet is one of the largest networks in the spam email threat ecosystem which has infected more than 9 million computers globally with victims in almost every country the world over. Botnet refers to a network of computers that a cybercriminal has infected with malware or malicious software. Once infected, hackers can remotely control these computers and use them to commit cybercrimes like attacks on other computers online and steal account credentials and confidential private data. The security community first identified the Necurs botnet in 2012 and observed it distribute many forms of malware, one being the Zeus GameOver Banking Trojan. During a 58-day investigation, it was found that Necurs-infected computers sent a whopping 3.8 million spam emails to over 40 million possible victims. Necurs is understood to be operated by Russia-based cybercriminals and has accounted for a wide range of cybercrimes that include the pump-and-dump stock scam, fake pharmaceutical spam emails, as well as "Russian dating" scams. Microsoft has joined hands with the FBI in the fight against botnet and malicious networks which are a major threat.

CASE ANALYSIS

In today's world of cloud computing and eCommerce technology, companies have thrived by using software and infrastructure as a service to better their business and give them an upper hand in nascent markets. The world is now a global marketplace which makes business transactions easier from any part of the world. Although this global marketplace is an optimum opportunity for businesses to advance, this has also been seen as an ideal chance for cybercriminals to steal sensitive information or commit cybercrimes such as online fraud, identity theft, phishing scams, and ransomware attacks, all of which would be very difficult to track on the web. Security and privacy are some of the most important factors that influence e-commerce, online trade, and cloud computing services. Providers of services online must keep identifying threats to prevent such attacks which could have dire consequences if cybercriminals were successful in achieving their goals.

Phishing

Phishing is an online identity theft that aims to steal sensitive information such as user IDs, passwords, and online banking details from its victims (Arachchilage & Love, 2014; Arachchilage et al., 2016). Phishing gets users to disclose sensitive personal information on a fake website by making it look and feel like a legitimate site. Phishing is a type of attack where users are cheated by hackers using technical subterfuge as well as social engineering techniques to exploit weaknesses in human nature (Mansfield-Devine, 2018). This tempts users to use communications that are disguised to appear as trusted sources while doing transactions or business online. The most common types of phishing are spear phishing, whaling, catfishing, and clone phishing which uses techniques such as link manipulation, website forgery, filter evasion, social engineering, and voice phishing, among others.

Bulk Phishing

As the name suggests, it is an approach that does not specifically target and is not tailored toward the recipient and the success of the scam depends upon large-scale bulk mailings reaching enough recipients who think these emails look important.

Spear Phishing

This attempt sent by email message is meant to target specific individuals or businesses and is tailor-made accordingly (Steer, 2017). In spear phishing cases, attackers are more likely to have done some prior research to enable them to frame the message to suit the individual or business and convince recipients more readily. As pointed out by Mansfield-Devine (2018), the more the attacker knows about the target company, the more a BEC scam is likely to succeed.

Clone Phishing or Link Manipulation

In Clone Phishing, scammers replace the link or attachment in a legitimate email with a link that is malicious while retaining the message contents and then spoof the sender with a claim that this message is an updated version of the email message.

Whaling

This type of phishing is used by hackers to target high-value individuals or higher-level management in an organization. Email continues to be the mode of communication, but they also use other mediums of communication to send threats like SMS (Smshing) or voice phishing or (vishing) that refer to equivalent threats using SMS and VOIP messages (Furnell et al., 2019).

The fact that Microsoft 365 has been a victim of these kinds of phishing attacks is evidence that companies of all scales are potential targets of this type of attack (Mansfield-Devine, 2016).

Business Email Compromise

As explored by Mansfield-Devine (2016), the US Federal Bureau of Investigation has categorized BEC, into five types.

Bogus Invoice Scam

The attacker exploits this business relationship between a business and a supplier by sending fake payment requests from the supplier to the client with the attacker's banking details.

Classic CEO Fraud Scam

A business executive gets an initiation request for a wire transfer of funds coming from the senior executive's compromised email account or through proofing requesting that the employee makes a transfer of funds.

Account Compromise

Business associates receive fraudulent email correspondence through compromised email accounts.

Business Executive and Attorney Impersonation

Frauds pretending to be lawyers of the firm who handle confidential and time-sensitive issues ask an employee of the target firm to make a payment while requesting it to be kept secret due to the sensitive nature of the matter.

Data Theft

Executives of the organization fall victim to a compromised email account which is then used to obtain sensitive data from employees (Mansfield-Devine, 2016).

Microsoft 365 is a cloud computing service provider which gives SaaS to clients. The phishing case and BEC scam targeted Microsoft 365 cloud clients and succeeded in phishing using spoofing and manipulating users to download a seemingly legitimate and harmless link. This brought them to the original Microsoft 365 site before redirecting them to a malicious website to steal confidential emails and orchestrate a bigger attack for financial gain.

Cloud Computing

Cloud computing which became part of the computing vocabulary of servers and users in the 1990s is now considered one of the most used technologies. It is now considered the fifth utility after water, electric power, gas, and telephony since most of our utility services today are hosted in server farms on the cloud. Cloud computing today is the most used alternative to the traditional in-house IT computing service which houses hardware and software services on-demand on the internet (Sahinoglu & Cueva-Parra, 2011).

Cloud computing has been the go-to technology with all the big players like Google, Microsoft, and Amazon joining the race to be the most used service provider in the market today. At first, people viewed cloud computing with a lot of skepticism, but many companies slowly started using it to get the cutting-edge technology key to advancing their business. Today, cloud computing services offer three different categories of services to home and enterprise users such as Software as a service (SaaS) like Microsoft 365, where the application is delivered to the user on-demand over the internet, Infrastructure as a service (IaaS) such as Google App Engine, and Platform as a service (PaaS) like Amazon EC2 (Gupta & Badve, 2016). One of the biggest advantages of cloud services is that they can be accessed from anywhere in the world using devices that connect to the internet and then use the service.

Many businesses around the world and individual users have invested in cloud computing and enhanced the efficiency of their processes while decreasing IT costs. Since cloud technology usage is rising fast, the need for security is becoming more and more critical (Kumar et al., 2018). Because cloud technology is still developing, security is very hard to implement due to the diverse types of users using the services who need different security requirements. A cloud user can be an individual, a software development team, a university, or a large business – all of which need different levels of security.

Data Security in the Cloud

Gupta and Badve (2016) discuss the importance of data security while using cloud services and indicate that cloud service providers must make sure that the data stored are secure to gain the trust of customers to ensure continued use of services. There is a threat of breaches, vulnerabilities in the security of the applications, and denial of service which would result in the loss of sensitive information and financial loss to the organization or user.

The main goals of cybercriminals against cloud computing are getting access to user data and preventing access to cloud services. Both can cause serious harm to cloud users and shatter confidence in the security of cloud services.

Hackers while assembling attacks in the cloud generally intrude into communications between cloud users, services, or applications by taking advantage of the weaknesses or vulnerabilities in cloud computing, stealing user-specific credentials somewhere outside the cloud, using users' cracked passwords to gain legitimate access, and acting as a malicious insider.

Although a known fact that data security is vital when using any technology SaaS user's dependence on cloud service providers for the proper security of their data has become a major challenge (Hashizume et al., 2013). Cloud users store data that contains sensitive personal or business information on the cloud. This data is vulnerable to attacks, loss, or damage due to application weaknesses, calamities, or emergencies all of which cannot be prevented by the service provider. However, cloud design engineers should apply the latest encryption algorithms to secure the reliability of data shared and stored by the user in the cloud.

Susceptibility of Cloud Services

Most cloud computing systems are designed to integrate multiple cloud services and these services have little protection against each other. Hackers can make use of the weakness in any of the cloud services to get into the other systems and steal the data of legitimate users. Since cloud services are still developing, weaknesses in cloud services are still being identified. The need is paramount to create a robust architecture that can help insulate a user's actions on the cloud.

API Vulnerabilities

Vulnerabilities in the Application Programming Interface (API) on the cloud may have a significant effect on the security of cloud service infrastructure and hence cloud developers need to look for, create, and implement stronger controls to manage, provision, and monitor APIs.

Cloud environments are susceptible to attacks by legitimate cloud users (malicious insiders) who orchestrate attacks or data breaches (Modi et al., 2012). Identity and access management (IAM) technologies could be used by cloud developers to minimize this threat and avoid loss of confidentiality and integrity (Kumar et al., 2018).

Shared Technology Vulnerabilities

Using cloud technology involves sharing technologies like cloud orchestration and a virtualization attacker can cause considerable damage by exploiting these weaknesses in any part of these shared computing technologies. The Virtual Machine Monitor (VMM) or hypervisor is responsible for virtual machine isolation. If vulnerabilities in a VMM or hypervisor can be compromised, hackers can potentially break in and control the host or virtual machines as well as execute cross-VM side-channel attacks and DoS attacks. Using virtual machine escape, the hackers can fully access and control the host through the shared resources. It is therefore important to choose a cloud provider whose security is strong (Hashizume et al., 2013).

Provider Lock-in

High switching cost is one of the ways cloud service providers make their customers dependent and stick with them. Cloud users cannot change easily even if they don't get all the services from their existing cloud provider. Customers should make sure that their business can easily switch or can migrate and import data from their providers seamlessly, supporting the import of data in multiple formats.

Weak Cryptography

Cloud developers should pay attention to data security before data is moved onto the cloud. Cloud providers use cryptographic algorithms that automatically generate random numbers used to encrypt and secure data, but these mechanisms are not enough as attackers also use sophisticated decoders to decrypt and hack the system. Thus, more thought should be given as to how to give data the strongest possible security on the cloud.

Although Microsoft is one of the biggest players in terms of software and cloud computing this case of a phishing attack on Microsoft 365 users shows that many security policies or frameworks are not foolproof solutions.

Phishing Challenges and Solutions

Vayansky and Kumar (2018) discuss a combination of steps that can be used to control and keep phishers at bay.

Step 1: Prevent Phishing

Prevention is better than cure. Prevent phishing before it reaches the user, by blacklisting, blocking, or filtering phishing emails. Use machine learning methods, statistical classifiers, and spam filtering techniques to identify potential phishing messages with varying degrees of effectiveness to block phishing sites and filter phishing emails (Sahingoz et al., 2019). Binks (2019) stresses the need for the right technology to be employed such as updated antivirus software and robust firewalls since they have a dual role in protecting emails and keeping the IT unfractured and guarded against threats and potential attacks by cybercriminals. Automated detection should be the first line of detection against phishing that way the attacks do not reach the users. Automatic phishing detectors are built into mail servers and clients, internet service providers (ISP), and web browser tools. These tools detect, blacklist, and block access to a phishing site and send a request to the ISP to take down the website (Alsharnouby et al., 2015).

Using an open authentication standard such as Domain-based Message Authentication Reporting, and Conformance (DMARC) aids in blocking malicious emails by spoofing your organization's domain before it touches the user's mailboxes (Derouet, 2016). DMARK is built on the backbone of DomainKeys Identified Mail (DKIM) and Sender Policy Framework (SPF) which are popular email authentication standards that validate domains which are registered with email and ensure that messages from a particular domain get delivered to the email inbox. Hackers sometimes register and authenticate their domains with these standards to ensure the emails get delivered to the intended recipients.

Palo Alto Networks, a cybersecurity service provider company, has developed a preventative to phishing, Next-Generation Firewalls (NGFW), which use Artificial intelligence and Machine Learning to detect malicious behaviors.

Step 2: Detect Phishing

Use active or passive browser indicators to alert the user that a site may be a phishing website by displaying a pop-up or warning message that the site is not potentially safe or suspected fraud (Moghimi & Varjani, 2016). Also applying a verification system for sites that are trusted and secure can be helpful. Alsharnouby et al. (2015) also mention some of the techniques used to detect phishing websites: machine learning, url feature classification, domain name analysis, contextual analysis, visual similarity assessment, behavioral prediction, and crowdsourcing. Google uses automated machine learning techniques, while Phishtank, an anti-phishing site owned by OpenDNS offers a blacklist for use by other tools employing an API. Crowdsourcing volunteers submit potential phishing websites, vote on whether they are legitimate or phishing websites, and thereby populate their blacklist. Peterson, (2018) points out that although subtle, there exist behavioral signs that indicate the presence of a malicious imposter, however, identifying these signs with certainty requires data analysis that can only be possible through machine learning.

Step 3: Stake Holder Training

Greene et al. (2018) discuss that technology solutions are far from secure and that phishing requires a multi-pronged defense. They stress the important part that users play in the security of an organization. Mandatory training sessions at organizations to spread security awareness are required. Phishing awareness training is particularly important when simulated phishing emails mimicking real-world phishing emails are sent to the employees. In using embedded phishing training techniques, organizations should understand its scope, limitations, and benefits since user awareness is the best way to combat phishing. Using training techniques such as embedded mock emails and game-based anti-phishing training should be used to make users aware of the indicators and be aware of the techniques that hackers use.

Binks (2019) emphasizes the need for company-wide education for it is crucial in fighting or thwarting these attacks. Security is a shared responsibility of the entire organization. Training may seem like a simple idea, but it is the most effective way to plan for a phishing attack. It needs to spread further than the IT department, making all employees aware of what phishing attacks look like, the risks they pose, and steps to take in case of a breach (Figure 2.1.1).

A LIST OF LESSONS LEARNED AND SUGGESTIONS

- A strict security policy should be put in place and enforced.
- Antivirus and firewalls should be installed and updated regularly.
- Train users to recognize phishing emails/messages.
- BEC scams are more likely to be successful if the attackers have added knowledge of the target organization and its structure.
- A significant belief of attackers is that they will be able to deceive users through websites with visual characteristics sufficiently believable to be accepted as legitimate. This holds good in the case of the Microsoft 365 incident (Alsharnouby et al., 2015).
- Palo Alto Networks a cybersecurity service provider company has developed Next-Generation Firewalls (NGFW) which use Artificial intelligence and Machine Learning to detect malicious behaviors.
- Simulated attack training is by far the most effective form of training, as it helps humans recognize the subtle clues to identify phishing attempts and gives employees a baseline understanding of the latest techniques that attackers are using.

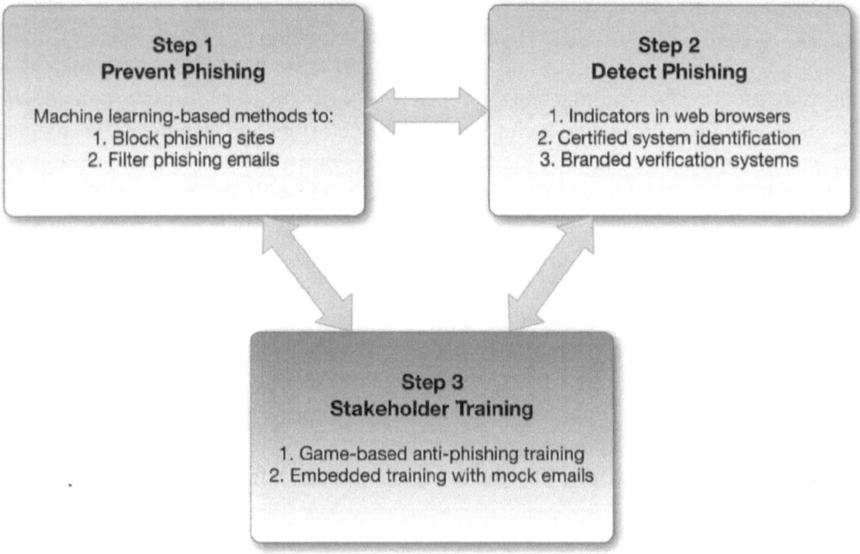

FIGURE 2.1.1 Proposed solution to combat phishing.

Source: Vayansky and Kumar (2018).

- Never click on links within emails.
- Look into blockchain security infrastructure security as a means of security.
- Organizations should hold mock drills to educate employees against phishing.
- Use two or more step verifications to secure password authentication.
- Use different passwords for different accounts.
- Cybercriminals are constantly evolving, rediscovering new ideas, and plotting new plans and are one step ahead. Therefore, we will need to keep re-evaluating our security, conduct constant audits, and plan ahead to mitigate these attacks (Binks, 2019).
- Information Security (Infosec) cybersecurity policies must be enforced in organizations and reviewed and monitored periodically.
- Use cutting-edge technologies such as blockchain AI and Machine learning techniques to detect and prevent attacks.

CONCLUSION

The Microsoft 365 Phishing case is a prime example that demonstrates that with the advancement in technology, even hackers are becoming more advanced and using new methods and levels of sophistication to achieve their goals. Cybersecurity professionals are finding it harder to identify phishing as it is growing in complexity (Vayansky & Kumar, 2018).

Most researchers and security experts have identified that users are the weakest link in IT security. The user aspect cannot be ignored when discussing security measures against threats; companies and providers need to work with users to make systems more secure. User training and appropriate security measures must be considered when utilizing IT services and technology to protect users and organizations against data hackers. User awareness is the best defense against phishing. Users should be aware of the threats and techniques used by hackers (Mansfield-Devine, 2018).

DISCUSSION QUESTIONS

1. How do we effectively guard against phishing?
2. What are the best ways to train a user against phishing?
3. How can organizations employ a security culture to avoid malicious attacks?
4. What are the latest technologies that be used to prevent phishing?
5. Whose responsibility is security in a cloud?
6. What can organizations do when they have been victims of a phishing data breach?

REFERENCES

Alsharnouby, M., Alaca, F., & Chiasson, S. (2015). Why phishing still works: User strategies for combating phishing attacks. *International Journal of Human-Computer Studies*, *82*, 69–82. https://doi.org/10.1016/j.ijhcs.2015.05.005

Arachchilage, N. A. G., & Love, S. (2014). Security awareness of computer users: A phishing threat avoidance perspective. *Computers in Human Behavior*, *38*, 304–312. https://doi.org/10.1016/j.chb.2014.05.046

Arachchilage, N. A. G., Love, S., & Beznosov, K. (2016). Phishing threat avoidance behaviour: An empirical investigation. *Computers in Human Behavior*, *60*, 185–197. https://doi.org/10.1016/j.chb.2016.02.065

Binks, A. (2019). The art of phishing: Past, present and future. *Computer Fraud & Security*, *2019*(4), 9–11. https://doi.org/10.1016/S1361-3723(19)30040-5

Derouet, E. (2016). Fighting phishing and securing data with email authentication. *Computer Fraud & Security*, *2016*(10), 5–8. https://doi.org/10.1016/S1361-3723(16)30079-3

Furnell, S., Millet, K., & Papadaki, M. (2019). Fifteen years of phishing: Can technology save us? *Computer Fraud & Security*, *2019*(7), 11–16. https://doi.org/10.1016/S1361-3723(19)30074-0

Greene, K., Steves, M., & Theofanos, M. (2018). No phishing beyond this point. *Computer*, *51*(6), 86–89. https://doi.org/10.1109/MC.2018.2701632

Gupta, B. B., & Badve, O. P. (2016). Taxonomy of DoS and DDoS attacks and desirable defense mechanism in a Cloud computing environment. *Neural Computing & Applications*, *28*(12), 3655–3682. https://doi.org/10.1007/s00521-016-2317-5

Hashizume, K., Rosado, D. G., Fernández-Medina, E., & Fernandez, E. B. (2013). An analysis of security issues for cloud computing. *Journal of Internet Services and Applications*, *4*(1), 5–13. https://doi.org/10.1186/1869-0238-4-5

Kumar, P. R., Raj, P. H., & Jelciana, P. (2018). Exploring Data Security Issues and Solutions in Cloud Computing. *Procedia Computer Science*, *125*, 691–697.

Mansfield-Devine, S. (2018). The ever-changing face of phishing. *Computer Fraud & Security*, *2018*(11), 17–19. https://doi.org/10.1016/S1361-3723(18)30111-8

Modi, C., Patel, D., Borisaniya, B., Patel, A., & Rajarajan, M. (2012). A survey on security issues and solutions at different layers of Cloud computing. *The Journal of Supercomputing*, *63*(2), 561–592. https://doi.org/10.1007/s11227-012-0831-5

Moghimi, M., & Varjani, A. Y. (2016). New rule-based phishing detection method. *Expert Systems With Applications*, *53*, 231–242. https://doi.org/10.1016/j.eswa.2016.01.028

Peterson, P. (2018). Unmasking deceptive attacks with machine learning. *Computer Fraud & Security*, *2018*(11), 15–17. https://doi.org/10.1016/s1361-3723(18)30110-6

Sahinoglu, M., & Cueva-Parra, L. (2011). CLOUD computing. *Wiley Interdisciplinary Reviews. Computational Statistics*, *3*(1), 47–68. https://doi.org/10.1002/wics.139

Sahingoz, O. K., Buber, E., Demir, O., & Diri, B. (2019). Machine learning based phishing detection from URLs. *Expert Systems With Applications*, *117*, 345–357. https://doi.org/10.1016/j.eswa.2018.09.029

Steer, J. (2017). Defending against spear-phishing. *Computer Fraud & Security*, *2017*(8), 18–20. https://doi.org/10.1016/s1361-3723(17)30074-x

Vayansky, I., & Kumar, S. (2018). Phishing – challenges and solutions. *Computer Fraud & Security*, *2018*(1), 15–20. https://doi.org/10.1016/S1361-3723(18)30007-1

2.2 Internet of Things Security

Trend Micro Experiment

Roydon Dominic Correya
MYOB, Christchurch, New Zealand

Tushar Ashok Sagathia
ASB Bank, Auckland, New Zealand

IOT AND CASE SUMMARY

In the article, *How the term 'Internet of Things' was invented* (Rayome, 2018) 'Kevin Ashton' explains how he coined the term Internet of Things (IoT) in the 1990s while he was working for Procter & Gamble and made a presentation to tag products with radio frequency chips to track them in their supply chain.

IoT terminology has come a long way after almost two decades and it is now being used as jargon by many people. IoT is an umbrella term for devices that use internetworking technology to transmit information to and from each other, in other words, any device that can communicate using the internet can be called an IoT device. These devices are connected to the internet using silicon chips that contain microprocessors and can be programmed to perform computing tasks such as sending and receiving information in a network. These devices are also connected with 'sensors' which are other hardware components that can be connected to the chipsets for 'sensing' physical variables such as temperature and pressure. Actuators are also a part of the device which can be used to create outputs or perform actions in the environment of the IoT device. A robotic arm used in a factory is an example of an actuator. Combining all these components we get what is a modern IoT device.

IoT technology wherein devices share information and perform tasks based on that information has changed the world as we know it. Factories are using more and more of these IoT devices for safer and efficient production, and individuals use smartphones which are a collection of sensors and actuators programmed to perform specific tasks. These smartphones might have much more computing power compared to other IoT devices and much less compared to Industrial IoT (IIoT) devices. IoT technology has been revolutionary and is being adopted on a wide scale. According to Klaus Schwab, the author of the book *The Fourth Industrial Revolution*, IoT is one of the technologies that is leading this new revolution. These IoT devices

DOI: 10.1201/9781003264415-18

are being used in modern households (smart homes), factories (smart factories), and entire cities (smart cities); hence, anything that contains the use of IoT devices is referred to as 'smart'.

The use of this technology is very promising and is expected to change many aspects in which our world functions creating more efficiencies, but like any technology, IoT has its downsides too. These IoT devices create a big security risk where, if proper measures aren't followed, their use can lead to disastrous consequences. What is more concerning is that this technology exists all around us penetrating our homes, workspaces, and public spaces. There is no escape from this, and misuse of such infrastructures can result in the compromise of privacy of individuals, loss of business data, and even cause extreme scenarios where people are harmed in the case of malfunction of devices caused by an intruder. Such extreme scenarios can be a path to lawsuits which can hurt businesses.

The Cybersecurity firm, Trend Micro, conducted an experiment where they mimicked the infrastructure of a real-time smart factory to study how vulnerable these IoT infrastructures used in smart factories can be (Onag, 2020). Trend Micro designed a '*honeypot*' where they set up an environment of 'Operational Technology' used in smart factories leaving vulnerabilities often observed in real factories. Set up to entice hackers, a honeypot does not contain actual business data and is created with some vulnerabilities to lure hackers inside the network, to determine the behavior of the hackers, and to find out ways in which the system can be compromised. This experiment shows how the technologies used in smart factories are vulnerable to common threats and can result in interruption of business activities having time cost and financial costs for the enterprise.

To better understand the attacks targeting Industrial Control Systems (ICS) environments Trend Micro created a highly realistic Industrial Prototyping company. The honeypot included real ICS hardware, physical and virtual machines to run the factory, and Programmable Logic Controllers (PLCs). The environment even included a Human–Machine Interface (HMI), Robotic and Engineering Workstations, and a File Server. The honeypot was made to be highly realistic mimicking a real-life smart factory.

Throughout 7 plus months of operation, the honeypot was under various attacks including installation of cryptocurrency mining software known as 'crypto-jacking', Ransomware attacks, and consumer fraud. The general belief is that industries are at the risk of attack on a nation-state level to disrupt key processes, but this experiment proves that commonplace threats are more likely to happen with the use of these technologies.

CASE ANALYSIS

IoT technology is at the forefront of Industry 4.0, is crucial in creating new efficiencies in the world, and is potentially connecting every individual globally. This can have some really good results where people in the remotest parts of the world have access to electricity, financial services, and the internet which will raise their overall standard of living. This is progressive and needs to be adopted rapidly to create more viable services for people, creating an all-inclusive environment. With having such noble use cases, the implementation of this technology must be done very carefully considering the major security risks in having such infrastructures throughout global societies and making them a crucial part of human life. The Trend Micro case study helps us understand the security perspective of IoT technology of how it can be exploited, what are the major and more common threats to ICS, and what safeguards must be taken to make it secure resulting in benefits for individuals and businesses. A systems perspective has been adopted for performing the analysis on this case and the parts of the system identified are given in Table 2.2.1.

Table 2.2.1 gives us the components of a systems perspective that was developed for carefully imitating a real-life smart factory environment. In this case, the researchers at Trend Micro (Hilt et al., 2020) describe how they created the honeypot and what components were used to build the system of this fake factory. The paper discusses how the researchers set up a honeypot by first building their ICS (PLC hardware) and HMI which was used to control the PLC hardware. A separate file server was set up using the Windows 7 Professional Operating System. Trend Micro also set up an engineering workstation and robotic workstation as these are core parts of modern smart factories. The file server was populated with random files generated using a script. The design of the system was kept as realistic as possible where the file server was connected to open Virtual Network Computing (VNC) and the robotic workstation was not kept open as generally the robotic workstations are connected to a separate closed network. The design of the system was crucial to guarantee that hackers not identify this to be a honeypot but rather continue their hacking of a presumed, actual company. The researchers even created a website and named their company 'MeTech' which provided rapid prototyping solutions to a list of anonymous clients which would be big companies. They even use Artificial Intelligence (AI) to generate fake images of people and upload them to the company website to make it look as realistic as possible. Real images of people were not used as images can be searched by 'reverse image lookup' services and so compromised the project. For the same reason, social media profiles for the company were avoided as they could lead to an increased margin of error. The systems were also connected to monitoring software tools such as Moloch, a network traffic analyzer. The monitoring was done by capturing the data which would be saved on a pocket-sized computer called Raspberry Pi 3, which could only be accessed by the virtual machine environments

TABLE 2.2.1
System Parts for Analysis

System Component	Examples
Hardware	Industrial Control Systems (ICS) Hardware, Human–Machine Interface (HMI), File Server, Raspberry Pi 3, and Cisco Firewall
Software	Operating system, virtual machines, and monitoring software tools
Data	Client lists, prototype design
Processes	Production of prototypes for clients using robotic arm and ICS hardware
People	Team members

which were set up alongside the File server. Mini computing boards like Raspberry Pi and Arduino are very popular for creating IoT devices, where these boards can be connected with sensors such as cameras and deployed on the field with low costs and space requirements. This is how the honeypot was set up by the researchers to understand the types of attacks that can be done on smart factories. The next section discusses the different attacks that took place in this system.

Attacks on the System

The honeypot went live on May 6, 2019. The researchers (Hilt et al., 2020) had exposed MeTech's HMI machine that controlled the ICS hardware by connecting it to an unsecured Virtual Network and used the same passwords for several workstations. This was done deliberately to lure attackers into the system, furthermore, the team even made it seem like the system had been compromised previously by leaking the information about it on online forums and databases. This made sure that the system was seen by attackers and the design of the system was done to make it look as real as possible.

The first attack that happened in the system is what is known as cryptojacking. According to Tuttle (2018), cryptojacking attacks siphon off victims' CPU processing power and electricity to mine cryptocurrencies. Cryptocurrencies are digital currencies; wherein certain currencies are created by solving mathematical problems on computers. For example, the most famous cryptocurrency, 'Bitcoin', uses the 'Proof of Work' (PoW) algorithm to create new Bitcoins. These Bitcoins are created when a participant of the blockchain solves these mathematical computations using their devices and publishes a new batch of Bitcoin transactions on the Bitcoin blockchain. In return, the individual/entity who does the work, also called a 'miner' is rewarded with 1 new Bitcoin. This process is referred to as mining Bitcoins. The network here requires the miner to show that they have done work by publishing new transactions on the blockchain and this is called the Proof of Work consensus mechanism. The distributed network agrees upon the current state of the blockchain, which is a list of immutable, cryptographic transactions. Similarly, other cryptocurrencies are mined but the consensus mechanism may be different for different cryptocurrencies. Cryptocurrencies are generally limited in supply so that their value doesn't go down with inflation; hence, the difficulty of mining these cryptocurrencies keeps on increasing with new coins being mined. This causes the cost of mining new coins to go up and getting good returns on investment in the mining business gets harder.

This has led to attackers using cryptojacking as a way of installing crypto-mining malware on victims' devices to mine these currencies and reap the profits. In this case, industrial devices can have a lot of computing power which can be utilised by the attackers bypassing the hardware and electricity cost on these victim devices. According to Tuttle (2018), McAfee Labs in their report saw a 629% increase in coin miner malware in the first quarter of 2018, jumping to more than 2.9 million total known samples from around 400,000 in Q4 2017. Hence it is important to make sure the smart factories are safeguarded against this common threat of crypto-jacking which could waste the resources and incur indirect financial costs and time costs in replacing the used-up hardware of the systems.

The second and third attacks on the MeTech system were ransomware attacks. According to Mansfield-Devine(2018), ransomware is a form of technological blackmail, where the malware encrypts files on the hard drive of your computer and then presents a message telling you how to get the documents unlocked again. That process usually involves making a transfer of funds to the cyber-criminals, most commonly using Bitcoins or other cryptocurrencies, in return for a decryption key. Modern operating systems are less prone to having easy-to-exploit vulnerabilities, where the error to compromise a system can be mostly based on social engineering attacks such as spear phishing attacks in which known and targeted individuals are picked for attacks. But this case focuses on IoT devices which may not run on these modern operating systems; organizations, on the other hand, do employ these modern operating systems on a large scale. There are more security-oriented best practices and policies in organizations than in factories, where the workforce also might be like the MeTech system which was hit with two ransomware attacks in consecutive months – the Crysis ransomware and a variant of Phobos ransomware, respectively.

Not being technically adept at understanding which loose ends are left vulnerable enables these attacks. but in the case of Trend Micro vulnerabilities were intentionally left open. The team even had many email exchanges with the attackers to make them believe that the data which was lost in encryption had value and that this was indeed a smart factory. In cases of ransomware attacks, attackers usually ask organizations to download the TOR browser which is an encrypted service that can be used to browse the web anonymously. Once the system had been set up the organizations were then directed to make payments to the mentioned darknet accounts. TOR browser is a great privacy tool but is not entirely secure. Used with a combination of other tools such as the Tails Operating System, the TOR browser might be more effective, otherwise using it from your local device just makes you more visible to the authorities and your Internet Service Providers (ISP).

There were other intrusions also recorded in the system over 7 months in 2019 by the team of researchers at Trend Micro. These other intrusions included attackers snooping around the system, renaming files, and later deleting files on the file server. Some attackers even left notes that advised users to change the passwords on the devices and on not doing so, the attackers returned and opened pornographic websites on the systems to make a point. These attacks put the security of IoT devices in perspective, as it has become more profitable for attackers to breach organizations for mining cryptocurrencies, blackmailing the organizations by encrypting their data, and so on. This also opens many vulnerabilities and commonly made mistakes in cloud computing environments.

These vulnerabilities affect the Confidentiality, Integrity, and Availability of data, also known as the 'CIA Triad'. The authors Kumar et al. (2017) discuss how Confidentiality relates to the privacy of data while Integrity is the confidence that the data served from the cloud is untampered and in its original state free of malicious code. Availability refers to the assurance that important business data is available when required. As discussed in the case study, the vulnerabilities found by (Hilt et al., 2020) in their honeypot reveal that all three aspects of the CIA triad are compromised by common attacks which can be avoided. Cryptojacking affects the

Integrity of data, Ransomware affects the Availability of data, and other snooping attacks impact the Confidentiality of data. Hence, the essential quality of the CIA Triad of data is lost in these IoT systems.

LESSONS LEARNED

There are many learnings from the MeTech case study, but the most important takeaway is to know that IoT technology which is being deployed extensively is not fully secure. Nothing is 100% secure, but comparatively, IoT devices lack the security protocols normally observed in bigger organizations and are mostly overlooked.

The Clark–Wilson Integrity Model as discussed by De Capitani di Vimercati & Samarati (2011) protects the integrity of commercial information by allowing only certified actions by explicitly authorized users on resources.

The Clark–Wilson integrity model, which is built upon principles of change control rather than integrity levels, was designed for the commercial environment. The change control principles upon which it operates are

- unauthorized users cannot make changes
- authorized users cannot make unauthorized changes, and
- internal and external consistency is expressly maintained.

These change control principles can make sure the integrity of the system is maintained and can be a safeguard against crypto-jacking malware. This model can also be used to tackle the control access vulnerability that was left intentionally open by the team at Trend Micro and is a common occurrence in such factories. The password management system of the routers and all other data can have integrity by following the Clark–Wilson Integrity Model.

Another strategy to combat data integrity issues can be to use deep learning as shown in the research by Darabian et al. (2020), where LSTM deep learning models are used to identify crypto-mining malware. The use of machine learning techniques such as deep learning is getting very popular and there are various physical and virtual infrastructures available to execute it and hence can be a very effective strategy to combat data integrity issues caused by attacks such as cryptojacking.

The issue of availability of data that can be caused by ransomware can be countered by creating micro-segments of the data in the organization as suggested by Mansfield-Devine (2018) where the organizational managers can create hundreds or thousands of encrypted packets which segment the various aspects of organizational data. Here, the data between each segment can be processed by not outside the segment, and this can prove to be effective against the spread of the ransomware malware. Another easier way to protect against ransomware threats that affect the availability of business data in the organization is by taking regular backups of data. But this is still vulnerable if the backup databases are connected to the primary database; the malware can copy itself and encrypt that too, like a worm malware. A combination of these strategies can be used to tackle ransomware attacks. Containment of the malware by micro-segmentation and using cold storage or securely connected backup databases which are regularly updated can help keep business continuity.

DISCUSSION QUESTIONS

1. Should you pay the ransom?
2. What are the advantages and disadvantages of IoT devices in businesses and daily life?
3. Is cloud computing and virtual computing technology worth it?
4. Are you aware of the data and information that IoT devices are collecting from us every day?
5. How secure is my data?
6. What are the implications of data breaches in IOT environments?
7. How can they be secure?

REFERENCES

Darabian, H., Homayounoot, S., Dehghantanha, A., Hashemi, S., Karimipour, H., Parizi, R. M., & Choo, K.-K. R. (2020). Detecting cryptomining malware: A Deep learning approach for static and dynamic analysis. *Journal of Grid Computing*, *18*(2), 293–303. https://doi.org/10.1007/s10723-020-09510-6

De Capitani di Vimercati, S., & Samarati, P. (2011). Clark and Wilson model. In H. C. A. van Tilborg & S. Jajodia (Eds.), *Encyclopedia of cryptography and security* (pp. 208–209). Springer US. https://doi.org/10.1007/978-1-4419-5906-5_814

Hilt, S., Maggi, F., Perine, C., Remorin, L., Rösler, M., & Vosseler, R. (2020). *Caught in the act: Running a realistic factory honeypot to capture real threats*. Trend Micro.

Kumar, P. R., Raj, P. H., & Jelciana, P. (2017). Exploring security issues and solutions in cloud computing services – A survey. *Cybernetics and Information Technologies*, *17*(4), 3–31. https://doi.org/10.1515/cait-2017-0039

Mansfield-Devine, S. (2018). Ransomware: Taking businesses hostage. *Network Security*, *2016*(10), 8. https://doi.org/10.1016/s1353-4858(16)30096-4

Onag, G. (2020). *Trend Micro: Honeypot reveals threats to smart factories*. FUTUREiOT. https://futureiot.tech/trend-micro-honeypot-reveals-threats-to-smart-factories/

Rayome, A. D. (2018). *How the term 'Internet of Things' was invented*. TechRepublic. https://www.techrepublic.com/article/how-the-term-internet-of-things-was-invented/

Tuttle, H. (2018). *CRYPTOJACKING how hackers steal resources to mine digital gold*. *Risk Management*. https://www.rmmagazine.com/articles/article/2018/08/01/-Cryptojacking-How-Hackers-Steal-Resources-to-Mine-Digital-Gold-

2.3 2014 Cyber-attack on eBay Case Study Analysis

Areeb Amanulla
Kainga Ora – Homes and Communities, New Zealand

Saadik Niyaz
Waste Management NZ, Ltd, New Zealand

CASE SUMMARY

eBay is one of the world's largest online shopping sites, popularly known for its consumer-to-consumer sales and its auctions. eBay is a sales channel that operates in more than 180 different countries across the world. The e-commerce giant operates a platform over the internet where buyers and sellers can interact with each other across the world to transact goods such as clothing, accessories, jewelry, watches, computers, networking, antiques, boats, and video games. eBay comprises 4.3 million new product listings daily; further, eBay has 167 million active users and 25 million sellers. Moreover, eBay announced that its annual revenue for 2019 was above $10.5 billion.

In early 2014, eBay announced to all its customers to change their passwords since the company had just become aware of a possible data breach taking place in its system. They stated that hackers had been able to acquire the login credentials of several employees, through this they were able to gain access to the system and access the usernames, addresses, contact information, email addresses, and passwords of more than 128 million users. However, eBay further made public that the credit card details of none of these users had been accessed by the hackers.

The details of how the attack occurred had not been completely made aware to the public. However, the general understanding is that this occurred due to a phishing attack done by the hackers. Through this, the hackers were able to access employee records and attack the system. Evidently, a failure of the employees of eBay was that they reuse the same credentials for passwords, and hence when they are compromised once, the breach gives hackers access for a very long time since the company has had no policy in changing credentials. Furthermore, investigators identified that eBay passwords did not have a two-factor authentication system. This allowed

DOI: 10.1201/9781003264415-19

anyone who obtained the password credentials of an employee to access the system with no further security protocol to breach. These failures were key to the data breach at eBay.

Among all these failures, that the credit card information of users was not breached by the hackers is astonishing. The security protocols applied by eBay to their payment portal and the personal portal were two different controls. Since the information consisted of two different classes of data, they needed different controls. The payment portal information comes under the regulation of Payment Card Industry Data Security Standard (PCI DSS), and eBay had to ensure they complied with these regulations. Therefore, information about credit cards had strict access controls with two-factor authorizations and hierarchies access points.

Other than the obvious failure in not having tighter security protocols for its employee information, eBay was largely criticized for delaying the announcement of the breach. The public was deprived of this information for 2–3 months from when it had occurred. However, eBay's response to the criticism was that the hackers had used their own employee credentials to access the system; furthermore, they had also managed to exfiltrate the data without any detection. Due to this, the team at eBay did not identify the breach as soon as it had occurred, and hence the delay in reporting the shortfall of it.

CRITICAL ANALYSIS

An investigation identified that eBay was an unfortunate victim of a phishing attack. According to Forcepoint (2019), a phishing attack is a deceptive method used by hackers to breach servers and gain access to confidential and sensitive information. The attackers deceptively present themselves as legitimate individuals/organizations and manipulate users to perform certain specific actions which unbeknownst to the users gives these attackers access. There are many types of phishing attacks (Newman, 2017), some such methods are summarized below.

CEO Fraud/Business Email Compromise

CEO fraud is a typical case where an attacker sends a lower-level employee an email claiming to be the CEO of the firm. In most cases, this email is typically sent to an individual operating in the accounts or finance divisions. The hacker's motive for sending this email is to get the victim to disclose sensitive financial information or in some cases transfer funds to a fake account. Even though it seems basic to avoid, according to the FBI, such attacks cost companies millions of dollars annually (Crane, 2019).

Clone Phishing

The motive of a hacker using Clone phishing is to take advantage of genuine messages a user could have already received and create a deceptive version of it to gain access to servers and systems. The attacker does this by creating a replica of the original message. This message is sent to the user through a separate email address which looks like the legitimate message the user had received. This email will be sent by the attackers containing malicious links and attachments (Sander, 2019).

Domain Spoofing

This is a method used by an attacker with the motive of spoofing an organization. Commonly, spoofing is done either to make recipients view emails received by the organization through a legitimate domain or to make a fake website look legitimate. The email spoofing is done by creating a replica of the email header of the organization to look legitimate. Website spoofing is done by creating a fake website which contains a replica domain which is similar to the legitimate organization (Crane, 2019).

Evil Twin

Evil twin phishing is a different kind of phishing compared to the other methods. This method focuses on capitalizing on the Wi-Fi networks of an organization. An attacker conducts Evil twin phishing by creating a similar wireless access point to a legitimate Wi-Fi access point. The motive of the hacker is to get a user to mistakenly access the rogue Wi-Fi access point, which allows the attacker to gain access to the user's servers. This method of phishing is commonly experienced in coffee shops (Sander, 2019).

HTTPS Phishing

This is statistically the most common method used by attackers. This method focuses on creating fake advertisements and emails with offers and vouchers with links to them. The motive of an attacker here is to get the user to access the link. The attacker uses a deceptive URL as a link which is similar to a legitimate link. Therefore such an attack is extremely hard to identify even for careful users (Bisson, 2019).

With the different types of major phishing types identified above, the discussion now moves to understanding the techniques which one can use to avoid being a victim of phishing attacks.

Keep Informed

The majority of individuals who succumb to phishing attacks are unfortunately victimized due to their lack of knowledge about these kinds of attacks. Induction training to educate all employees of an organization regarding this should be made mandatory. Furthermore, workshops should be conducted often updating employees regarding new forms of phishing attacks (Berdine, 2019).

Verify a Site's Security

Individuals commonly enter various sites across the internet and if a secure website is entered into, there should not be a problem. The most common way of identifying a secure website is first to ensure that the URL begins with "https" while also checking if the "closed lock" icon near the address bar is visible. A further method of verifying a website's security is to check if the site issues a security certificate.

Keep the Browser Up to Date

A key advantage of keeping the browser up to date is that this ensures that any security loopholes that phishers utilize to attack are being covered. Whenever a new loophole is discovered by phishing attackers, the security teams initiating the browser send across updated security patches to ensure this loophole cannot be used by an attacker to breach another user (Berdine, 2019).

Be Wary of Pop-Ups

Pop-up windows in the present context are a very common experience among users. Users should be aware of how to handle pop-ups, which often lead to a separate window with options of "proceed" and "cancel"; users should not click either. Rather, a user should look for an "x" button to close the pop-up. A simpler way however is to change the browser settings to block pop-ups.

A further failure that the team at eBay encompassed was their failure to implement a two-factor authentication to their system. This enabled anyone who had access to a password to gain access to the system. According to Coco-Scotts (2019), a two-factor authentication is a multi-factor authentication security measure which acts as a cross-verification when users want to log into their system, commonly generating a PIN number to their personal mobile or Email address. However, two-factor authentications also have their downsides. According to Imperva (2019), one of the main cons of two-factor authentication is that the additional step increases the login time of users, requiring them to go through an extra process which can sometimes take long. Furthermore, a two-factor authentication also requires dependence on third parties such as mobile service providers or email providers. Finally, maintaining two-factor authentications can also become an issue since it could disrupt the efficiency of users.

LESSONS LEARNED FROM THE EBAY DATA BREACH

Until now, eBay has not released any technical details of how the hackers obtained employee credentials. History teaches us that a majority of data breaches occur as a result of a vulnerability exploited in a web application or a social engineering attack. One single application which is vulnerable within an organization could pave the way to a full breach of database servers and web servers. The eBay data breach experience teaches us many lessons to counteract such attacks in the future. Some such important lessons to be noted are summarized below.

Keep Web Server Operating Systems and All Software Updated

Considered to be the first layer of security, updating with up-to-date security patches is critical to ensure that the operating system on which the database servers, web servers, and the other network servers run smoothly and safely. A very common mistake for most organizations is to have unpatched software or operating systems.

Always Follow the Principle of Least Privilege

The principle of least privilege is to have the webserver be granted the least access privilege when interacting with the database. The web server should not have the authority to perform any administrative functions in making structural changes to databases. In such an instance when the server is hacked the principle limits the damage which would be caused to the system.

Limit Remote Access

A major learning from the eBay data breach is to limit remote access. Ideally, in the process of maintaining tight security protocols, database servers and web servers should only have access authorization via the local network, rather than the open internet. In a situation where employees would require remote access, such employees should be given permission to access the system using whitelisted virtual private networks. Furthermore, ensuring that employees are educated on the risks of accessing the system through external internet networks is important.

Do Not Use Shared Hosting

Even though shared hosting is one of the cheapest solutions it also has its fair share of risks. A hacker would be able to directly access the database server or the webserver if he manages to gain access to another site on the same server. Avoiding shared hosting is one of the basic steps to take to protect servers. eBay however could not avoid this.

Frequently Monitor and Audit

Making appropriate use of log files is highly important for organizations. These files should be analyzed frequently to identify any possible suspicious activity. A process of monitoring and auditing should be formulated and practiced. Log files should also be stored in separate areas to ensure malicious users do not gain access to and are then able to tamper with them. How frequently the management of eBay audited and monitored these best practices is unclear.

Keep Informed

Finally, all organizations should ensure that they are being kept updated on the latest matters related to security in the present. Formalized processes should be introduced to ensure employees are kept informed of these current security matters; furthermore, employers should encourage employees to share their findings with the organization and also create a mechanism of how their ideas can be implemented across the organization.

DISCUSSION QUESTIONS

1. Was eBay wise in not disclosing details of the breach to the public?
2. Do you believe the security policies of eBay have been taken seriously by the company? Or, do you believe eBay has been focusing more on the business end of things to the detriment of security?

REFERENCES

Berdine, M. (2019). 10 tips to avoid phishing scams. Emailonacid. Retrieved from https://www.emailonacid.com/blog/article/industry-news/how-to-prevent-phishing/

Bisson, D. (2019). 6 common phishing attacks and how to protect against them. TipWire. Retrieved from https://www.tripwire.com/state-of-security/security-awareness/6-common-phishing-attacks-and-how-to-protect-against-them/

Coco-Scotts, K. (2019). Multi factor authentication pros and cons. Security Boulevard. Retrieved from https://securityboulevard.com/2019/10/multi-factor-authentication-pros-cons/

Crane, K. (2019). Types of phishing attacks. HashedOut. Retrieved from https://www.thesslstore.com/blog/10-types-of-phishing-attacks-and-phishing-scams/

Forcepoint (2019). What is phishing. Forcepoint. Retrieved from https://www.forcepoint.com/cyber-edu/phishing-attack

Imperva. (2019). Two factor authentications. Imperva. Retrieved from https://www.imperva.com/learn/application-security/2fa-two-factor-authentication/

Newman, L. H. (2017). Phishing scams even fool tech nerds, here's how to avoid them. Wired. Retrieved from https://www.wired.com/2017/03/phishing-scams-fool-even-tech-nerds-heres-avoid/

Sander, R. (2019). 10 types of phishing attacks and phishing scams. ITportal. Retrieved from https://www.itproportal.com/features/ten-types-of-phishing-attacks-and-phishing-scams/

2.4 Internal Revenue Service Scams

Nishanth Chakkere Ramesh
Kapiti Coast District Council, New Zealand

Ying-Pin Lan
Hato Hone St. John, Auckland, New Zealand

CASE STORYLINE

The Internal Revenue Service (IRS) is a federal agency in the United States that is responsible for collecting federal taxes, aiding citizens in matters relating to taxation, and resolving errors in taxation. The scams related to the IRS can be broadly classified into three categories – monetary scams, scams relating to information/data of citizens, and a combination of both. Most of these scams play upon the vulnerability of citizens who do not understand the policies, laws, and practices of the IRS. Moreover, with the increase in internet penetration and with the role of information systems in the management of taxation rising, scammers have found targeting citizens in multiple ways is more opportunistic ("Tax Scams/Consumer Alerts," 2020).

The majority of scams are still carried out over the telephone, especially by perpetrators who are outside the United States. Traditional scams targeted individuals by warning them that they may be arrested or fined for not filling the tax forms in the correct manner or failing to pay the correct amount of taxes. The callers then asked for the credit/debit card numbers or directed the individuals to transfer money to a given bank account. Most of these scams play with fear tactics because ordinary citizens know that the consequences of not having a correct and up-to-date tax record can be serious (Fontinelle, 2019).

The key question behind these scams is – how did the scammers access the personal information of the victims? The information is usually available through unencrypted hard drives or the cloud, where information is stored by various acquaintances – employers, financial advisors, tax agents, or in the victim's own devices. Obtaining access to the Social Security Numbers (SSNs) of citizens will make the victim feel like the calls are legitimate and originating from the IRS. Moreover, it is increasingly difficult to keep up with the different tactics and media used by the scammers to contact victims. Although telephonic scams are popular, scammers are now using emails to contact the victims (Fontinelle, 2019).

DOI: 10.1201/9781003264415-20

Email scams are more difficult to uncover because the email addresses, the appearance of the email, and the details in the email seem very professional, and this tricks the victims into thinking that the emails are being sent by the IRS itself. The emails may also compel the victim to act quickly as they may provide links to online payment portals to rectify the mistakes or shortcomings in the filing of taxes (Fontinelle, 2019).

The onset of the Covid-19 global pandemic which has particularly affected the United States in a serious manner can increase the cases of IRS scams. These scams target the stimulus packages initiative of the government. Some scams purport that there is a processing fee required to be paid by the beneficiary before they receive the stimulus package. Other scams use emails to target victims by spreading misinformation regarding additional stimulus packages or other concerns stemming from the pandemic. The IRS has also identified scams which target victims by proposing that they invest in false firms which are on the verge of finding a vaccine for the Covid-19 virus (Erb, 2020).

Even without contacting the victims, the scammers can obtain the personal information and act as proxies of the victims. Fraudulent refunds can be claimed by scammers who use the SSNs of the victims, without the victim ever realizing it. The IRS may send a letter to the victim regarding a suspicious activity with their account, and only then would the victims realize the scam (Morgan, 2016).

The IRS has not been immune to data hacks and data breaches. In May 2015, it was reported that between 615,000 and 724,000 taxpayers had their data compromised by scammers who had obtained the data from elsewhere. The hackers then used this information to obtain the tax transcripts of the taxpayers through impersonation. The tax transcripts show the tax history of the taxpayer. The IRS notified the victims about this scam through email. This scam is related to identity theft, and the consequences may be more severe if the stolen information is used elsewhere such as payment for electronic purchases (Morgan, 2016).

The major questions surrounding these varied scams involving the IRS are, first, how to ensure that only the right information is being shared with the stakeholders, so as to eliminate parties that manipulate taxpayers through phone calls, emails, and other means. Second, it is crucial to ensure that only legitimate parties to the taxation information have access to it. This two-pronged problem is not unique to the IRS, but as a government entity, the consequences of data breaches can be more severe in comparison to those of private players. This case will provide insights into how public entities can manage information security to tackle varied challenges to the privacy and integrity of stakeholder data.

CRITICAL ANALYSIS

The data for this case analysis was gathered through the official IRS website, as well as online news publications. The IRS website provides comprehensive educational material for citizens to learn about the various types of scams and the scam reporting mechanism. The news articles cover topical issues related to scams and serve as the first point of reference for gathering preliminary information by the citizens. Since the nature of IRS scams are opportunistic, they keep changing with the changing times. The Covid-19 pandemic highlighted that scammers are always on the lookout for recent developments that may serve as a weak point for stakeholders, so that they may succumb to the traps laid by the scammers (Erb, 2020). Therefore, news articles serve as a good primary source of information about recent developments related to IRS scams.

The social context behind the IRS scams is transnational. While some scammers are based in the United States, many are based in South Asia. While India has built a reputation for hosting numerous call centers which provide customer service and technical assistance, investigations also show that many organized fraud cases are also traceable to India. On the surface, it is difficult to distinguish legal call centers from illegal ones. Since there is a large source of English-speaking graduates, they are hired to make scam calls by modifying their accents to be clear of foreign language influence. The U.S. Department of Justice, along with the Indian authorities, have made numerous indictments charging illicit call centers who scam U.S. residents. However, new illegal call centers spring up when the old ones are closed because the revenue generated through these activities is high – estimated at around 200,000 USD per day per call center (Bengali, 2016).

The large number of graduates in India, and the consequent strain on the job market, is also to blame for the existence of these illegal call centers. This means that candidates are willing to work for such illicit call centers because they know that their victims are mostly overseas, and this reduces their accountability under the law. They also earn a modest amount of salary, about 300 USD per month. Most of these call centers are based in metropolitan cities and their suburban towns, which provides access to a large number of graduates proficient in English (Bengali, 2016).

The scams are particularly targeted at the elderly who may not be tech savvy to identify that the calls or emails were illegitimate. Elders are also vulnerable because they are concerned about losing their government-funded financial support such as pension funds or medical rebates if they fail to pay their taxes or if there are any discrepancies with their tax information. Therefore, it is easier to convince them to make payments so as to sort out the issues with their tax accounts. Even though elders fall prey to IRS scams more than younger people, the rate of their reporting the matter to the authorities is significantly lower. Elderly victims of IRS scams can also face depression and premature death because they are unable to re-earn their retirement savings if they are lost to scammers (Gray, n.d.).

IRS scams have existed in a variety of ways for decades; however, their prevalence has increased after the information technology revolution which made the use of computers for communication more accessible. This removed the geographic barriers and gave the scammers access to a wide variety of information about their targets.

Related to scams are news about the leak of personal information. These two activities (scams and data leaks) complement each other, as the purpose of one is to bolster the other. There is routine news about the leak of SSNs by different parties and for different reasons. For example, in 2017, credit reporting company Equifax confirmed that there was a risk of identity theft of more than 140 million U.S. customers because unauthorized data access occurred from mid-May to July 2017. The personal identification information that was accessed were name, SSN, birthday, address, and driver's license number. The application vulnerability served as the access point for criminals (Symanovich, 2017).

In 2020, amidst the global Covid-19 pandemic, data leaks related to Pandemic Unemployment Assistance databases across some U.S. states were reported. Unemployment assistance applicants could see the information of other applicants when they were registering for the programs. The information included were names, SSNs, banking details, addresses, number of dependents, and, in some cases, correspondence between the unemployment office and unemployment applicants (Allen, 2020).

The similarity between both these cases and many other IRS scams is the access of SSNs. The nine-digit number is issued to U.S. citizens and residents and serves as a means of identification for a wide variety of government services, from taxation to public welfare schemes. This number is more sought after than other information because it is sort of a metadata that links various other data under one number. With this number, the hacker can access information relating to address, date of birth, employer details, etc. This information can be used to manipulate other information such as credit scores and health record details. Obtaining a new SSN is extremely difficult, and for most people, this number is for life. Therefore, securing the SSN is most critical (Horaczek, 2018).

The ability to secure against identity theft is also crucial to preventing identity theft. However, most authorities focus on recovering from an identity theft after it has occurred, rather than trying to prevent it in the first place. Identity theft can happen in the following ways.

Losing physical wallets, purses, and mail: although focus is provided on the data breaches online, physical loss of materials containing important information is perhaps more common. These personal belongings may have bank and credit card statements, tax information, credit offers, etc.

Stealing personal information through unsecured online sites and business records: phishing, malware attacks, malicious mobile apps, and other techniques may be used to elicit information through electronic means (details in succeeding paragraphs).

1. Contacting through phone or email for personal information: imposters may pretend to be employers or landlords and obtain the required information directly from the individual. This is a balancing act because it may be easy to identify the imposters or it may be easy for the imposters to obtain their target information, depending on the awareness of the target.
2. Buying personal information from third parties: For instance, information obtained through photocopying ID cards or scanning tax documents may be retained in the system unless they are routinely deleted by the organizations. This information may be sold to identity thieves. (Social Security Administration, 2018)

The following paragraphs will focus on internet-based information theft and the preventive measures that may be undertaken, in relation to the IRS scams.

Personal data can be obtained through the following ways online:

1. Phishing: Fake emails are sent to the targets, which look like legitimate emails from the IRS authorities. These emails may communicate that the situation is urgent, and that immediate action must be taken to prevent losses. A link in the email may open an illegitimate portal where the taxpayer's personal information can be retrieved once they have finished typing it in.
2. Malware: Malicious software can steal data from the system without active intervention from the user. Installation of pirated software or not updating the operating system may cause the loss of taxation information. For example, spyware and trojans can monitor the activities on the infected computers. Information related to tax filing, tax refunds and Social Security Number may be obtained through such monitoring.
3. Malicious mobile applications: Since most organizations, including government departments and banks have mobile applications, it is important to distinguish between the legitimate ones and the illegitimate ones. Microphone access through such apps may record calls and steal information when the taxpayer is calling a legitimate bank or IRS call center. Access to Gmail account through login may create a loophole for scammers to access contact and mailing information.
4. Smishing: Similar to phishing, smishing obtains information through calls or SMS that look legitimate but are fake. This involves building a certain level of trust with the taxpayer, and eventually obtaining their personal information by convincing them that it is for a legitimate purpose.
5. Insecure networks: Connecting to a free public network may provide the scammer access to personal banking information, and scammers in this case may be shopkeepers or employees of the organization that is providing the free public Wi-Fi network (Johar, 2017).

Taxpayers must take the following precautions against the above-mentioned means of compromising personal data:

1. Looking for inconsistencies in the email addresses may prove that the email is actually a phishing email, and not a legitimate one. Therefore, caution must be given to spelling mistakes or layout that looks like it is informal, as these may reveal the email to be fake.
2. Using legitimate anti-virus software may help prevent malware attacks. Downloading fake anti-virus may also cause malware attacks. Therefore, legitimate anti-virus from well-known brands may be installed, in addition to constantly updating the operating system.
3. Malicious apps can be prevented by checking the reviews for the app and understanding the permissions being requested by the app. Apps may be downloaded from iOS stores and Google Play Store as these companies constantly monitor the apps listed on their portals.

4. Critical information must never be shard through a call or SMS. Verify the identity of the caller or texter before information is sent to them.
5. Using personal data for internet access may be the best option as it is difficult to evaluate the safety of public Wi-Fi. Strong encryption may be used for home Wi-Fi (Johar, 2017).

LEARNINGS FROM THE CASE

The case highlighted the varied methods by which data may be compromised by a government entity. It also showed how global affairs such as the Coronavirus pandemic can be used to create newer means of scamming victims. The best practice of the IRS would be to be constantly vigilant and adopt public education measures to ensure that fraud detection information is widely available to taxpayers.

The IRS website has information about and a comprehensive list of scams that are prevalent. The webpage has classified the scams into many different categories. Scams targeting taxpayers, scams targeting tax professionals, soliciting information from payroll and human resources professionals, phishing and malware scams, and fraudsters impersonating the Taxpayer Advocacy Panel are categories identified by the IRS. A section titled 'recent scams' provides taxpayers with updated information about new scams that have arisen due to changing global and local situations. Publishing such information is a constructive way of ensuring that all stakeholders are aware of the scams and misrepresentations ("Tax Scams/Consumer Alerts," 2020).

The IRS also has policies and protocols for victims and the general public to report scams and other suspicious information. The Treasury Inspector General for Tax Administration oversees such reports about fraud and takes necessary measures to investigate complaints. An easy-to-use chart provides information about which authorities to contact to report specific types of scams. Identity Theft Affidavits can also be filed if the crimes are more serious and involve unauthorized impersonation. Such measures instill confidence in the general public about the measures being adopted by the authorities. Streamlining the scam reporting procedures is important in maintaining the trust of citizens, as well as consolidating information about scams and data breaches ("Tax Scams/Consumer Alerts," 2020).

A security plan, which includes the entire set of personnel, policies, plans, and initiatives adopted by the organization for managing information security, is a good starting point for the IRS to deal with scams. Since scams form a large part of concerns raised by the general public, a security plan integral to the operations of the IRS will act as a holistic measure to address these concerns. The IRS must classify information security as a key organizational priority to ensure that the required level of emphasis is afforded to information security across formal and informal actions of the organization.

Security in large organizations such as the IRS must concentrate on different functions within the larger information security umbrella. Legal compliance will be a high-priority function since taxation and its disputes are highly regulated through various legislations of the U.S. Congress. Incident response is another important function because responding to particular data breaches/scams is crucial for damage control and for maintaining trust among the general public. Effective training of IRS staff ensures that they are proactive in their responses to the complaints of citizens and victims (Whitman & Mattord, 2018).

Since the data breaches occur mostly through third-party sources such as hard drives and cloud storage of employers, organizations, and the victims themselves, it is difficult for the IRS to ensure the safety of the data which is not under its purview. Unlike other data breaches where the security threats can be resolved by improving the security of the internal system, it is not possible to do the same in the case of the IRS. The scammers will still be able to access sensitive personal information through the above-mentioned unencrypted sources and will continue to harass taxpayers through impersonation. Therefore, there are a few measures which may reduce the incidence and impact of such third-party data breaches. Wide publicity about personal and organizational information security may help mitigate the issues regarding data theft. Organizations not prioritizing the data privacy of associates and acquaintances must be penalized according to the law. Individuals must be educated about maintaining their personal data in a secure manner. Such measures may be the only defense against data theft from third-party data breaches where the IRS systems have not been compromised in any manner (Fontinelle, 2019).

The IRS commemorates the National Tax Awareness Week in December every year to disseminate information regarding data security and taxation policies. It targets a wide range of stakeholders including security experts, tax professionals, and ordinary citizens. Audio visual tools such as YouTube videos and radio public service announcements are made to keep citizens up to date about dealing with scams targeting the IRS. Such measures can aid in the enhancement of information security knowledge among the general public ("Taxes. Security. Together. We all have a role to play in protecting your data," 2020).

The case of IRS scams helps information security professionals understand the challenges in maintaining information security in two scenarios – information security of the internal systems of the organization, and security against scammers who obtain personal data of taxpayers from third-party sources and use this data to manipulate taxpayers. A holistic vision for security which looks beyond the security of internal systems is essential to ensure that information security is maintained.

DISCUSSION QUESTIONS

1. What is the role of public education in the prevention of Internal-Revenue-Service-related scams?
2. How has the Covid-19 pandemic affected Internal-Revenue-Service-related scams?
3. How can the Internal Revenue Service safeguard the data of stakeholders when the data breach occurs indirectly through third-party websites or services?
4. What is the difference between data security issues for government and private sector organizations?

REFERENCES

Allen, T. (2020). Unemployment Data Leaked: Names, Full SSN, Addresses. Retrieved from https://www.forbes.com/sites/terinaallen/2020/05/21/unemployment-data-leaked-names-full-ssn-addresses/#3f2091235a3a

Bengali, S. (2016). Inside the Indian IRS scam that cheated U.S. taxpayers out of millions. Retrieved from https://www.latimes.com/world/la-fg-india-irs-scam-20161027-story.html

Erb, K. P. (2020). IRS Warns On COVID-19 Scams Involving Stimulus Checks, Fake Cures & More. Retrieved from https://www.forbes.com/sites/kellyphillipserb/2020/06/08/irs-warns-on-covid-19-scams-involving-stimulus-checks-fake-cures--more/#56f5aefa37a9

Fontinelle, A. (2019). Know the Sneakiest IRS Scams. Retrieved from https://www.investopedia.com/taxes/know-latest-irs-scams/

Gray, J. (n.d.). Tax Scams Targeting the Elderly. Retrieved from https://www.agewisekingcounty.org/ill_pubs_articles/tax-scams-targeting-elderly/

Horaczek, S. (2018). Your Social Security Number Probably Got Leaked and That's Very, Very Bad. Retrieved from https://www.popsci.com/social-security-number-equifax-leak/#:~:text=According%20to%20Equifax%2C%20the%20breach,for%20much%20more%20than%20that.

Johar, A. (2017). Internet Security 101: Six Ways Hackers Can Attack You and How to Stay Safe. Retrieved from https://economictimes.indiatimes.com/tech/internet/internet-security-101-six-ways-hackers-can-attack-you-and-how-to-stay-safe/articleshow/61342742.cms?from=mdr

Morgan, S. (2016). IRS Reports 700,000 U.S. Taxpayers Hacked And 47 Million 'Get Transcripts' Ordered. Retrieved from https://www.forbes.com/sites/stevemorgan/2016/02/28/irs-reports-700000-u-s-taxpayers-hacked-and-47-million-get-transcripts-ordered/#638e18d7b93a

Social Security Administration. (2018). Identity Theft and Your Social Security Number. Retrieved from https://www.ssa.gov/pubs/EN-05-10064.pdf

Symanovich, S. (2017). Equifax Data Breach Affects Millions of Consumers. Here's What to Do. Retrieved from https://www.lifelock.com/learn-data-breaches-equifax-data-breach-2017.html

Tax Scams/Consumer Alerts. (2020). Retrieved from https://www.irs.gov/newsroom/tax-scams-consumer-alerts

Taxes. Security. Together. (2020). We All Have a Role to Play in Protecting Your Data. Retrieved from https://www.irs.gov/individuals/taxes-security-together

Whitman, M. E., & Mattord, H. J. (2018). *Management of Information Security* 6th Edition. Cengage.

2.5 International Student Scams

Nishanth Chakkere Ramesh
Kapiti Coast District Council, New Zealand

Ying-Pin Lan
Hato Hone St. John, Auckland, New Zealand

CASE STORYLINE

In the case study (Bidgoli & Grossklags, 2017), several international students at Penn State University, U.S.A., received a phone call which claimed that their visa status had encountered some problems and needed further verification. The people who contacted international students wsaid that they were from an official authority or the U.S. Foreign Affairs Department. Moreover, when students asked the callers for evidence of their identification, the callers could correctly tell the students' personal information, including address, email address, date of birth, major, year of study, and nationality. Because the callers could precisely list the international students' personal information, it was difficult for the students to tell whether this was a scam or an actual event. If students asked for a solution, the callers provided several ways of payment such as through a recharge card, bank account transfer, or online payment. Also, the callers promised the victims that their problems would be solved after the students completed the payment. If the students did not follow the instructions and arrange payment, the callers threatened the students that they might face deportation, law violation charges, and jail terms. To avoid any trouble, some students did transfer the money by purchasing point cards or doing a bank transfer. After the students completed the payment, the callers disappeared and were unable to be contacted again. Eventually, students realized that this was a phone scam and that the caller had deceived them.

Besides the phone call scam, another common scam at Penn State University was scam emails. Students received an email from the so-called "immigration authority" or "foreign affair department" which claimed that the students' visas will be canceled due to unpaid tax or for violation of specific laws. The email contained the logo of the government department and contact information which looked like a formal document. Also, the email addresses were listed as custom@usa.gov or Visa@foreignaffair.usa.gov, which confused the students, especially people who arrived in the U.S. for the first time. If students replied to the email and asked for help, the sender would

DOI: 10.1201/9781003264415-21

convince them to transfer the unpaid tax with penalty fee to a specific account and promised that everything would be fine once the students completed the transfer. After the students arranged the payment, the senders disappeared.

The case study brings up two concerns. First, most of the student scam cases seldom get reported. Victims think that reporting such a scam is a waste of time and do not believe the government can do anything about it. Also, they tend not to report due to the sense of shame for being deceived. Some of the students did report, but all the police officers did was record the case instead of launching an investigation or providing any physical help. Therefore, students rather keep this issue to themselves, and the number of recorded scams is far fewer than the scams that actually happened.

Second, the main victims are first-year students and international students (Broadhurst, Skinner, Sifniotis, Matamoros-Macias, & Ipsen, 2020). The number of cases deceived was significantly more than domestic students. The reason is that international students are vulnerable to the scam scheme due to the different cultures, unfamiliar regulations, and language barriers. Visa status and tax issues are the primary concerns for all international students since no one wants to take the risk of being deported or expelled from the school. If the students lack support and awareness, they will fall into these scam schemes and become victims eventually.

CRITICAL ANALYSIS

According to the IC3's (Internet Crime Complaint Center, U.S.) 2015 Internet Crime Report, it has received 288,012 cases in total concerning scam complaints, reaching a total loss of $1,070,711,522 ("2015 Internet Crime Report," 2015). The average financial loss is around $3,717.60 for each case in 2015 alone. As a result, organizations must have a comprehensive program to tackle different types of scams and help their staff and employees to avoid the potential risks. A 2018 study by Fraud.org confirms the persistence of the problem and provides statistics by category of fraud (Fraud.org, 2018).

SETA Program

From the case, although the university did have an existing information security policy, it did not work as expected because some students have become victims of the scams. An organization needs to establish a SETA (Security Education, Training, and Awareness Program) program to reduce accidental information security breaches and promote awareness of personal information security. This program does not only apply to companies but also to schools and universities. Through a comprehensive Security Education, Training, and Awareness Program, all the staff and students will benefit from this well-planned procedure. Therefore, the chief information security officer (CISO) needs to come up with an effective method to promote the sense of information security awareness through a top-down approach (Chart 2.5.1).

There are several factors that are important while building the program. First, focus on people. In the case of the university, the staff and students are the most critical elements of an effective information security program. If no one follows the guidelines, the policy will eventually be a failed project. Second, do not overload the users.

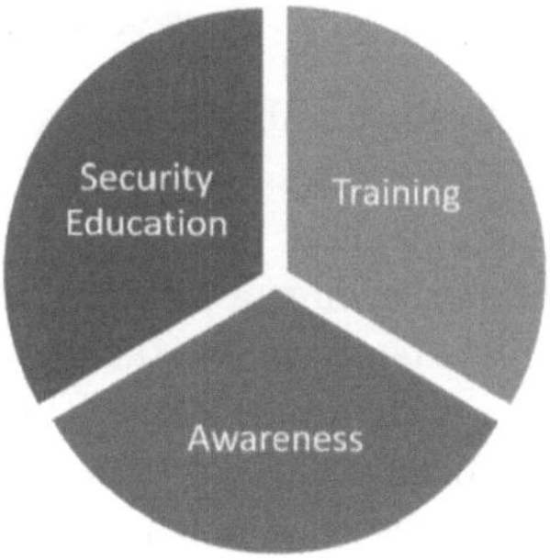

CHART 2.5.1 Security Education, Training, and Awareness (SETA).

In China, the government tries to remind its citizens to follow the rules by using posters and banners all over the major roads and streets. It might look helpful, but people seldom read the posters. As a result, the security program needs to reach a so-called "middle way" that uses every available venue but keeps things light. Third, help users to understand their responsibilities in the entire program. Without knowing their roles, even a perfect policy might not be as helpful as an ordinary one. Since the scammers can target everyone, the program needs to educate people to understand the risk and obey the regulations thoroughly.

As for the security awareness components, the officer can consider the following items, including videos, newsletters, brochures and flyers, and trinkets. Video and newsletter will be formal ways to promote awareness, while trinkets (such as coffee cups, pens, and T-shirts) help to focus the attention of the crowd. By combining these methods, a successful security program will be established.

The Evolvement of Scams

Scams toward students such as phone scams and email scams are common around the world. Regardless of the region, scammers are trying to gather the victims' personal information and use it against the victims themselves. Long before the existence of the internet, scammers used different ways such as questionnaires to gather personal information and send scam letters or make phone calls to victims. Since victims could not verify the information in time, they had no choice but to follow the instructions of the scammers to transfer the payment. Besides, people used to have a strong sense of mutual trust which was used by the scammers to gain the trust of victims. The cases of scams had gradually increased because the scammers realized that victims had no efficient tools to verify the information and chose to believe the scam.

Fortunately, nowadays, people can check information in time by using a smartphone. It does help to reduce the percentage of those being a victim of this type of scam. However, the techniques of scams also improved and are now involved in different types of schemes. Besides the common types such as phone scams and email scams, scammers try to obtain personal information by using a virus and dummy software (Isacenkova, Thonnard, Costin, Balzarotti, & Francillon, 2014a). As a result, it is gradually getting more challenging to protect students from being targeted.

How Did Scammers Collect Student's Personal Information?

One of the methods to collect personal information, especially of students, is to use Facebook promotion or lucky draw advertisements to gather data such as name, school name, country of origin, telephone number, and address. Since students, in general, have a limited budget, the scammers will launch a campaign and ask people to sign up for a chance to win free burgers, for example, to attract students to fill out the form and give away their personal information. This method does not require a programming background or excellent hacking skills, so this is also the most common way to obtain personal information. Once the data is collected, they will filter the data and select the target victims for the scam.

Another way is to purchase or hack the database of student clubs or associations. Most of the students' organizations keep the members' information list. The scammers will try to contact the senior members for the datasheet of contact information and offer a price to buy the resource. The scammer might use some excuses such as NGO activities or marketing research companies to gain the trust of the senior members. Also, scammers can hack into the database since most of the clubs do not apply reliable information security measures. In contrast, they mainly use free sources, including Google Drive and Drop Box to keep the members' contact information. Once hackers gain access to the database, they can copy personal information without being noticed.

Besides, a scammer with a tech background might use third-party software to gather personal information. The typical case is that of free-to-use software but with codes that will transmit personal information to the third-party's server. For example, the hackers might hack a paid application to make it become a free-to-use app and publish it to a private website. However, once users install it and agree to the terms and conditions, the app will collect personal information from the phone and send it to the hacker's server such as contacts, credit card numbers, and messages. Therefore, it is highly recommended not to install a hacked program even though it is free.

Finally, online game registration is also a way to gather information from students. The hackers might copy an existing web page game and insert the code into it. Once a user registers for a free play session, they also give away their personal information to the hackers as well. In conclusion, there are several ways to collect students' information without their permission.

The Evolvement of Scam Technique: Email

McCoy, Park, Shi, and Jakobsson (2016) stated in their report that spam-like and non-targeted scams are decreasing gradually while targeted scams with specific victims are increasing over the past ten years. This might explain that instead of calling students randomly, the scammers tend to collect enough information and do the background checks in advance to guarantee a higher successful victimization rate. The research analyzed thousands of scam emails and messages between 2006 and 2014 and categorized the type of message into non-targeted and targeted scams. In early 2006, the major types of scam emails were non-targeted scams, such as lottery scams, money transfer scams, and loan scams, all traditional forms of fraud that targeted a mass crowd and sent messages to anonymous users. The message usually claimed the authority of a certain department to gain the trust of victims. After 2008, most of the email system companies started cooperating with anti-virus software and were able to filter suspicious emails. Once the system found a specific text or email address in the message, the system would automatically remove the message from the inbox and place a warning sign with that email. This measure dramatically reduced the chances for a spam email to reach the potential victims. Nonetheless, the targeted scams, including rental scams and romance scams, were picking up and the cases have increased significantly. Since these types of messages state the victims' personal information, the scammers were able to exploit this knowledge and compose the message in a believable way. In addition, scammers developed a way to escape the checkpoint of the anti-virus system by using substitute email addresses to disguise the real address. This technique is hard to detect by the software because it shows that the sender is from an authority instead of from an address in the blacklist. Fortunately, with the development of machine learning and text mining, the software can detect more than 82% of the disguised emails and block them (Isacenkova, Thonnard, Costin, Francillon, & Balzarotti, 2014b).

LEARNINGS FROM THE CASE

How to Make the Reporting Process Easier

The study shows that after students have encountered the scams, only less than 10% of them will report it to the authorities. As a result, the security office is unable to collect the correct data to reveal the actual situation, and the number of cases is often underestimated. To solve this problem, the security program needs to actively encourage everyone to report the cases by providing incentives to do so. The reason that most of the cases do not get reported is that students think that it is not helpful and a waste of their time to report. Even if students did inform the authorities regarding the case, they receive a "thank you" note without further information such as where the scam came from or how long it takes to track and arrest the scammers. As a result, a student might choose to ignore a scam if they did not become a victim. This reasoning helps scammers to continue playing the same tricks at different universities without worrying about being caught. Therefore, the authority can consider providing incentives such as trinkets and rewards to encourage students to report scam cases. It can not only help the officers to tackle scams faster but also keep a precise record of the scam details and raise awareness among the general public.

Do Not Reveal Personal Information

As discussed above, there are several ways to get students' personal information, including third-party software, FB campaigns, and online free games. In some cases, students might choose to give it away for free in exchange for some benefits. However, the side effects usually cost more than the convenience they obtained. Therefore, the information security program can focus on these perspectives and develop strategic plans to help students not to fall into the scams and become victims.

Provide Support to Students

Besides the precautionary measures, the information security program should also develop consultation sessions for victims. The case indicates that victims tend not to report to university due to a sense of shame. This embarrassment and scam can also lead to severe mental problems and depression because of money loss and possible financial difficulties. As a result, a supportive consultation program is essential to help students step out from the shadow of being deceived.

Raise Awareness of Suspicious Bait

To establish an effective information security program, the officer should focus mainly on the protection of personal information and the transfer of money. Since the scammers' target is to obtain money by whatever means possible, the awareness of suspicious messages and offers is a key factor in reducing scams (Cross & Kelly, 2016). If students understand how their personal information will be used against them, it can help them to distinguish between a scam and a real message from an authority.

CONCLUSION

In conclusion, student visa scams and email scams are common tricks that scammers use to obtain money illegally. If students accidentally give away personal information through FB campaigns, promotions, or market surveys, they may be the target of a phone scam or mail scam. To tackle this problem, the information security program in the university needs to establish the SETA program, first to raise the awareness of possible scams and promote the understanding of cybersecurity policies. Second, SETA will provide supportive programs to students in the case of reporting a scam or consultation. Students also need to be careful when dealing with strange messages or emails that ask for personal information.

This case highlighted the importance of the SETA program in information security management. With the improvement of scam techniques, people sometimes cannot distinguish the real message from the phishing emails. Although the anti-virus software helps users filter most of the suspicious messages, users still need to be always aware to avoid scams. Therefore, with the awareness of scams and a comprehensive SETA program, organizations can overcome the scams.

DISCUSSION QUESTIONS

1. Are there any other ways for scammers to obtain a student's personal information?
2. With the Deepfake app, a scammer can change the face and voice and gain so trust of a victim. What kinds of scams will use this technology?

3. Besides developed countries, will it be different in developing countries for this type of scam?
4. Could technologies be used to help track the scammers and find their location?

REFERENCES

2015 Internet Crime Report. (2015). Retrieved from https://www.ic3.gov/Media/PDF/AnnualReport/2015_IC3Report.pdf

Bidgoli, M., & Grossklags, J. (2017). *"Hello. This is the IRS calling.": A case study on scams, extortion, impersonation, and phone spoofing*. https://ieeexplore.ieee.org/document/7945055

Broadhurst, R., Skinner, K., Sifniotis, N., Matamoros-Macias, B., & Ipsen, Y. (2020). Phishing risks in a university student community. *Trends and Issues in Crime and Criminal Justice* (587), 1–16. Retrieved from https://www.aic.gov.au/publications/tandi/tandi587

Cross, C., & Kelly, M. (2016). The problem of "white noise": Examining current prevention approaches to online fraud. *Journal of Financial Crime*, *23*(4), 806–818. doi:10.1108/JFC-12-2015-0069

Isacenkova, J., Thonnard, O., Costin, A., Balzarotti, D., & Francillon, A. (2014a, 2013). *Inside the SCAM jungle: A closer look at 419 scam email operations*.

Isacenkova, J., Thonnard, O., Costin, A., Francillon, A., & Balzarotti, D. (2014b). Inside the scam jungle: A closer look at 419 scam email operations. *EURASIP Journal on Information Security*, *2014*(1), 1–18. doi:10.1186/1687-417X-2014-4

McCoy, D., Park, Y., Shi, E., & Jakobsson, M. (2016). Identifying scams and trends. In M. Jakobsson (Ed.), *Understanding social engineering based scams* (pp. 7–19). Springer New York.

Top ten scams of 2018: Imposter scams, romance frauds, and bogus loans showed big gains in 2018 (2019). Retrieved from https://nclnet.org/wp-content/uploads/pdf/2018_top_scams_report.pdf

2.6 Security and Privacy Risk with Social Robotics

Satish Prabhu
Capgemini, Christchurch, New Zealand

Abhishek Kumar
University of Canterbury, Christchurch, New Zealand

INTRODUCTION

In our daily lives, robots are becoming an integral part of living. Apart from industrial use, they have an important role in medical, household, social, military, and space applications. Several types of research have shown substantial innovation in various fields as well, such as field robotics, social robotics, intelligent robotics, and sex robotics. The widespread use of robots in social and personal life could be possible because of recent advancements in sensor development, artificial intelligence, robot learning, motion planning, decision-making, computer vision, and other relevant areas.

Robots have their potential benefits which are clearly and widely acknowledged. The use of technological advancement, social robots, is gradually becoming more interactive and more involved with people in their daily and personal lives. However, despite several advantages of such close human–machine interaction, some of its perceived drawbacks have raised an issue of public debate. The key concern that evolved from this human–machine interaction with these social robots is correlated to cybersecurity and user privacy. The function of these robots is constructed on data analysis which consists of greater mobility and autonomy than static devices. They are built on a computational platform, being bonded with actuators, sensors, and other hardware such as cameras. Therefore, new security and privacy issues have been introduced by these robots. As they are networked and connected to the internet, their presence not only raises the concern of cyberattacks but also the development of a completely new set of security concerns that can cause privacy issues if hacked or even dangerous, physical harm. Yet, there is scarce or no empirical evidence on cybersecurity and privacy issues from such robots, but one could say that people are not largely familiar with privacy implications from them.

Security and privacy of one's data is a fundamental human right, and individuals have the right to have control over their data. Below is a critical analysis associated with security and privacy concerns of social robotics which is a privacy-sensitive recent technological advancement.

DOI: 10.1201/9781003264415-22

CRITICAL ANALYSIS

In society, privacy is believed as a fundamental right. People having the freedom to act without tracing and surveillance is considered an important basis for self-development. Privacy is also theorized as the protection of liberal selfhood. Privacy facilitates intellectual endeavors and protects social values by imparting speaking, reading, and exploring ideas. This value-based privacy may be distinguished by a cognate-based understanding, where privacy is considered as the individual's perception, mind, and cognition instead of absolute moral norms or values. With this cognate-based understanding, privacy is knitted with individuals' control of their personal space and information. Therefore, privacy is claimed by individuals, institutions, or groups to determine how, when, and to what extent, data or information can be shared with others by themselves.

To discuss privacy, one must consider the distinction between private and public spaces. For the last few years, virtual elements have been increasingly invading the physical world resulting in fuzziness when separating public and private spaces. This blurring of boundaries is more conspicuous with the use of autonomous vehicles and the Internet of Things (IoT). Social robots theorize the same concept. They have their own mobility, use the IoT, and even can access personal and private spaces, such as bathrooms and bedrooms, more easily than any other technology. Thus, with the advancement in technologies, privacy challenges are increasingly more complex and can be differentiated into the following four dimensions:

1. Psychological privacy: values and thoughts
2. Physical privacy: personal space and territory
3. Informational privacy: personal information
4. Social privacy: social contacts and influence

Every individual consists of his own personality which is considered predominantly sensitive, and people get uncomfortable when such information is gathered. In addition, some view that a social robot can collect more information about what an individual can perceive, capturing the forces and changes of electromagnetic waves of a brain. These robots can identify places that are even not accessible to users. Thus, robots may "know" more about the users and their living conditions than what the users know about themselves. They can collect more sensitive data than perceived by any other technologies before.

The use of any online technology depends on the mental calculus that weighs the benefits and risks of any newer technology. If the risks are outweighed by its benefits, people start using newer technology. Similarly, despite privacy concerns, if the use of social robots is recognized as extremely useful and valuable for people's personal lives, they will be more likely to develop the intention to use it. A person's familiar concerns are more vulnerable to prevent the use of any newer technology than unfamiliar concerns. Therefore, critical analysis of social robots is increasingly important and divided into three categories – Institutional Privacy Concern, Social Informational Privacy, and Physical Privacy. Mostly, people are more familiar with information privacy concerns (both social and institutional) from social media and the internet but, they have limited knowledge about physical privacy risks from social robots. Physical Privacy risks are associated with the interaction of social robots with physical spaces and physical objects like using personal objects or by entering private rooms. Institutional Informational Privacy concerns data protection and information control from different institutions like government agencies, social robot manufacturers, cloud storage providers, or data brokers. Social Informational Privacy is more related to personal data protection of an individual by hackers or other users. Therefore, social robots are not only challenging users' informational privacy but also influencing their psychological and physical privacy due to their potential social bonding and autonomy.

First, with the technological advancement in robotics, there is increasingly more interaction between humans and robots. Extensive use of a sophisticated array of sensors makes the robots quite different and powerful from mobile phones. Users underestimate their capabilities and misunderstand how and what kind of data robots are copying. Second, the autonomy and mobility of robots can often fade the traditional boundaries between public and private spaces. As an example, video surveillance is more agreeable in public places, but not in private spaces. However, what rules and regulations should be appropriately implemented for robots is not clear, for these human-like machines can roam in both private and public spaces (e.g., social robots that navigate in public places or sex robots that sail across personal spaces). These robots have brought also an entirely new challenge as they even can move to private rooms and can access confidential documents to scan. Third, social robots use human-like conversational dialects to interact with users, and thus, they are often sentimentalized by their owners. With these interactions, attachment and trust may build up between robots and users, which, in turn, could become an opportunity for robots to exploit, influence, and manipulate their users to impulsively disclose sensitive and confidential information to robots. This interaction and association are diverse in different users and depend on how users commonly integrate with other devices like mobile phones.

The social robots that are found in the workplace, have high operating costs and are likely to be shared between different users. Such robots have only a single interface to interact with many users and are collecting data in a single database which makes it more vulnerable to unauthorized data sharing with others. Social robots usually share a common physical space with many people, and they could accidentally gather data from bystanders while they take on their core users.

Furthermore, the storage of collected data by social robots on the cloud is also counted as problematic. Social robots are connected to the internet and among each other. Such networking makes it more susceptible to hacking. This risk potential is confirmed by the experimental studies and warned that household robots are easily infiltrated by intruders without having direct contact with them. In other words, hackers can get much deeper insights into the daily lives of the users of social robots through handheld devices such as mobile phones, tablets, or laptops. Recognizably, these social robots may become the focal point of hackers. A small group of experts advocate the use of social robots for better security in larger companies compared to small start-up companies. A few state that social robots are not at all security threats because of their low adoption at this point in time. However, it also poses a security threat to society.

Some social robots are meant to monitor and surveil buildings and facilities to keep away intruders prohibiting the misuse of premises. They alert the security in a timely manner if any unknown intruder infiltrates or violates the rules. A robot communicates with another robot over Wi-Fi but they interact with humans using the HTML5 web GUI (graphical user interface). A cyberattack penetrates the network of robots and gets access to sensitive information, infiltrating the premises and escaping without being noticed by human surveillance and security. Similarly, in ROS-based systems (Robot Operating Systems), an attacker can easily intrude into the network and can send shutdown commands to obstruct or kill any ROS node and release distorted messages on important functions, e.g., velocity commands to change the direction of a robot. However, detecting malicious attacks can be very challenging, as manipulations can be masked with simple bugs, when the ROS may be under attack. Thus, it is important to ensure secure and preventive design principles for social robotic applications. Similar to ROS, Yet Another Robot Platform (YARP) is another robotics-oriented architecture. Processors, actuators, and sensors are linked to the software in the YARP. Messages are transmitted using the transmission control protocol (TCP). An intruder can connect with TCP and YARP, making it more vulnerable to attack or hack into and corrupt the data and disrupt confidential information.

Security is another aspect of the risk of social robotics. In medical hospitals, some robots are involved in operating surgeries on patients. Many times, the commands given by doctors to robots are not encrypted and the system is more prone to cyberattacks. An unauthorized person can take control of the surgical robot and the consequence may be disastrous or even fatal to the patient. Similarly, in defense and space, drones are being used as aerial vehicles for surveillance and combat. Although the communication between robots and humans is encrypted, sometimes it may not be. The intruder may snoop into a drone by hacking its control system. The drone may then either be crashed into populated areas, take all sensitive data, or sabotage confidential records. Like defense robots, disaster robots are responsible for accessing, repairing, breaking, and disrupting harmful systems. An intruder can take control of such robots which has been deployed for disaster management or any nuclear platform. This may cause a hindrance by the intruder who disconnects the robot from the commanding system.

For the last few years, the automotive industry has been one of the rapidly growing industries devising intelligent robotic cars. These robotic cars are equipped with intelligent keys, digital instrumentations, and hands-free door locks, which warn and help prevent collisions and the automatic generation of various signals. Audi AG, Google, Toyota, and Hyundai continue innovating and improving such robotic cars making them fully autonomous. These robotic cars rely on in-built internal wired networks. Wheels have also sensors prompting wireless communication. Researchers are working on the invasion of both small-range wireless networks and long-range cellular networks. Such cars receive various radio signals and connect to various IoT devices. Hackers can get access to the internal network, disable the breaks affecting the sensor breaks, make variations in the speedometer, or even install malware resulting in a compromised autonomous system. In this way, hackers can do spoofing of vehicles, people, and obstacles remotely causing severe threats to persons' privacy.

Household robots are a type of social robots that are used as domestic helpers or assistants. They assist in the daily chores of housework. These robots are equipped with cameras, microphones, and sensors which can gather a vast amount of information. Many times, they may collect sensitive information related to the health status of the family members. This information can be shared by hackers who may take control of the robots and gain access to confidential information. These robots develop emotional reactions by making eye contact, using natural voice, assuming facial expressions, affecting memories, and changing attitudes. These facilitate the building of personal relationships with humans. Therefore, they are dealt with politely, perceived as a companion, and can evoke gender stereotypes. They can influence physical cues and behavior to gain confidential information and then, hackers obtain and copy this information.

LESSONS LEARNED

To avert the privacy breach and security threat from the use of social robots, "privacy by design" is one of the identified concepts. The fortification of an individual's privacy and social security needs to be taken from the very beginning of the development process of social robots. However, the concept of "privacy by design" is not measured to be influential in all circumstances. Many times, social robots are configured by the users themselves, and such pre-programming is not considered sufficient and becomes ineffective. Thus, instead of designers, manufacturers, or regulators, users are primarily responsible for privacy breaches and security threats. Users are encouraged to protect from privacy breaches themselves. It is obvious that in private spaces, jurisdiction tends to be ineffective therefore, the technological aspect plays an important role in protecting user privacy. Technological solutions to prevent the breach of privacy of an individual includes the facility to switch off the robot as and when required, restricting its movement into private spaces, designing the robots in such a way that they can generate signal through ears or eyes letting the users know that they are collecting and storing the information, and do the data anonymization in which personal data can be irreversibly altered so that subject can no longer be recognized. On the contrary, some researchers depicted that privacy-enhancing behaviors (PEBs) are more observed for surveillance cameras instead of social robots because the anthropomorphic shape of social robots concealed the probability of data gathering and storing in their storehouse.

Finally, it is an opportunity to replace the written terms of use with verbal interactions from social robots.

The specific interactions like the physical and psychological behavior of the users can't be protected from being captured by social robots. Therefore, understanding privacy concerns goes ahead of informational privacy. This understanding and experience of users with advanced technologies of social robots have given the linkage of informational privacy with the functions and roles of robots. It is hard to access psychological and social implications causing new challenges. Users themselves have the primary responsibility to protect themselves from such challenges. Thus, experts have proposed different means to construct user awareness such as discussing privacy and security concerns from robots, educating about the subject in schools and other institutions, advertising on television and on social media platforms, and exchanging personal experiences via publishing articles and via interpersonal communication.

Although there are laws to control privacy and security breach through robotic technologies, several legal gray areas still exist that provides the unequal distribution of power to states and companies against users. Re-construction of legal frameworks is required for new technologies. Strengthening user rights needs to improve. With time, social robots have several specific features posing various novel threats and ask for further development of new and updated data protection laws to sustain individual privacy and security. The purpose of social robots should be limited and data should be minimized. Data minimization in the context of the General Data Protection Regulation should be restructured. However, several legal means are pre-existing, and they need to be implemented consistently. The ethical code of conduct for social robots should be constructed in accordance with cyberethics, technological ethics, and the ethics of cyberspace. Together, they will make social robots more humane to fit into society.

To secure social robots from hijacking is also a prime concern. Social robots work on the traditional computer system platform. Web browsers need to be secured with firewalls to prevent unauthorized access. Need to define a rigid interface separating the main components of the framework that enhance the security of the robotic system. In autonomous social robots, data security services – Secure Dispatching Service (SDS) and Key Management Service (KMS) should be used. SDS empowers the confidentiality and authenticity services by encrypting and decrypting the messages and verifying the proof of authenticity. Similarly, KMS generates the new access keys either periodically or upon a vehicle leaving. These measures make the task tough for the hijackers or intruders thus, privacy and security can be maintained effectively.

CONCLUSION

In conclusion, social robots have different privacy and security risks that go ahead of other newer devices like smartphones, smartwatches, etc. The psychological impact of such robots is more critical due to emotional bonding with individuals and further in an open and intimate way. It tends the individual to share more confidential data causing privacy and security risks. However, overcoming these risks is not an easy task and several recommendations are advised such as securing robotic software systems, keeping their software updated, educating the users and awareness in society, encryption and decryption of robotic communication, realizing cybersecurity best practices, performing security audits, and avoiding granting access to an unauthorized individual. Such interdisciplinary effort empowers the users to protect themselves from privacy and security breaches.

DISCUSSION QUESTIONS

1. What shall be the extent of IS privacy and security when the U.S. implements autonomous robot soldiers by 2035?

2. What can be the security leaks in future endeavors of Nanorobots?
3. Can AI Robotics dominate human society in the future?
4. Does AI Robotics have more pros than cons?

REFERENCES

Lee, M. K., Tang, K. P., Forlizzi, J., & Kiesler, S. (2011). Understanding users! perception of privacy in human-robot interaction. Paper presented at the 2011 6th ACM/IEEE International Conference on Human-Robot Interaction (HRI).

Lutz, C., Schöttler, M., & Hoffmann, C. P. (2019). The privacy implications of social robots: Scoping review and expert interviews. *Mobile Media & Communication*, *7*(3), 412–434.

Lutz, C., & Tamó-Larrieux, A. (2020). *The Robot Privacy Paradox: Understanding How Privacy Concerns Shape Intentions to Use Social Robots*. https://doaj.org/article/2e166 3d3a32f405bba3951369734e707

Portugal, D., Pereira, S., & Couceiro, M. S. (2017). The role of security in human-robot shared environments: A case study in ROS-based surveillance robots. Paper presented at the 2017 26th IEEE International Symposium on Robot and Human Interactive Communication (RO-MAN).

Priyadarshini, I. (2018). Cyber security risks in robotics. In *Cyber Security and Threats: Concepts, Methodologies, Tools, and Applications* (pp. 1235–1250): IGI Global. https://www.igi-global.com/pdf.aspx?tid=203489&ptid=190539&ctid=15&t=Editorial%20Advisory%20Board&isxn=9781522556343

2.7 Financial Fraud (IS System Controls) at NCL Ltd

Satish Prabhu
Capgemini, Christchurch, New Zealand

Abhishek Kumar
University of Canterbury, Christchurch, New Zealand

SUMMARY

NCL Ltd was established in 1950 and is the fifth-largest construction company in India with an annual turnover of 100 million NZD. They have offices spanning all the metro cities in India such as Bangalore, Chennai, Hyderabad, and Noida with its head office in Mumbai and an employee base of around 1200 people strong across the country.

The company's mission statement is "We are committed to the Growth and Development of all our People (NCL Family) upon whom we rely on to make this happen" and their vision statement is "To be one of the top 3 construction companies in India in terms of Turnover, Type of Projects handled, Pan India Geographical Presence, Quality and Safety."

NCL Ltd implemented one of the best Enterprise Resource Planning (ERP) systems in the world, Systems Applications Products (SAP) in 2014 for which it received the award for the best implementation in the construction industry category in 2014. The IT system was centralized in Mumbai with 200 concurrent users across the country. The Financial Statements were audited every year by Big 4 Audit firms along with the mandatory Information System Audit as per the statutory regulations.

During one of the Project review meetings with the CFO in which the profitability and cash flow analysis were presented by the Project Account Manager for their respective construction sites, the CFO did a random check of expense vouchers which were attached in SAP and identified a few payment vouchers that did not have acknowledgments (signatures) by sub-contractors. A policy existed in NCL that sub-contractors were issued bank cheques against their invoices and got the acknowledgments (signatures) on the payment vouchers. The Project Account Manager failed to explain this since he had delegated all this work to his Account Assistant. He quickly called his Account Assistant who sits on the same floor where the review was going

DOI: 10.1201/9781003264415-23

on. The Account Assistant failed to give a satisfactory explanation to the query of the CFO and started avoiding them. He further requested to leave the office early since he had to attend a family function which led to more suspicious activity.

Since all the of them present in the meeting sensed that there was something wrong with the transaction, the CFO immediately ordered that all the transactions had to be investigated thoroughly. An SAP team was assigned to get on the job quickly.

As a first step, they prepared a file in which all the attachments from SAP in the Document Management system were downloaded against each of the payments made. All these vouchers were checked thoroughly as to whether they had the acknowledgment (signature) of the sub-contractor; the SAP team found that there were two vendors in which the acknowledgment was missing. The vendors were created by one of the members of the SAP team who was the friend of the Account Assistant. The Account Assistant had mailed the SAP member without keeping cc to anyone for approvals giving reasons that the vendor needs to be created for urgent payment.

An audit trail was done to check the source and application of the funds for these vendors. NCL Ltd had a policy to make online payments in case of urgency. It was found that there were few online payments made to these vendors. On further investigation, it was found that the bank account number to whom online payments were made was of the Account Assistant's mother. The Project Account Manager then revealed that he had shared the password with the Account Assistant for posting payments since he trusted him the most and found him sincere and hardworking. Fraud of 5 million Rupees was unearthed after a thorough investigation. Security is not about how well you know your enemy, but how well your enemy knows you. If your enemy knows nothing about you, then there's nothing for him to attack (Wingreen, 2020).

Immediate steps were taken by management to check whether there was any further leakage in revenues such as obtaining new cheques or bank statements from the vendor to verify with the bank account number in SAP Vendor Master. Internal Controls of the maker and the checker as two different persons were strictly enforced. The bank cheque which earlier had only names printed now had the bank account number too from Vendor Master as an additional check. Since the vendor creation process involved large numbers of vendors created daily, it was streamlined by including the Site Project Manager and the Project Account Manager to approve the vendor creation. The Site Project Manager could validate the onsite sub-contractors and the Project Account Manager of the Zonal office was responsible for the profitability of the Project.

SAP processes were revisited to check for lapses if any with new responsibility assigned to the SAP Project Head to monitor for suspicious transactions with Exception Reports. Legal action was taken against the Account Assistant, and he was imprisoned. Also, the Project Account Manager was given a warning since he had a very good work record in the past. There were important lessons to be learned such as thorough background checks to be conducted by the HR department while recruiting new employees and mandatory holidays for 1–2 weeks need to be given to the employees as this shall help in detecting frauds. NCL Ltd should implement the Harrison–Ruzzo–Ullman (HRU) model which defines the method to make changes to the access rights.

CRITICAL ANALYSIS

NCL Ltd had its mission statement, vision statement, and also values for its organization, and despite all these, the fraud took place because of a lack of strategic planning to lay out the long-term direction to take the organization to the next level.

The Information Security Policy remained only on paper and in most organizations, it is not implemented at all. NCL Ltd limited management and may not have spent adequate time and effort with its employees to practically implement the Info Security Policy by integrating it with their day-to-day work. This led to an opportunity for the Assistant Account to use this to his advantage.

Security in the organization is constructed by a robust Governance System which is achieved by adopting various strategies. In this case, there was a flaw in the New Vendor Creation process as well as payment approval processes. The Governance System didn't establish a multi-layered approval system which not only makes the system more secure but also prevents the business transactions from becoming more vulnerable to any such threats. Therefore, such a feeble approval process had put the Operational Security at stake which couldn't maintain the integrity of information.

In this case study a traditional method has been followed in the organization which includes planning, organizing, staffing, directing, and controlling (POSDC) but in this approach, the strategies have been missing in planning which has led to a vulnerable system.

SAP applications were used to provide several layers of approval processes which had not been used in the configuration system. Either it was not required by the Governance Team or the IT manager didn't inform such use of multi-layered approval processes to organizations. Therefore, understanding of functionalities of a technology is also an important factor for any successful implementation, the criterion of success is not just implementing SAP as they were awarded as one of the best implementations in the construction industry – assuming the construction standards were carried out for the organization. Basic workflows which are available in SAP were not included in the scope of work during implementation which implies that the organization was not harnessing the technology at its atomic level. Also, it is possible that the priority was not given to important features of workflow or any adequate risk measures undertaken.

Authentication is provided with the intent of the availability of some documentation rights. The Project Account Manager shared his password which not only breached the security policy but also there was a loss of control over authentications which were provided to the Project Account Manager by the organization. As we know, the availability of certain documents should be given to authorized persons. But, in this case, password sharing violated the intent of availability to access all the documents. Also, the SAP Manager should have tracked this from audit logs which give the data about different instances against a single user ID. Password sharing weakens the defensive control system of security & confidentiality. The organization should have included not sharing passwords as one of the directives in the job description. The code of conduct should have been established by the organization, which would act as a deterrence in preventing an illegal activity.

Accountability is the most associated with system audit logs. After losing the authenticated control over the system and then business processes, the Project Account Manager as well as the CFO who was signing authority for payments shall be equally accountable for such loss of security and fraud. The CFO relied on the information he received from his subordinates and approved payments without checking whether his subordinates had taken substantial measures to check the authenticity of transactions.

Human error or failure can be considered as a threat since there is no deliberate intent of the person to harm the organization. Such events happen by mistake or due to inadequate knowledge of the user. This cannot be considered as a threat since the password had been shared by the Project Account Manager to the Account Assistant. This act of sharing passwords implies that the system was left loose for an attack and that the internal control of the system was not tightly knit; the Accountant Assistant could steal or damage the confidential financial information. He might also perform some intentional or unintentional financial issues apart from this fraudulent act that can damage or otherwise compromise financial information and thus affect the credibility of the organization.

The Assistant Accountant exploited the financial system for his personal gain. He befriended his SAP Master Data team, finessed the creation of a fake vendor account, used the authorization control of his boss, and credited the payment.

The SAP Project Manager had failed to implement controls in the SAP system where a single user could directly create a vendor. The authorization roles should have been defined in which the single user, "Master Data Team", should not be allowed to create the vendor. Also, exception reports should have been developed for monitoring suspicious transactions. Management pointed to the SAP Project Manager for not alerting them about the non-implementation of workflow in the payment process in SAP.

LESSONS LEARNED

Steps were immediately taken by NCL Ltd upon learning that fraud had taken place in the organization.

- All Vendors were communicated to send their bank statements and canceled cheques for verification with SAP Vendor Master records.
- Vendor Accounts would not be created on urgency in the future and the authentication process implemented in SAP with two-level authorizations.
- Authorization of roles in SAP was audited with the introduction of daily exception reports and daily log analysis to find any exception activity.
- The Vendor Master form would be signed by the Project Manager of the Site, the Project Site Accountant, and the Head Office Accountant.
- The workflow was implemented in SAP for the payment process with a division of roles and responsibilities.
- Password was reset by the System periodically where user had to create a new one.

- IS Security Continuous Review team would review monthly all the governance policies at a granular level with the introduction of various training programs at all levels.
- Periodic meetings with the business and IT teams were scheduled to discuss the various issues which shall help to bridge the gap analysis between the organization and IS.

NCL Ltd needs to review its Information system framework to include ISO/IEC 27001: 2005 Information Technology Code of Practice for InfoSec Management, which is designed to give recommendations for InfoSec management. Access Controls need to be reviewed thoroughly and Mandatory Access Control (MAC) needs to be introduced. From the various models which are available, NCL Ltd should implement the HRU model which defines the method to make changes to the access rights and the addition and removal of subjects and objects. The HRU model involves a specific set of commands and a set of generic rights which is used in the access control matrix.

As a self-assessment for recommended security practices, Gartner group has offered 12 questions for people, processes, and technology. NCL Ltd needs to apply several questions such as whether the HR department conducts the background checks of all its employees, whether the enterprise is annually updated, and whether the networks and systems are periodically reviewed to prevent malicious software attacks.

Since this fraudulent event has already occurred, steps need to be taken by NCL Ltd so that such types of events do not happen again or in the near future. Therefore, they need to implement the NIST Risk Management Framework (Kohnke et al., 2016) in which the following steps such as framing the risk, assessing the risk, responding to the determined risk, and finally, monitoring risk on an ongoing basis are done.

The HR policy should include compulsory leave of 1–2 weeks for all employees in the calendar year. This would help organizations to expose the employees who continue to play fraudulent activities with the company. Major frauds are unearthed when employees go on leave therefore banks have preventive vigilance leave for at least 10 days every financial year (News, 2011).

NCL Ltd's mission statement reads "We are committed to the Growth and Development of all our People (NCL Family) upon whom we rely to make this happen." NCL Ltd also had the vision statement to be one of the top three construction companies in India in terms of Turnover, Type of Projects handled, Pan India Geographical Presence, Quality, and Safety. NCL Ltd acted against the fraudster Assistant Accountant by prosecuting him but at the same time did not fire the Project Account Manager due to his excellent previous records. The company also considered his honesty in admitting his mistakes and helping the company to be a witness against the fraudster's Assistant Accountant. Management of NCL Ltd acted as per the mission statement.

NCL Ltd needs to also implement the above policies using the Security System Development Life Cycle (SecSDLC) method by adopting a knowledge-based security system (Alowibdi, 2011) with the help of a security expert team who will be able to identify, monitor, and organize all kinds of security functions and issues.

DISCUSSION QUESTIONS

1. To What extent should we delegate our work to subordinates?
2. Sharing passwords with our colleagues shows how much we trust them but at the same time, it is against the policy of the organization; how many times do we come across such scenarios?
3. NCL Ltd did not fire the Project Account Manager despite his blunder, giving reasons for his previous excellent records. Is that ethically correct for any organization to do? What message does the employee get from this act?
4. Is SAP ERP software somehow responsible for this fraud? Explain your reasons.

REFERENCES

Alowibdi, J. (2011). Adopting Knowledge Based Security System for Software Development Life Cycle. Athens. https://www.cs.uic.edu/~jalowibd/uploads/SER4014.pdf

Kohnke, A., Sigler, K., & Shoemaker, D. (2016). Strategic Risk Management Using the NIST Risk Management Framework. *EDPACS*, *53*(5), 1–6. https://doi.org/10.1080/07366981.2016.1160713

News, G. B. (2011). *SBI to ask staff to go on compulsory leave to limit fraud.* https://economictimes.indiatimes.com/industry/banking/finance/banking/sbi-makes-offs-compulsory-for-staff-to-clean-out-frauds/articleshow/7315285.cms

Wingreen, S.C. (2020). Lecture notes, "Risk Management". From a Lecture Delivered at the University of Canterbury, 2019.

2.8 When You Can't Believe What You See or Hear

Anjali Krishnan
University of Canterbury, Christchurch, New Zealand

Shameer Thottoli
Acclaim Group Limited, Auckland, New Zealand

CASE SUMMARY

As per a report in The Wall Street Journal [1] (WSJ), the CEO of an unconfirmed UK-based energy company presumed he had been on the phone with his boss, the German parent company's chief executive, and followed orders to transfer €220,000 (approx. $243,000) to a Hungarian provider's bank account instantly. In reality, the voice belonged to a fraudster who spoofed the German chief executive utilizing AI software algorithms. The details were shared with WSJ by Rüdiger Kirsch of Euler Hermes Group SA, the insurance provider at the bank. He clarified that the CEO recognized the obvious German accent in the voice of his boss – and in addition, that it held the "melody" of the man's voice as well. This was the first acknowledged incident of a deep-fake voice created by artificial intelligence used for a scam.

As per Kirsch, the organization was contacted three times by the anonymous fraudster: the first time to facilitate the payment, the second to wrongly say it had been refunded, and a third time to request a follow-up payment. The victim became suspicious at this point; he could see that the alleged repayment had not gone through, and he realized that the call was made from an Austrian telephone number. Although he had not sent a second payment, the first payment had already been transferred from the Hungarian bank account to the one in Mexico and then reallocated to several other locations. Kirsch told WSJ that he suspects that a computer program was used to imitate the German executives' voice in the manner that is used by Dessa, an AI company based in Toronto that successfully created a recreation of the voice of popular podcast host and comedian Joe Rogan. If this is indeed the case, this incident will have been the first documented example of AI voice manipulation to be used for fraud.

Although AI has been marketed by the cybersecurity community as a way for developers to simplify processes and for businesses to identify abnormalities, this event also demonstrates how the system could be effectively used fraudulently. Security analysts have cautioned that AI is indeed a blessing for fraudsters who may

DOI: 10.1201/9781003264415-24

use the technology to automate phishing. There are authentication methods that could recognize these attempts. However, companies need to be cautious and use appropriate authentication methods – such as multi-factor detection, biometrics, and robust identity proofing – to combat emerging AI challenges.

CASE ANALYSIS

Deepfake innovation has acquired a new degree of sophistication, enabling the illegal incorporation of audio and visual imagery into social media sites and making it more difficult than ever to separate real from fake content. Deepfake technology not only addresses individualized dangers to citizens but also destroys trust in institutions and impacts national security. As individuals choose to believe everything they want, instead of what they hear and see on online platforms, deepfakes continue to progress with their knowledge warfare. Deepfake videos have grown quickly in the political arena and have been used on social media to embarrass politicians. Nevertheless, the greatest distorted content has been mostly porn that has been available over the Internet for years. According to a poll published [2] by Amsterdam-based deep learning company Deeptrace Labs in December 2018, almost 14,678 deepfake videos exist on the Internet.

Deepfake films, audio, or photographs are usually created by combining new pictures, audio, and sound into an existing film, audio, or photographs to create the appearance of speech. This combining process is created using adversarial generative networks, or GAN, a group of machine learning techniques. A GAN harnesses [3] two algorithms of artificial intelligence one against the other. The first technique, defined as the generator, simulates an image or audio. This simulated picture or sound is then applied to a pool of actual photographs or voices – of people, say – loaded into the second technique, called the discriminator. The digital images or audio recordings may look nothing like faces or original voices at first. However, repeating the procedure numerous times with performance reviews enhances the discriminator and the generator. With plenty of loops and reviews, the generator will begin to produce realistically attainable faces or voices of non-existent people.

The four major types [4] of deepfake creators are as follows:

1. Deepfake hobbyist groups
2. Political actors, such as governments or protesters
3. Malicious actors, such as fraudsters
4. Legal actors, such as broadcasting networks

Hobbyists prefer to see AI-designed content as a novel model of online humor and approach contributing to the development of this technology as tackling a rational riddle, rather than a means for tricking or threatening individuals. Their deepfakes are intended to be humorous, amusing, or politically provocative, and often are created to attract social media followers. Many hobbyists could be searching for further tangible rewards, such as promoting education on the possibilities of deepfake technologies or getting deepfake-related jobs (e.g. making music videos or TV programs). Different political actors, like political provocateurs, hackers, extremists, and foreign nations can use deepfakes in fake news campaigns to influence public sentiment and reduce trust in other governments. In these times of modern warfare, deepfakes can be armed deception intended to interfere with elections and cause civil unrest. Scammers can use deepfakes for asset fraud or other financial crimes.

The most threatening repercussion of deepfakes is their ability to get us to question what we see. The more common deepfake innovation becomes, the less we can believe our own vision. Deepfakes are a big danger to our culture, political environment, and industry as they can [4]:

- Place pressure on journalists trying to root out true news from false reports
- Endanger a country and its future by distributing misinformation and intervening in elections
- Dismantle citizens' confidence in government
- Pose issues of cybersecurity for individuals and organizations

It is highly likely that, due to deepfakes, the journalism industry will have to confront a huge customer confidence issue. Deepfakes constitute a bigger problem than "ordinary" false news because they are harder to identify, making them a threat to the integrity of news reporters and the press in general. A video clip, filmed by an event eyewitness, could give a news organization strategic gain. With deepfakes, there is a risk that the video clip provided is false. Reuters, a global news company owned by Thomson Reuters, found 30 hoax videos concerning events during the surge of conflicts between India and Pakistan in 2019. In most cases, old videos from several other events were uploaded with new annotations to make them appear current. Video evidence, such as a genuine street protest or vicious scuffle, that is posted and falsely attributed to happenings elsewhere is a significant problem that will be magnified by the rise of deepfakes. When digging for witness footage of the gun massacre in Christchurch, New Zealand, Reuters stumbled upon a video that appeared to depict N.Z. police shooting the gunman. They soon learned the footage came from a different attack in the U.S.A. and that the gunman was not involved in the Christchurch attack.

Intelligence agencies are worried about the use of deepfakes to propagate electoral deception, manipulate election processes, and endanger national security. U.S. intelligence officers have regularly cautioned regarding the possibility of international intervention in affairs concerning the U.S., particularly in the lead-up to election day. Speech and action falsely attributed to a person in a viral deepfake is an effective tool in today's wars of deception, as these distorted videos will quickly distort the opinions of voters. Consider a recording of a politician utilizing a racist slur or accepting a bribe, a political contestant revealing an involvement in an offense or cautioning another nation of an impending conflict, a government executive in an apparently precarious position revealing a clandestine plot to conduct a coup, or U.S. troops performing war crimes like murdering citizens. Although these fake videos would undoubtedly trigger domestic instability, protests, and political disturbance, some nation-states might also opt to enforce foreign policies based on these deepfakes, contributing to disputes between countries.

Deepfakes are anticipated to hinder computer literacy and the trust of civilians in data presented by the authority because false videos of government executives engender public distrust in government. Most people are influenced by misleading spam created by AI and by false news that grows on hateful content, forged images, and a myriad of conspiracies. Nevertheless, the most destructive feature of deepfakes may not be misinformation per se, but how repeated interaction with disinformation causes people to believe that all information, even images, can not necessarily be trusted. This will lead to a trend called the "truth crisis" or "fact ignorance". Moreover, people may also reject real videos as false, merely because they've been rooted in the idea that something they hesitate to hear must be false. The biggest danger is not that the public will be fooled, but that they will start to see everything as manipulation.

The business environment has also used deepfakes, for example, by depicting a CEO saying racial or bigoted words, declaring a bogus takeover, issuing false claims about monetary failure or bankruptcy, or presenting them as doing a felony. In this way, deepfakes can be used for customer and asset exploitation. Additionally, deepfake technology can impersonate and portray an executive in real time to persuade a worker to make an immediate money payment or to provide classified details. Last but not least, deepfakes can help disperse malicious files. Experts recently discovered that an internet site dedicated to deepfakes clandestinely utilized the machines of their users to mine cryptocurrencies.

Deepfake innovation also has beneficial applications [5] in other markets, including films, digital networks, electronic media, gaming, sports, medical care, social media, material sciences, textiles, and e-commerce.

Deepfake technologies can help the movie industry in many ways. For starters, they can help to render digital sounds for movie actors who have vocal issues (i.e. due to sickness) or to repair film clips digitally rather than scheduling a reshoot. Film creators would be able to replicate classic movie scenes, create modern movies starring long-dead stars, use effects on visuals and advanced post-production facial editing, and create high-quality amateur pictures. Deepfake technology even facilitates programmed and original voice dubbing of films in different languages, allowing diverse audiences to understand films and academic material better.

Deepfake technology facilitates online gaming and interactive chat windows with improved telepresence, natural sounds, and visual replication. It can assist people who have lost their dear ones by virtually bringing back into existence a dead relative, allowing them to bid goodbye. It could enable those in the transgender community to view themselves in their true gender identity. These technologies may also aid people with Alzheimer's, by allowing them to connect with youthful faces that they might recognize. Experts are also researching the usage of GANs for detecting X-ray anomalies.

Companies are interested in deepfakes, as they can dramatically change online shopping and advertisement. For instance, brands can employ digital models to present modeling looks, selecting models with varying skin colors, weights, and heights. Furthermore, deepfakes could be created that transform shoppers into the model to simulate how an ensemble looks on them before buying.

In the present, it is relatively easy to say if a video is a deepfake or not. Slightly irregular motions in the head, ambiguous shadows, and a lack of eye movement are typical signs that a video is not genuine [6]. However, GANs are improving with every passing day and as the images that we watch gradually converge with visual realism, it may be up to software engineers to build forensic recognition systems. Fortunately, there are steps being taken to create classifiers for deep learning. Such classifiers will examine the raw video features to demonstrate validity via biometric video watermarks. This isn't an indefinite solution, particularly if GANs can be designed to counteract this type of forensic analysis. Cybersecurity leaders must continue to find novel ways to counter deepfakes.

LESSONS LEARNED

Accessing an online universe riddled with deepfakes is like walking into a dark room; terrifying, because you do not know what's there. Instead of closing the door of that dark room, locking it, and chaining it, we need to turn on the lights and determine if anything there could be helpful. Anxiety surrounding deepfakes and what they can do is a prevalent concern and can be addressed by understanding the benefits and risks of deepfakes.

The methods for creating "synthetic media" are becoming faster, more popular, and simpler to use when compared with previous years. Today, anybody can create hyperrealistic deepfakes, even with no machine learning expertise. A search for free applications to build deepfakes on Github reveals over 100 results. Some of them are a variation of the technique called face switch, but others offer different ways to exploit film, audio, and photographs.

Thus far, deepfakes have mainly been used to threaten famous people. This is because datasets are accessible containing photographs of these famous people on digital platforms. These datasets are used to create the deepfakes. However, as photographic data source content is readily accessible on social media, anybody can turn into a possible threat. For deepfake developers, it is just a matter of accessing these photos and wanting to use them for identity fraud, bribery or to propagate false news. Therefore, deepfakes can potentially affect anyone, not just political leaders, CEOs, and other powerful people.

Many deepfake methodologies emerged in scholarly communities and were then embraced by companies utilizing these tools and methods for profit. An issue that is central to many of today's arguments is why students and professionals developed deepfake methodologies if they could be harmful. It is important to remember that, aside from harmful uses, there are many beneficial implications of deepfakes.

No widely viable devices are yet available to support deepfake detection. Blockchain might become part of the answer. Perhaps the most realistic solution to rebuilding trust in videos could be the use of blockchain technologies [7]. Blockchain is a radically different method of storing online data that has the potential to provide unchangeable archives of data and transactions in a global open ledger.

All these technical approaches look promising, but it is necessary to adhere to a series of guidelines in the meantime. These include:

- Maintaining skepticism and being mindful that exploiting content is normal.
- Investigating information before distributing it.
- Checking multiple sources.
- Teaching others how to assess and process the reliability of information.

DISCUSSION QUESTIONS

1. What steps can social media giants like Facebook, Twitter, WeChat, and YouTube take to detect deepfake content and stop promoting it?
2. Should the government regulate what content can be posted on social media as a means to control deepfakes?
3. What legal frameworks currently available can be used to address deepfakes?
4. What can organizations do to avoid falling for deepfake scams?
5. How prepared are humans to survive AI threats like deepfakes?

REFERENCES

[1] https://www.wsj.com/articles/fraudsters-use-ai-to-mimic-ceos-voice-in-unusual-cybercrime-case-11567157402

[2] https://deeptracelabs.com/mapping-the-deepfake-landscape/

[3] Hasan, H. R., & Salah, K. (2019). Combating deepfake videos using blockchain and smart contracts. *IEEE Access*, *7*, 41596–41606.

[4] Westerlund, M. (2019). The emergence of deepfake technology: A review. *Technology Innovation Management Review*, *9*(11), pp. 40–53. https://doaj.org/article/6cf7440c848d4870b8642344c5b301e7

[5] Caporusso, N. (2020, July). Deepfakes for the good: A beneficial application of contentious artificial intelligence technology. In *International Conference on Applied Human Factors and Ergonomics* (pp. 235–241). Springer, Cham.

[6] https://www.uscybersecurity.net/deepfake/

[7] Hasan, H. R., & Salah, K. (2019b). Combating deepfake videos using blockchain and smart contracts. *IEEE Access*, *7*, 41596–41606.

2.9 Robots
Attack Vectors and Safeguards

Alisha Gilbert
Canterbury District Health Board, Christchurch, New Zealand

Zhigang Gong
University of Canterbury, Christchurch, New Zealand

CASE SUMMARY

A shopping mall in Singapore started to use a robot to fight COVID-19 on 21 May 2020. The robot uses ultraviolet light instead of chemical substances to kill the virus. The firm manufacturing the robot claims that the role played by the robot is dangerous, dull, and dirty. It's dangerous because ultraviolet light must be deployed in an environment in which no person gets involved, dull because the job is repetitive, and dirty because a virus exists in the environment. Therefore, it follows that a robot is the most suitable substitute for a person in this situation. The robot has a certain level of intelligence: it turns off the ultraviolet light when it detects a person nearby, it can perform cleaning work autonomously if its route has been fully mapped, and it recharges without human intervention.

Fast-food chain Kentucky Fried Chicken (KFC) built a new store in Moscow, Russia in 2020. This store is different from other stores in the world. KFC deployed robots in the storefront. The workers in the kitchen put the food on the conveyor belt. The robot takes the food from the conveyor belt and puts it in the food locker. Customers get their orders from the food locker. There is little to no contact between the customer and the cashier. KFC claimed that the deployment of robots in restaurants makes the restaurants more sanitary. On the other hand, robots cause unemployment ("Restaurant Of The Future" – KFC Unveils Automated Store With Robots And Food Lockers, 2020).

A bar in Spain uses humanoid robots to serve drinks to customers during the COVID-19 pandemic. The reason to use robots is for enforcing social distancing and limiting contact. These robots look friendly. The interaction between the customer and the robot seems to be not that difficult. When a robot comes to a table, the customer takes away the drink from the tray and presses a button on the robot, then the

DOI: 10.1201/9781003264415-25

robot leaves. In addition, customers can ask questions of a robot, in the same way they do with a real waiter, and the robot can search the Internet for the answers. It is said that robots intrigue a large amount of enthusiasm from people (Lo, 2020).

Spot, a robot dog created by Boston Dynamics, was tested to do the patrolling work in one of Singapore's parks during the COVID-19 pandemic. The robot broadcasts a pre-recorded message that tells visitors the importance of keeping to social distancing requirements as it patrols the park. It captures photos of the surroundings and sends them to officials to help them analyze the extent of people's gatherings ("Meet Spot, the Robot Dog that Enforces Social Distancing in Singapore," n.d.).

The camera-based eyes currently used on robots can be replaced with artificial eyes. An artificial eye looks like a biological eye because it mimics the iris and retina of the human eye and is particularly important for humanoid robots. Because it resembles the human eye, human-robot interaction could become more amicable. Technically, the artificial eye can provide more capabilities than the human eye. It could have a resolution that is higher than that of the human eye. It distinguishes between visual light and infrared radiation. It can see smaller objects and further. So, the artificial eye has the potential to transform the robot's vision system (Cuthbertson, 2020).

Price is an important factor when people consider using robots. At present, the price of buying a robot is quite expensive, and robot rental services are available as a cheaper alternative (Thomson Reuters Foundation, n.d.).

Robots that operate in the physical world and have little to no interaction with humans can be classified as physical robots. A robotic arm that is used to perform a particular job in a factory production line is a classical physical robot. Physical robots usually can work on their own. As the networked technologies of robots develop, physical robots can connect to the company network and the Internet. Although the connection can bring huge benefits, it also increases the risk of cyberattacks ("Overcoming Industrial Robot Security Threats," n.d.).

Robots that operate in the social world and have frequent interactions with humans can be classified as social robots. The robot acting as a waiter in a bar or restaurant, or as a receptionist in a hotel, is a classical social robot. People communicate with social robots through speech or non-verbally. Social robots are usually humanoid robots and are often designed with features that make people proactively engage in communication with them. To mitigate human frustration during communication, social robots often respond quickly to environmental cues such as sound or movement. In efforts to humanize social robots, facial recognition, and gaze-locking features are not uncommon. Social robots can even speak in the accent used by the person they are talking to. Research indicates that the cultivated appearance of social robots, where they appear pretty, innocent, and willing to serve, makes people feel relaxed when talking to them (IDLab – imec, Ghent University, Belgium, n.d.).

CASE ANALYSIS

While there are security models and standards designed to protect robot systems, it is possible for a hacker to exploit a flaw in the robot's operating system to gain root access. Therefore, robot manufacturers need to have a clear understanding of the possible security threats to their products ("Overcoming Industrial Robot Security Threats," n.d.).

The consequences of implementing a cyberattack on a physical robot are diverse and destructive. Consider a robot being used in a manufacturing plant. An attacker can gain access to a robot's operation instructions and modify them, resulting in the robot acting in a way that causes defects in part production. In this scenario, the whole production line is compromised, and the company will incur great losses. Additionally, attackers can launch ransomware attacks on network-connected robots. This could prevent the robots' ability to receive operation-critical data or to access the production system. Any security vulnerabilities in a robot can become attack vectors that hackers can exploit to gain access to and steal confidential information from the robot, or from the wider network and system ("Overcoming Industrial Robot Security Threats," n.d.).

Currently, most people don't think social robots are a threat to them, often considering social robots to be advanced mechanical dolls: pretty, simple, kind, helpful, and trustworthy because they won't tell what you say to them to anyone else. On the contrary, social robots are much more complicated. First, social robots usually don't operate in isolation. They are usually connected to a network to receive the information they need. Social robots can also upload data collected through sensing and interacting with their environment to the network. Second, although social robots seem to behave autonomously, this may not be true for all social robots. Some may be controlled by a remote human operator. For example, the voice that people hear from the social robot may not be spoken by the social robot itself, but by a remote operator whose voice has been distorted to sound like a robot voice. In such a case, it is possible that a conversation conducted by a remote operator through a social robot could be perceived by a person as confidential, affording that person a false sense of security that can lower inhibitions when sharing private information, while in fact, it is not. Furthermore, people, especially children, can be influenced by social robots (IDLab – imec, Ghent University, Belgium, n.d.). When people make decisions, they tend to follow others for two reasons. The first reason is uncertainty. For example, if the menu of a restaurant is written in a foreign language, people tend to order the same food as the neighboring table. The second reason is that people don't want to be different from the collective group. For example, if a participant knows that others completing a test have the same answer that is different from theirs, the participant is likely to change their answer, even if their original answer is the correct one. It is for all these reasons that unconditionally believing in social robots is risky (IDLab – imec, Ghent University, Belgium, n.d.).

The world continues to develop toward higher digitalization and mobility. Industry 4.0 (4IR) which depends on automation and the use of robots plays an important role in this process. It brings efficiency to production processes and improvement to user experience. We can see robots working in various places, such as car factories, restaurants, and hospitals. Robots often work in highly networked environments. They communicate with various services and systems. Willow Garage introduced the Robot Operating System (ROS) in 2007 which is a unified and standard platform used by robots. Software used by robots is based on the middleware frameworks in the ROS. Nodes running ROS form a distributed network. In addition, ROS provides libraries for solving problems relating to the recognition of objects in image and space mapping. However, there are few security mechanisms in ROS. The designers of the ROS intentionally excluded security mechanisms from it because they didn't have a clear security threat model. At present the ROS is mainly used by researchers, for the sake of convenience, they prefer an environment without security constraints. People expect to use the ROS in mass markets besides the laboratories. However, a system without strong security mechanisms can easily be the attack target. By conducting a cyberattack on such a robotic system, a robot could leak sensitive information captured through sensors and cameras about its environment or be manipulated to cause security risks (Galov, 2019).

LESSONS LEARNED

The deployment of robots in an organization has an impact on its security management. First, the security modules in the robot must be enabled. For example, the security module in the anti-virus robot deployed in the Singapore mall turns off its ultraviolet lights automatically when it detects there are people around it. Second, the suggestions relating to the secure use of the robot provided by the manufacturer must be respected. For example, the robot arm manufacturer suggests installing a fence around the robot arm. Third, the organization's security strategy and security policies must be revised. For example, a person employed by a park to do patrolling work, when encountering a danger during their patrolling period, can make a call to park security to ask for help. If we consider Spot, the robot dog, a 'call for help' functionality would be of paramount importance. Spot costs around $75,000, which could make it a target of theft. The challenge in implementing this 'call for help' functionality is in determining the situational or environmental markers that would trigger a call to park security.

The access control security model is used in the information system to manage the availability of information assets. People are usually authenticated and authorized to access a certain kind of information. The social robots have the same requirement as the information system: to keep the security of the information asset. As social robots have a certain level of artificial intelligence (AI), they contain a significant volume of information assets and, as such, must be programmed to limit user access to those information assets. A robot waiter, for example, has at least two kinds of users: the customers it serves and the manager it is controlled by. The information obtained by customers (e.g. the menu) is public, while sensitive information (e.g. business data) is private and only available to the manager. Therefore, the access control security model must be enforced during the interactions. Several techniques could be used to authenticate users, such as speech recognition, system and user passwords, and biometric (fingerprint or face) scanning.

When parents buy a companion robot for their children, they often focus on the robot's functionalities. Playing music and video, singing, and storytelling are basic functionalities. Advanced functionalities we can see in the market include teaching as well as video and phone calls. A versatile robot with a reasonable price is a rather attractive choice. Additionally, security is a significant concern for parents when choosing a companion robot. Companion robots for children are usually small and light, with limited mobility, making them unlikely to be a physical threat to children. However, most robots have connections to the Internet through Wi-Fi or cellular networks. Therefore, the robots can download content from the Internet and children can make calls to their parents or friends through the robot. The combination of the camera and the connection to the Internet is a vulnerability that could be exploited by attackers for illegal monitoring. In addition, research shows that children can be influenced by robots because of their relatively low level of judgment. If a companion robot has been manipulated, this could compromise how it performs tasks, such as manipulating the originally programmed Internet content safety rules. This change in robot behavior can negatively impact children. Since hackers can find attack vectors to compromise systems that are connected to the Internet, it is important for parents to periodically check their child's companion robot to confirm that it works as intended.

DISCUSSION QUESTIONS

1. What are the security threats confronted by physical robots?
2. What are the security threats confronted by social robots?
3. How do people get along with social robots?
4. How to prove that a robot is not a threat to people?
5. How should security strategies, policies, and programs be revised when robots are involved?

REFERENCES

"Restaurant Of The Future" – KFC Unveils Automated Store With Robots And Food Lockers. (2020, June 27). Retrieved July 20, 2020, from https://www.zerohedge.com/technology/restaurant-future-kfc-unveils-automated-store-robots-and-food-lockers

Cuthbertson, A. (2020, May 21). Artificial eye that "sees" like a human could transform robotics. Retrieved July 20, 2020, from https://www.independent.co.uk/life-style/gadgets-and-tech/news/robot-artificial-eye-bionic-retina-blind-a9525151.html

Galov, D. (2019, October 24). A glimpse into the present state of security in robotics. Retrieved July 20, 2020, from https://securelist.com/robots-social-impact/94431/

IDLab – imec, Ghent University, Belgium. (n.d.). The potential of social robots for persuasion and manipulation: a proof of concept study. Retrieved July 20, 2020, from https://media.kasperskycontenthub.com/wp-content/uploads/sites/43/2019/10/14081257/Robots_social_impact_eng.pdf

Lo, T. (2020, June 2). Bar employs humanoid robot waiter to serve and take orders during Covid-19 pandemic. Retrieved July 20, 2020, from https://www.dailystar.co.uk/news/world-news/bar-employs-humanoid-robot-waiter-22126862

Meet Spot, the Robot Dog that Enforces Social Distancing in Singapore. (n.d.). Retrieved July 20, 2020, from https://www.cndenglish.com/others/travel-technology/meet-spot-robot-dog-enforces-social-distancing-singapore

Overcoming Industrial Robot Security Threats. (n.d.). Retrieved July 20, 2020, from https://www.robotics.org/blog-article.cfm/Overcoming-Industrial-Robot-Security-Threats/200

Thomson Reuters Foundation. (n.d.). Exterminate! UV robot sent to Singapore mall to zap coronavirus. Retrieved July 20, 2020, from https://news.trust.org/item/20200521093707-bzkae

2.10 Ransomware *Hollywood Presbyterian Medical Center*

Tridhawan Choudhary
AFFCO New Zealand Limited, Hamilton, New Zealand

Priyanka Jagre
UNIFY Solutions, Wellington, New Zealand

CASE SUMMARY

Ransomware continues to be a major threat to organizations, including medical facilities. Data and systems can be kept on hold until the ransom amount is paid. This is what happened at Hollywood Presbyterian Medical Centre (HPMC), a well-established healthcare facility that opened in 1924. This ransomware attack significantly affected the Hollywood community hospital and became an international headline because HPMC's data was blocked by the attacker, prompting HPMC to immediately pay $17,000. WIRED magazine has revealed that HPMC was attacked by the Locky virus, which blocks user access to system data. The Locky virus is sent as an email attachment requiring the user to enable macros. In HPMC, an employee launched the virus by mistakenly opening an email attachment disguised as an invoice.

Once Locky has been launched on the system, a countdown prompt is displayed on the screen instructing the user to pay a specified amount to regain access to the system. Because of the ransomware attack, hospital authorities and staff were not able to access critical documents such as patient records. Loss of data access is of particular concern in the healthcare sector. For example, patients need regular check-ups and treatments, for which hospital staff needs a treatment and medication history. Because of the ransomware attack, patients needing immediate care were shifted to a nearby hospital.

Once HPMC staff were made aware of the ransomware attack on their hospital system, an immediate state of emergency was announced within the hospital, which limited the use of computers and electronic devices. Computer systems were taken offline and online services such as cloud-based storage could not be used. Some departments like "Radiation Oncology" which simply could not operate without computers had to pause all services until the attack had been resolved.

DOI: 10.1201/9781003264415-26

The attackers demanded payment of 40 bitcoin – roughly equivalent to $17,000 – to reinstate HPMC's data access. Although the FBI and LAPD conducted an investigation concerning the source of the ransomware attack, HPMC needed to pay the ransom because of the time-sensitive and critical nature of the data being ransomed. Since payment was made in bitcoin, the identity of the attacker receiving the payment could be kept anonymous.

This attack lasted for more than a week, during which HPMC remained offline and kept pen-and-paper records of the medical services they were able to provide without system access. Fortunately, HPMC was able to modify its operations, in response to the ransomware attack, in a manner that did not compromise patient health outcomes. Once the ransom was paid, data access was restored and, after a thorough sweep to ensure there were no traces of the malware remaining, HPMC's systems could be brought back online. After the Locky ransomware attack, HPMC started looking more into its security policy. HPMC has since developed a security infrastructure and no further security breach has been reported by HPMC (Baraniuk, 2016).

CASE ANALYSIS

One of the most prolific criminal businesses is ransomware. It's malicious software and this software aims to obtain money from its victims. The aim of ransomware is not to hack the data but to make data inaccessible and ask for ransom in the form of bitcoin. Users often feel complicit and foolish, because the ransomware was launched as a result of them opening a document containing the virus. The only way to protect systems from ransomware is to raise awareness of the method used by attackers: in this case, by sending emails containing affected MS Word documents disguised as an invoice or any other important document. Once the system is affected by the virus the screen is locked and no further activities can be done. The only thing is to let the data go or to pay the ransom to get the decryption key which can reinstate access. To pay the ransom, instructions are shown on the screen on how to pay the amount to get the decryption key. Ransom costs can range from hundreds of dollars to thousands of dollars and can only be paid in bitcoin to the ransom criminals (Jareth, 2019).

There are various methods used to launch ransomware attacks by which the whole system can be locked (e.g. Grimes, 2017). A few methods are as follows:

- Social Engineering
- Malvertising
- NotPetya
- Drive-by-Downloads
- USB drives and portable computers
- Pirated software
- Network propagation

Social engineering includes ways to distribute the virus. Distribution usually occurs by duping a victim into opening an email, instant message, or text message. An attachment or link is provided through mail or message which can lead to the installation of malware into the system. As mentioned earlier, ransomware is not targeted and can be launched without the recipient knowing that the link or the message may contain the virus and affect the whole system. Attackers also use social media platforms and emails and insert URLs that contain the virus.

Malvertising is a popular and fast-growing method of ransomware attacks these days. Attackers purchase online ad space to disseminate advertisements on the web that contain the virus. The advertisements promote random products, sales and discount deals, or free software installation. Once the ad is clicked, it installs malware on the system.

Drive-by-Downloads launch malware on systems that visit an infected website. The system user doesn't need to download or click on anything on the website to fall victim to the attack.

NotPetya is malware that can introduce a virus into your system. This virus appears in infected computers only and corrupts data in such a way that it cannot be recovered. This is not a widely used attack but is very dangerous.

USB drives are a common way for malware to infect a system. USB drives are ubiquitous all over the world for all age groups. Simply connecting an infected USB drive to your system can infect your device.

Pirated software is of particular concern because bad code can be inserted into software that has been illegally copied. When a user tries to install the software, they instead install malware hidden inside it.

Network propagation is another method to launch ransomware attacks. Early ransomware only infected the local machine. The advanced malware variants currently being used have self-propagating mechanisms that allow the malware to infect the host machine and then propagate to other devices that are connected to the same network (Fruhlinger, 2020).

Methods to prevent ransomware attacks are as follows:

- Only open attachments from trusted senders and try to build strong network security.
- Do not open attachments, emails, messages, etc., that require modifications to system or application settings (e.g. enabling location tracking).
- Systems, computers, browsers, and apps that are used often should be up to date.
- Install security patches and ad-blocking software.
- Establish strong network security in conjunction with the network operator and security manager.
- Never use pirated software or websites.
- Do not plug unknown devices into your system.

LESSONS LEARNED

HPMC needs to make its security infrastructure stronger than ever before. Every year there are many new cases of devices and systems that are affected by ransomware attacks. Ransomware attacks are generally opportunistic, and victims are not usually targeted. Ransomware can be designed to attack any device or platform on a mobile phone, laptop, tablet, or personal computer. Many organizations that share a lot of data across a network have a weak security infrastructure. Even with strong security infrastructure, some organizations rely so heavily on their networked data that they readily pay the ransom when they fall victim to a ransomware attack. In the end, the main way to protect against ransomware attacks is by maintaining a strong security infrastructure that includes employee education.

HPMC is an example of an established organization that has fallen victim to ransomware. While there are many methods to launch ransomware attacks, there are also many ways to protect systems from ransomware. Organizations and their employees need to understand how ransomware works, its scope of effect, and best-practice security measures that can be taken to mitigate the threat of ransomware.

The most effective and important step an organization can take to protect against ransomware is having a strong information security infrastructure. Making regular backups, downloading software updates, and installing anti-malware and ad-blocking software are digital safeguards that must be integrated into an organization's information security. Education and corporate policies around digital safety and vigilance are also critical.

DISCUSSION QUESTIONS

1. What is ransomware?
2. How was the ransomware attack against HPMC launched?
3. What methods can organizations use to protect their systems from ransomware?

REFERENCES

Baraniuk, C. (2016, February 18). *BBC News*. Retrieved from Hollywood hospital pays ransom to hackers: https://www.bbc.com/news/technology-35602527#:~:text=A%20Los%20Angeles%20hospital%20has,out%20some%20tasks%20on%20paper

Fruhlinger, J. (2020, June 19). *CSO*. Retrieved from Ransomware explained: How it works and how to remove it: https://www.csoonline.com/article/3236183/what-is-ransomware-how-it-works-and-how-to-remove-it.html

Grimes, R. A. (2017, January 24). *CSO*. Retrieved from How to prevent ransomware infection and recover if you're hit: https://www.csoonline.com/article/3160766/the-evolution-of-and-solution-to-ransomware.html

Jareth. (2019, December 19). *EMSISOFT*. Retrieved from How ransomware spreads: 9 most common infection methods and how to stop them: https://blog.emsisoft.com/en/35083/how-ransomware-spreads-9-most-common-infection-methods-and-how-to-stop-them/

2.11 Fast Food Phishing
Cyber Espionage

Alastair Blackett
M2X, Auckland, New Zealand

Cherie Chun Mei Mak
LivePerson, Singapore, Singapore

CASE SUMMARY

Hackers are having a field day with the COVID-19 crisis. According to a report released by Google's Threat Analysis Group, over 18 million COVID-related phishing and malware Gmail messages and 240 million COVID-related spam messages are detected by Google each day (Huntley, 2020). Government-backed attacker groups are using COVID-19 themes as bait for phishing and malware scams, targeting government officials on every continent except Antarctica (Huntley, 2020; Ng, 2020). One notable scheme was used only in the US and targeted personal accounts of US government employees (Huntley, 2020). The ploy was to entice US government employees to click on emails that directed them to phishing pages; disguised as meal delivery websites and designed to steal their personal Google account credentials (Huntley, 2020). The bait: coupons and free meals from fast-food chains in response to COVID-19 (Huntley, 2020). According to Google, most of these emails were preemptively intercepted and sent to spam, so no user accounts were compromised by this campaign.

Fortunately, there is a silver lining: there was no overall increase in government-backed phishing attacks, merely a change in tactics. Rather there was a slight decrease in accounts that received "government-backed attacker" warnings, as seen in Figure 2.11.1. Perhaps, even hackers were not immune to productivity lags amidst the COVID-19 pandemic (Huntley, 2020).

CASE ANALYSIS

To tackle the analysis of the case, it is important to identify who these government agencies are using as their cybersecurity partners.

Google's Threat Analysis Group, LLC (TAG) is experienced in security risk reduction and evidence-based outcomes, where they utilize performance-based, industry-specific risk assessment methodologies that incorporate effective deter, detect,

 DOI: 10.1201/9781003264415-27

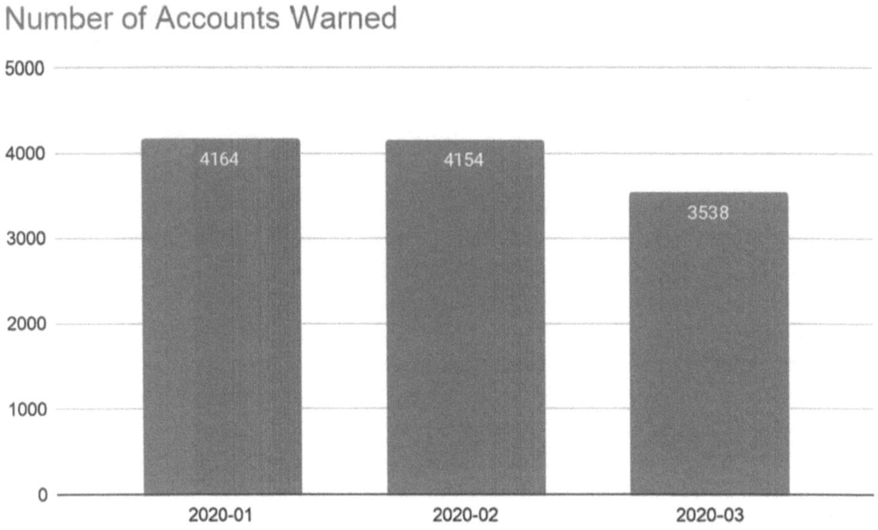

FIGURE 2.11.1 Google accounts that received attack warnings (Huntley, 2020).

delay, and response criteria for protecting assets (Threat Analysis Group, 2015). While offering its technical services, the group CrimeAnalysis™, a web-based software application, identifies the threat level at each organization's facility (Threat Analysis Group, 2015). The application uses actual crime records to generate security metrics to help benchmark threats and help select appropriate countermeasures to reduce risk (Threat Analysis Group, 2015). The group offers a strong set of expertise in managing risk and cybersecurity, though there is very little research analyzing the failures of the organization. However, this may be subjective due to the group being a subsidiary of Google. It is possible the company is preventing the publication of negative material on one of its agencies.

The US government requires a strong business partnership with TAG due to its history of poor cybersecurity practices. The most infamous example of the government's failure in maintaining cybersecurity is the Hillary Clinton private email case. It was revealed that Clinton had exclusively used her private email address and server for government business while serving as Secretary of State (Jacobson, 2016). The congresswoman was aware of the US transparency laws that are designed to inform the public of government work and create public accountability for government officials (Jacobson, 2016). Therefore, the Secretary was not allowing for disclosure of government business to the people of the US (Jacobson, 2016). However, despite this incident and subsequent changes, government employees still use private emails for non-classified information. Because of this incident, Clinton had 31,000 emails deleted by her staff and a couple thousand emails submitted to Congress. During her time as Secretary of State, Clinton's private email was reportedly involved in at least five phishing attempts. Though the hackers were not successful, the persistence of attacks underlines the damage that a careless click could have caused by releasing a large volume of sensitive information (Jacobson, 2016).

This does not suggest that Clinton's private email practices were better or worse compared to her fellow employees. Indeed, the previous Secretary of State, Colin Powell, relied on personal email accounts for communications with American officials and foreign leaders. By communicating outside the proper channels, government officials undermine public and institutional oversight and create an incomplete historical record (Jacobson, 2016).

This sort of scrutiny by the media revealed the significant shortcomings of government information technology (IT) departments in implementing strong information security policies and measures. (Jacobson, 2016). The government's IT departments do have a classification system that prevents the disclosure of sensitive information that could injure national security. However, the government needs to follow best practices and intertwine its email and classification systems. It is largely acknowledged that the government's IT capabilities must be improved. A substantial investment in IT would enhance cybersecurity practices and the convenience of using government email accounts (Jacobson, 2016).

While the US government has looked for partnerships to improve its IT cybersecurity, there have been many discussions about partnering with the private sector. The federal government's record on cybersecurity does not inspire confidence that it can provide a solution to the cybersecurity threats faced in the private sector (Rosenzweig, n.d.). Former US president Barack Obama announced in 2009 that securing cyberspace cannot succeed if done in isolation (Rosenzweig, n.d.). This means the private and public sectors need to work together to fix the cybersecurity problems occurring throughout the country.

The government's role was initially thought to be the management of public sector resources, such as medical, education, and traffic services. This means these sectors are now more exposed to cyber threats due to the cyberattacks launched on the private sector and the unequal development of security strategies in the public and private sectors. This is evident in the Collins Cybersecurity Act of 2012 which illustrated that the private sector has done an inadequate job of protecting itself and that market incentives have not developed appropriately to promote better protection (Rosenzweig, n.d.). A statement released by former US president, Donald J Trump, showed that the 15-year National Cyber Strategy will continue to collaborate with the private sector and work to strengthen the ability of the US to deter and, if necessary, punish those who use cyber tools maliciously (Government of United States of America, 2018). It appears that the US government and the private sector will continue to work together despite the high risk of exposing each other to cyberattacks.

While online hacking occurs all around the world, it is different in this case involving the US government where hackers launch cyberattacks for espionage purposes. Cyber espionage can involve the theft of military, government, and trade secrets (Khosrow-Pour, 2020). The most typical method of gaining such information is using social engineering in the form of phishing, which plays to human psychology. Initiated by hackers on the dark web, phishing is an effective way to gather information without hacking into any of the US government's sophisticated computer systems that would be highly protected by more sophisticated cybersecurity systems (Khosrow-Pour, 2020). The dark web makes it difficult for the emails used to launch phishing attacks to be detected, as it allows hackers to mask not only their identities but their locations.

An example of a hacking agency is the Julian Assange case. Julian Assange founded WikiLeaks, an international non-profit organization dedicated to publishing documents and information for government transparency (Khosrow-Pour, 2020). Though WikiLeaks was unsuccessful, and Julian Assange was brought to justice, there have been cases involving other agencies. In February 2012, the US Department of Security website was hacked and taken down for days due to an attack from an anonymous group. In June 2007, the Secretary of Defense's email account was hacked by unknown foreign intruders. The list of such incidents is extensive (Khosrow-Pour, 2020). Espionage against a country's government is sometimes more valuable than financial information, as it can allow the attacking organization or country to have total power over the victim.

While phishing is a simple psychological social engineering scheme for hackers to gain the information they need, a lot of technically knowledgeable IT specialists often make the mistake of believing that strong security measures will succeed in protecting their stakeholders (Dhamija et al., 2006). Researchers have found that computer users fall victim to phishing attacks based on a lack of understanding of how computer systems work, a lack of attention to detail, and the visual deception perfected by phishers. Attackers use two key strategies to make recipients believe the phishing emails and websites are safe and authentic: third-party endorsements and personalization (Dhamija et al., 2006). Many hackers used well-established brands, like IT company giants Verisign, to make them look credible, or commercial brands, like Nike and Adidas, to encourage victims to provide personal details or entice them to click on a link (Dhamija et al., 2006). A high degree of personalization increases the trustworthiness of the consumer, whether it is through emails or web pages (Dhamija et al., 2006). The more personal information is presented to the user, the more likely the user is to believe the website or email is authentic. Therefore, phishing attacks can be effectively masked and launched without alarming victims or making them feel they are being hacked.

As this case has explored, US government employees were clicking on links and providing personal information to obtain free coupons and discounts on fast food items. The hackers had identified this ploy as an innocuous and enticing cover to gain personal information and conduct espionage. A study found that every food ad was coded for intended primary targets, based on the content of the ad and the nature of the appeal (Fleming-Milici & Harris, 2016). Table 2.11.1 depicts the average advertising time, per hour, of food advertising content. This has been categorized by demographic and shows, by and large, people who are 20–64 years old as the main target audience. Table 2.11.2 shows the average age of members of the US government subject to their hierarchical position, demonstrating that most government employees fall into that target market.

Even in today's world, where online hacking has become common knowledge, people still do not acknowledge its presence. This may be attributed to the fact that impulse buying has become a human psychological behavior. Combined with the convenience of online shopping, people are always intrigued to purchase or discover items online (Jakobson, 2019). Online impulse buying has received some attention

TABLE 2.11.1
The Average Minutes and Seconds Per Hour of Food Advertising Content by Target (Clinton, 2015)

Target Audience	Average Minutes Per Hour
Children and/or Teens Only	0:40
Adults and Teens	0:35
Adults Only (20–64)	0:46
Older Adults (65+)	<0:01
All Age Equally	0:06

TABLE 2.11.2
The Average Age of Congress Members in the US Government (Manning, 2018)

Congress	Representatives (years)	Newly Elected Representatives (years)	Senators (years)	Newly Elected Senators (years)
115th	57.8	50.8	61.8	57.8
114th	57.0	52.3	61.0	50.7
113th	57.0	49.2	62.0	53.0
112th	56.7	48.2	62.2	52.1

and considerable academic studies have attempted to describe online impulse buying behavior (Fleming-Milici & Harris, 2016). Researchers have concluded that emotions dominate the impulse buying process and impulse buying behavior is a direct result of hedonic motivation (Jakobson, 2019). It is found that consumers who make impulse buying decisions also make unintended, unreflective, and immediate purchases just to fulfill the urge to buy something. The reasons for this can be summarized as low prices, marginal need for items, emotions, and low cognitive control or spontaneous behavior; when there are discounts, these people cannot resist the allure of online impulsive shopping (Ozen & Engizek, 2014). This research has revealed that impulse buying satisfies several hedonic needs such as fun, novelty, and surprise. This is reflected in the case of government employees being offered cheap or free meals. One may suggest that due to the high pressures and long hours of working in Congress, there is a desire to obtain something quick, enjoyable, and convenient to eat. This demonstrates the positive correlation that exists between the three dimensions of hedonic shopping motivation – adventure, relaxation, and value – and online buying tendencies (Ozen & Engizek, 2014). While ignoring the effectiveness of security practices and hackers' capabilities, humans have developed a behavior that creates major cybersecurity risks, posing challenges for an IT department tasked with managing the organization's IT infrastructure and protecting it from attacks.

LESSONS LEARNED

Human, technological, and organizational factors all play integrative roles in an organization's information security (Safa, Solms, & Futcher, 2016). Employees are often implicated as a significant threat to information assets and the root cause of many information security incidents (Safa et al., 2016). A plausible reason concerns irresponsible human cybersecurity behavior when faced with time pressures and deadlines that can jeopardize security assets and increase vulnerability to threats (Chowdhury, Adam, & Skinner, 2019). For instance, when under pressure to meet deadlines, an employee may send confidential company data to personal email accounts so they can work from home. Other people-related security concerns also include social engineering attacks and lack of compliance with security policies (Frauenstein & von Solms, 2013). Nonetheless, people can play a significant role not just in risk creation but also in risk mitigation, where aggregation of employee efforts, abilities, and knowledge can prevent information security breaches (Safa et al., 2016; Wiley, McCormac, & Calic, 2020). Security knowledge sharing was posited to be the most important factor that mitigates information security incidents, as hackers often utilize social engineering techniques and phishing to target their victims. Knowledge sharing among employees serves as an effective approach to increasing awareness while also reducing the cost of organizational information security (Safa et al., 2016). In the context of this case, Google's Threat Analysis Group compiled a report on current COVID-19-themed phishing and malware schemes, publishing their findings on the Internet for community awareness. By identifying threats and sharing this knowledge, implicated persons can be alerted and reminded to be vigilant and on guard against the current "phishing-theme-of-the-season".

Another point of consideration is organizational culture. A practical recommendation in research, for organizations aiming to improve employee information security awareness, was to focus on understanding, developing, and strengthening their organization's security culture (Wiley et al., 2020). An ideal security culture involves awareness of both the organization's external security environment, to proactively mitigate new threats, and internal security environment, to identify the types and root causes of security breaches that occur (Ruighaver, Maynard, & Chang, 2007). An empirical study conducted across 12 countries illustrated how information security policy positively influences security culture. Organizations that ensured employees read the information security policy had stronger information security cultures (Da Veiga, 2016).

While there is no common consensus standardizing the cultivation of positive information security culture in an organization, information security culture is believed to be a subculture reflecting organizational culture, as enforcement and development of security policies require support from organizational culture (Nasir, Arshah, Hamid, & Fahmy, 2019; Ruighaver et al., 2007). As email remains a major form of communication within and between organizations, this case highlights the importance of organizational policies that can serve as a protective security measure. Organizational email policies not only outline specific guidelines for acceptable email behaviors – such as the use of personal emails for corporate communications – but also include policies aimed at equipping employees with the necessary tools to protect against email-related threats (Ertan, Crossland, Heath, Denney, & Jensen, 2018). It is suggested in academic literature that while security policies and procedures may be in place, enforcing employee compliance remains a key challenge (Ertan et al., 2018).

DISCUSSION QUESTIONS

1. Why have employees often been considered the weakest link in an organization?
2. What role does organizational culture play in information security?
3. How can we prevent employees from using personal email accounts for work matters?

REFERENCES

Chowdhury, N. H., Adam, M. T. P., & Skinner, G. (2019). The impact of time pressure on cybersecurity behaviour: A systematic literature review. *Behaviour & Information Technology*, *38*(12), 1290–1308. https://doi.org/10.1080/0144929X.2019.1583769

Clinton, L. (2015). Best practices for operating government-industry partnerships in cyber security. *Journal of Strategic Security*, *8*(4), 53–68. https://doi.org/10.5038/1944-0472.8.4.1456

Da Veiga, A. (2016). Comparing the information security culture of employees who had read the information security policy and those who had not. *Information and Computer Security*, *24*(2), 139–151. https://doi.org/10.1108/ICS-12-2015-0048

Dhamija, R., Tygar, J. D., & Hearst, M. (2006). Why phishing works. *Proceedings of the SIGCHI Conference on Human Factors in Computing Systems - CHI '06*, *4*(8), 754–788. https://doi.org/10.1145/1124772.1124861

Ertan, A., Crossland, G., Heath, C., Denney, D., & Jensen, R. B. (2018). *Everyday Cyber Security in Organisations: Literature Review*. https://arxiv.org/ftp/arxiv/papers/2004/2004.11768.pdf

Fleming-Milici, F., & Harris, J. L. (2016). Television food advertising viewed by preschoolers, children and adolescents: Contributors to differences in exposure for black and white youth in the United States. *Pediatric Obesity*, *13*(2), 103–110. https://doi.org/10.1111/ijpo.12203

Frauenstein, E. D., & von Solms, R. (2013). An enterprise anti-phishing framework. In *IFIP International Federation for Information Processing 2013* (pp. 196–203). https://link.springer.com/content/pdf/10.1007/978-3-642-39377-8_22.pdf

Government of United States of America. (2018). National cyber strategy of the United States of America (1). https://www.whitehouse.gov/wp-content/uploads/2018/09/National-Cyber-Strategy

Huntley, S. (2020). Findings on COVID-19 and online security threats. https://blog.google/technology/safety-security/threat-analysis-group/findings-covid-19-and-online-security-threats/

Jacobson, J. (2016). The secretary's emails: The intersection of transparency, security and technology. *Purpose of Transparency Laws*, *88*(2), 1442–1503.

Jakobson, M. (2019). The human factor in phishing. *School of Informatics*, *3*(4), 78–98. http://markus-jakobsson.com/papers/jakobsson-psci07.pdf

Khosrow-Pour, M. (2020). *Encyclopedia of criminal activities and the Deep Web*. Information Science Reference.

Manning, J. E. (2018). Membership of the 115th Congress: A Profile. https://www.senate.gov/CRSpubs/b8f6293e-c235-40fd-b895-6474d0f8e809.pdf

Nasir, A., Arshah, R. A., Hamid, M. R. A., & Fahmy, S. (2019). An analysis on the dimensions of information security culture concept: A review. *Journal of Information Security and Applications*, *44*, 12–22. https://doi.org/10.1016/j.jisa.2018.11.003

Ng, A. (2020). Hackers tempt federal workers with free fast food in COVID-19 scams. https://www.cnet.com/news/hackers-are-targeting-us-government-workers-with-covid-19-scams-offering-free-fast-food/?fbclid=IwAR08dr1TNYczHzNXng7Uiv7OCXjx-2ByvxMyRTnUX0um31t4Nr9SbADLOTI

Ozen, H., & Engizek, N. (2014). Shopping online without thinking: Being emotional or rational. *Asia Pacific Journal of Marketing and Logistics*, *26*(1), 78–99. https://doi.org/10.1108/APJML-06-2013-0066

Rosenzweig, P. (n.d.). The alarming trend of cybersecurity breaches and failures in the U.S. government (2). Backgrounder. https://www.heritage.org/defense/report/the-alarming-trend-cybersecurity-breaches-and-failures-the-us-government

Ruighaver, A. B., Maynard, S. B., & Chang, S. (2007). Organisational security culture: Extending the end-user perspective. *Computers & Security*, *26*(1), 56–62. https://doi.org/10.1016/j.cose.2006.10.008

Safa, N. S., von Solms, R., & Futcher, L. (2016). Human aspects of information security in organisations. *Computer Fraud & Security*, *2016*(2), 15–18. https://doi.org/10.1016/S1361-3723(16)30017-3

Threat Analysis Group. (2015, June 14). Security risk management. Independent Security Consultants. https://www.threatanalysis.com/security-risk-management/

Wiley, A., McCormac, A., & Calic, D. (2020). More than the individual: Examining the relationship between culture and information security awareness. *Computers & Security*, *88*(101640). https://doi.org/10.1016/j.cose.2019.101640

2.12 Workplace Robots

Alastair Blackett
M2X, Auckland, New Zealand

Cherie Chun Mei Mak
LivePerson, Singapore, Singapore

CASE SUMMARY

Katie Johnson explores the topic of workplace robots in her article 'Hey robot, is that report ready'. In her article, she quotes MIT graduate and chief executive of Eckhart, Andy Storm who states "automation is not going to be Armageddon for the American Worker" (Johnson, 2018). The idea of the robots entering the workplace is to "allow people to get out from underneath mind-numbing, repetitive tasks and move into roles where they're using their minds and their god-given talents" (Johnson, 2018). According to the World Economic Forum, 29% of the workforce is operated by robots, and by 2025 machines will carry out more than half of the workplace tasks (Johnson, 2018). The medical industry, warehouse operations, and manufacturing plants have all experienced the positive and negative effects of the introduction of robots. Northeastern University professor Joseph Aoun believes it essential that higher education prepare people for this change (Johnson, 2018). Education and training are the most important investments in human capital, and the traditional economic ideology is that investment in human capital improves business performance and therefore drives the economy upward (Beckler, 1992). Many studies in the United States demonstrate that investment in human capital raises a person's income and adjusts for better family backgrounds and greater abilities of more educated people (Beckler, 1992). However, they do not discuss how people, due to the unequal opportunities within many developed countries around the world, are often not afforded the same opportunities to achieve higher level education, and therefore workplace robots pose a significant threat to the livelihood of low-income and low-skill workers. The US educational system is one of the most unequal systems in the world, and students routinely receive dramatically different learning opportunities based on their social status (Schmidt & McKnight, 2015). Despite quantifiable differences in funding, teacher quality, curriculum, and class sizes, many still hold the view that a student's success, or failure, is solely determined by their abilities and scholastic aptitude. This places extreme pressure on, and can cause serious harm to, minorities and economically disadvantaged students (Schmidt & McKnight, 2015). Such an environment begs the question: Will robots be an asset to the wealthy and a threat to the poor?

DOI: 10.1201/9781003264415-28

The first robots introduced were used for simple tasks, like pick and place, and had no external sensing (Wallen, 2005). In 1956, Joseph F. Engelberger and George C. Devol, an engineer and inventor/entrepreneur, respectively, developed the Unimate robot, which extracted die castings from machines and carried out spot welding on vehicles manufactured by General Motors in the United States (Wallen, 2005). Since then, the field of robotics has grown immensely. The industrial growth of robots within the last 20 years has been significant, and research has shown that the use of robotic workers is likely to continue increasing. As the development and utilization of robots continued, researchers discussed that robots should only be implemented if they make work more meaningful (Rosso et al., 2010). Robots should only be implemented if their use leads to job satisfaction and worker well-being. More importantly, robots should not have a detrimental effect on individual lives by disrupting the way we live them, such as by reducing the hours individuals work in a week (Acemoglu & Restrepo, 2017). Indeed, a prevalent philosophical argument states that, as individuals, we must fulfill our duty to not restrict any individual's access to meaningful work.

A Boston-area hospital implemented an MIT-trained robot that learned to function as a nurse manager, managing workflows and making suggestions in real-time (Johnson, 2018). The nurses at the hospital mentioned that "this allowed people to focus on the parts of the job that truly require human judgment and experience and if we can offload many of the easier decisions on the machine, we can get the best of both worlds" (Johnson, 2018). In the healthcare sector, patient safety errors account for over 400,000 preventable deaths annually in the United States, where 70% of them are caused by team communication breakdowns stemming from hierarchical structures and an asymmetrical power dynamic between medical professionals and patients (Greco et al., 2006). Nurses are positioned to identify and prevent these errors but feel unempowered to do so. The introduction of a Robotic Team Support System (RoboTSS) helped prevent these errors by detecting them and allowing the nurses to demonstrate the problems. As a result, the nurses felt empowered to work with physicians to prevent errors from occurring again (ROESSLER, 2011). Recent research illustrates that empowering nurses has led to an increase in job commitment, positive behavior, and improved patient outcomes because nurses can focus more on human interaction with patients (Smids et al., 2019). Studies found that nurse empowerment issues originated in the 1990s with the healthcare reform in the United States (Taylor et al., 2019). Due to budget cuts, private organizations began managing healthcare systems that adversely affected nurses as part of the downsizing of labor forces in hospitals, and therefore nursing staff levels decreased and workloads increased (Taylor et al., 2019). These conditions to work in were unfavorable and undesirable and nurses experienced more emotional exhaustion and stress (Burke et al., 2014).

CASE ANALYSIS

Through the lens of agency theory, several principal–agent relationships are discussed in a robot-inclusive workplace. A classic issue in principal–agent relationships is the asymmetry of information, where the agent is often presumed to have valuable information or expertise unbeknownst to the principal who does not precisely grasp what is being performed by the agent (Baker, 2019; Thompson, 2018). Hence the importance of aligning agent actions with principal preferences (Thompson, 2018). Tricky situations can occur when relationships between effort and outcome are unclear. For example, when an agent's efforts or outcomes are not directly observable, or when outcomes rely on teamwork and input from multiple agents (Thompson, 2018). Straightforward contracts, comprised of a fixed wage along with a bonus that varies with exact output, were recommended to realize the best outcome in these tricky situations (Thompson, 2018).

We will explore the principal–agent relationship between shareholders and executives. Executives or management ought to act in the interest of the shareholders, where a business's social responsibility is to increase its profits. Executives are agents contractually bound to fulfill their fiduciary responsibilities to pursue the desires of the shareholders (Lampert, 2016). While automation possesses the potential to improve productivity and profits, the social impact of automation makes it a crucial reputational issue for businesses considering the development and use of automation (Archer, Writer-Davies, & McGeoghegan, 2018). In the UK, most parliamentarians were largely supportive of automation of work, but plans for government policy and interventions in response to potential negative impacts of automation were unclear. Public perceptions of automation fell between ignorance and suspicion (Archer et al., 2018). Nonetheless, should the public and/or parliamentary consensus shift, reputational risks exist for businesses that automate too much and lay off many workers since the business could lose both customers and potential employees to competitors who perceive hiring human employees as part of their social responsibility (Archer et al., 2018). While fulfilling fiduciary duty toward shareholders, executives may consider "automation [as] the ultimate exit from the costs and risks associated with human labor". However, they must consider that they are responsible not just for business revenue but also for reputation.

An implicit assumption of agency theory is that both the principal and agent are opportunistic and motivated by self-interest based on their divergent and independent interests (Renmans, Paul, & Dujardin, 2016). For instance, the executives in an organization wish to maximize output while paying employees as little as possible; contrastingly, employees may wish to expend minimal effort yet earn as much as possible. A level of moral hazard exists within these relationships as information asymmetries, created from a principal's inability to observe or monitor agent behavior, can be opportunistically exploited by the agent for self-interest at the expense of the principal (Baker, 2019; Bendickson, Muldoon, Liguori, & Davis, 2016). Moral hazard can be described as a situation in which an agent does not have to bear the negative consequences of his or her actions. Consequential shirking behavior occurs when the agent performs actions against the interests of the principal (Renmans et al., 2016). It has been acknowledged that guile is a human tendency and not the sole preserve of agents (Zardkoohi, Harrison, & Josefy, 2017). For example, human employees may exploit automation in their workplace, seizing the opportunity to reduce their efforts by passing their workload on to the robot while still being paid by their employer.

An area of future research is the potential reduction in human responsibility resulting from increased interaction and collaboration with intelligent, technological interfaces (Limerick, Coyle, & Moore, 2014). For instance, initial research demonstrates a diminished sense of agency in human flight operators who use highly automated flight control systems. This highlights an inherent risk in such situations (Limerick et al., 2014).

Occupational automation is said to be limited and driven by socioeconomic influences. Certain roles may be reinforced or restructured, with the same technologies theorized to create joblessness (Fleming, 2019). For example, the future of a robotics-enabled mining industry could see human miners deploying collaborative robots to perform dangerous, repetitive, or strenuous tasks. While techno-pessimists posit that automation will take over all jobs, automation advocates argue that the human is integrated into current automation initiatives. Given the pervasiveness of task automation by computers and machines, it has been highlighted that designers of such systems need to carefully consider the optimal balance between appropriate technical assistance and a human's sense of agency. This is especially true in safety-critical and semi-automated systems, where a human task supervisor is held responsible for task failures (Limerick et al., 2014).

As principals can also act opportunistically, management may still replace a human employee regardless of work performance (Zardkoohi et al., 2017). "Robotic mechanization is molded and constrained by the pricing of labor, organizational power relations, and the nature of the task itself, all of which are interrelated at various levels" (Fleming, 2019, p. 28). While techno-pessimists may claim that low-skilled workers are more vulnerable to robotic automation, realistically, companies are likely to automate a job only if it is economically worthwhile; weighing the availability of cheap labor with the capital and maintenance required for automation (Fleming, 2019).

Three broad scenarios for employment are proposed:

> a pessimistic scenario that emphasizes job losses from automation; an optimistic scenario that emphasizes job creation from technological disruptions; and a skeptical scenario that considers both the negative and positive implications of technology but emphasizes the potentially long and difficult transition period involved.
>
> (Blit, St. Amand, & Wajda, 2018, p. 4)

It is submitted that both blue-collar and white-collar jobs will be affected by automation; some individual roles will disappear altogether, and new types of roles will be created, but jobs will not be lost abruptly (Wisskirchen et al., 2017). Instead, a gradual transition is already underway, where disruptions will differ within industries and companies (Wisskirchen et al., 2017). Before the COVID-19 crisis, companies had been gradually replacing human work with automation to reduce labor costs and improve profitability. COVID-19-induced social-distancing directives, along with the need for increased manufacturing capability, could accelerate automation in various industries. Although companies are claiming that reliance on technology is temporary, reassigned employees may return to find themselves redundant, replaced in their role by a. Robots are costly to create and integrate, but cheaper than human employees when amalgamated into the organization. Once a business invests in a robot replacement, it is unlikely to rehire for the role.

Robotic automation may deprive human workers of their rights and responsibilities. Described as "the experience of controlling both one's body and the external environment" where "control is central to the experience of being an agent", a sense of agency is a vital consideration for assessing how people experience interactions with technology (Limerick et al., 2014, p. 2). Automation raises the issue of the potential loss of a user's sense of agency when interacting with automated technologies. This is particularly important considering the prominence that a user's sense of agency has in the designs of new interfaces or in situations where automated systems go awry (Moore, 2016). A key social function of a sense of agency is the attribution of responsibility: a means where people are held accountable for their actions and which resultantly allows legitimate implementation of behavioral management via punishment or reward (Moore, 2016). While some organizations might claim automation or collaborative robots will not replace human jobs, undistinguishable attribution of employee responsibility that could accompany automation might cause businesses to rework their performance evaluation system. Theoretically, the utilization of collaborative robots could blur distinctions between the robot–employee work output, despite the anticipation that automation improves productivity and performance. Consequentially, semi-automated occupations may face downward pressure on wages and conditions, especially if the technology can do most of the work and the required human component has been relatively deskilled such that anyone can perform it (Fleming, 2019).

As applications and accuracy of robotics continue to increase, security issues and risks have correspondingly insinuated themselves into the technology. While physical dangers posed by industrial robots to human workers are well appreciated, collaborative robots designed to work alongside humans can present significant safety issues if compromised (Schneider, 2017). Industrial robots, previously designed to be isolated, have now evolved to incorporate exposure to corporate networks and the Internet (Schneider, 2017), containing connections to control networks via heterogeneous control devices for production advantages and ease of maintenance – a trend that poses cybersecurity risks (Pogliani et al., 2019). Industrial robots deployed in a workplace can be vulnerable to network and physical attackers. Network attackers can exploit various entry points to communicate with and compromise a networked robot (Maggi et al., 2017). A common profile of a physical attacker is the robot operator who has access to the robot's manual interface. Malicious contractors or technicians can also plug devices into the robot controller's Ethernet port or other proprietary I/O interfaces, gaining full access to the robot controller's computer (Maggi et al., 2017).

We mentioned that opportunistic employees could shirk work responsibility by delegating their work to robots. One may argue that robots were created for such a purpose. In a principal–agent relationship, robots possess a duty to protect the interests of their human principals. A major hurdle in automation lies in ensuring humans and their robotic co-workers can safely work alongside each other (Johnson, 2018). Current research in modern robotics seeks to extend a robot's role beyond simply being a machine with defined functionalities and limited to operating in a confined workspace separated from humans (Salge & Polani, 2017). Human safety around robots often involves a rudimentary 'kill switch', a default approach that may be undesirable, especially if a robot performs "a vital function where an immediate shutdown would lead to harm of a human or itself, or when, to maintain safety and prevent damage, the robot is required to *act* rather than to *stop acting*" (Salge & Polani, 2017, p. 2).

LESSONS LEARNED

The application of agency theory typically hovers around two perspectives and their identified problems. Principal–agent research focuses on two possible problems, risk-sharing, and agent monitoring, and positivist agency theory focuses on governance mechanisms that limit agents' self-serving behaviors (Bendickson et al., 2016). It is posited that strong corporate governance mechanisms could induce self-interested executives or managers, who might otherwise adopt certain risky ventures without commensurate returns, to adjust their strategies to ensure alignment with shareholder financial returns (Chari, David, Duru, & Zhao, 2019). These governance mechanisms provide discipline by using a combination of monitoring and incentives to dissuade managers from pursuing risks that erode shareholder returns (Chari et al., 2019). According to NIST,

> information security governance can be defined as the process of establishing and maintaining a framework and supporting management structure and processes to assure that information security strategies are aligned with and support business objectives, are consistent with applicable laws and regulations through adherence to policies and internal controls, and provide assignment of responsibility, all in an effort to manage risk.

It is suggested that the governance of robots should be dynamic; treating robots as digital workers and enabling control while allowing for change and improvement (Deloitte, 2017). An emerging area of research lies in robotic governance, which aims to create a framework of voluntary guidelines in the form of the 'Robot Manifesto', a Code of Conduct that all stakeholders who purchase, use, or produce robots can commit to (Bösl & Bode, 2018).

The trend of human–robot collaboration promises a variety of advantages, such as increased productivity and efficiency, but also new risks and threats. By adopting robotic automation, organizations and businesses need to consider these risks and potential technological vulnerabilities. IT risk management targets three primary areas: the security of data and information, the integrity of hardware and systems, and the implementation of IT projects (Govindji, Peko, & Sundaram, 2018). The National Institute of Standards and Technology (NIST) outlines four components in risk management processes: framing, assessing, responding to, and monitoring risk (NIST, 2012). Risk models are used to assess risk factors and their relationships, where typical risk factors include threat, vulnerability, impact, likelihood, and predisposing conditions (NIST, 2012).

As discussed above, networked robots can be vulnerable to network and physical attacks. For instance, attackers could gain control over robots and thereby put the human operator at risk. Risk-sharing, in this case, may involve the alignment of principal–agent interests regarding robust cybersecurity behaviors to safeguard against such attacks. As a degree of information asymmetry exists, where employers cannot fully monitor employee security behavior all the time, organizations may incentivize or reward employees who practice good cybersecurity behaviors. Agency theory often attempts to mitigate moral hazard by offering remuneration schemes and agent contracts to align principal–agent interests. Theoretically, linking pay with a measure of performance aligns the agents' objective function (e.g. increased outcomes or outputs) with that of the principals' (e.g. performance-based financing) (Renmans et al., 2016).

DISCUSSION QUESTIONS

1. How can governance mechanisms regulate the deployment of robots in the workplace?
2. What are some risks that can accompany robotic automation?
3. How can organizations prevent information asymmetry within human–robot collaboration?

REFERENCES

Acemoglu, D., & Restrepo, P. (2017). Robots and jobs: Evidence from US labor markets. https://doi.org/10.3386/w23285

Archer, H., Writer-Davies, R., & McGeoghegan, M. (2018). AI, automation, and corporate reputation. Retrieved from https://www.ipsos.com/sites/default/files/2019-05/ai_automation_cr_web.pdf

Baker, R. M. (2019). The agency of the principal–agent relationship: An opportunity for HRD. *Advances in Developing Human Resources*, *21*(3), 303–318. Retrieved from https://doi.org/10.1177/1523422319851274

Beckler, G. S. (1992). Human Capital and the Economy. *Proceedings of the American Philosophical Society*, *136*(1), 85–92. https://www.jstor.org/stable/pdf/986801.pdf

Bendickson, J., Muldoon, J., Liguori, E. W., & Davis, P. E. (2016). Agency theory: Background and epistemology. *Journal of Management History*, *22*(4), 437–449. Retrieved from https://doi.org/10.1108/JMH-06-2016-0028

Blit, J., St. Amand, S., & Wajda, J. (2018). *Automation and the Future of Work: Scenarios and Policy Options* (CIGI Papers No. 174). Retrieved from Waterloo, Canada: https://www.cigionline.org/publications/automation-and-future-work-scenarios-and-policy-options

Bösl, D. B. O., & Bode, M. (2018). Roboethics and robotic governance – A literature review and research agenda. In *Advances in Intelligent Systems and Computing* (pp. 140–146). Springer, Cham. Retrieved from https://doi.org/10.1007/978-3-319-70833-1_12

Burke, R. J., Ng, E. S., & Wolpin, J. (2014). Economic austerity and healthcare restructuring: Correlates and consequences of nursing job insecurity. *The International Journal of Human Resource Management*, *26*(5), 640–656. https://doi.org/10.1080/09585192.2014.921634

Chari, M. D. R., David, P., Duru, A., & Zhao, Y. (2019). Bowman's risk-return paradox: An agency theory perspective. *Journal of Business Research*, *95*, 357–375. Retrieved from https://doi.org/10.1016/j.jbusres.2018.08.010

Deloitte. (2017). The robots are ready, are you? The 3rd Annual Global Robotics Survey. Retrieved from https://www2.deloitte.com/content/dam/Deloitte/de/Documents/Innovation/Global_Robotics_Survey_2017.pdf

Fleming, P. (2019). Robots and organization studies: Why robots might not want to steal your job. *Organization Studies*, *40*(1), 23–38. Retrieved from https://doi.org/10.1177/0170840618765568

Govindji, S., Peko, G., & Sundaram, D. (2018). A context adaptive framework for IT governance, risk, compliance and security. In *Context-Aware Systems and Applications, and Nature of Computation and Communication* (pp. 14–24). Retrieved from https://doi.org/10.1007/978-3-319-77818-1_2

Greco, P., Spence Laschinger, H., & Wong, C. (2006). Leader empowering behaviours, staff nurse empowerment and work engagement/Burnout. *Nursing Leadership*, *19*(4), 41–56. https://doi.org/10.12927/cjnl.2006.18599

Johnson, K. (2018, November 4). Hey robot, is that report ready? *The Boston Globe*. Retrieved from https://www.bostonglobe.com/business/2018/11/04/hey-robot-that-report-ready/LrGEKOLaBJB50MA8tzTXbP/story.html

Lampert, M. (2016). Corporate social responsibility and the supposed moral agency of corporations. *Ephemera: Theory & Politics in Organization*, *16*(1), 79–105. Retrieved from http://www.ephemerajournal.org/sites/default/files/pdfs/issue/16-1ephemera-feb16.pdf#page=83

Limerick, H., Coyle, D., & Moore, J. W. (2014). The experience of agency in human-computer interactions: A review. *Frontiers in Human Neuroscience*, *8*(643). Retrieved from https://doi.org/10.3389/fnhum.2014.00643

Maggi, F., Quarta, D., Pogliani, M., Polino, M., Zanchettin, A. M., & Zanero, S. (2017). Rogue robots: Testing the limits of an industrial Robot's security. Retrieved from https://documents.trendmicro.com/assets/wp/wp-industrial-robot-security.pdf

Moore, J. W. (2016). What is the sense of agency and why does it matter? *Frontiers in Psychology*, *7*(1272). Retrieved from https://doi.org/10.3389/fpsyg.2016.01272

National Institute of Standards and Technology (NIST). (2012). *Guide for Conducting Risk Assessments*. Retrieved from Gaithersburg, MD: https://nvlpubs.nist.gov/nistpubs/Legacy/SP/nistspecialpublication800-30r1.pdf

Pogliani, M., Quarta, D., Polino, M., Vittone, M., Maggi, F., & Zanero, S. (2019). Security of controlled manufacturing systems in the connected factory: The case of industrial robots. *Journal of Computer Virology and Hacking Techniques*, *15*(3), 161–175. Retrieved from https://doi.org/10.1007/s11416-019-00329-8

Renmans, D., Paul, E., & Dujardin, B. (2016). *Analysing performance-based financing through the lenses of the principal-agent theory*. Retrieved from Antwerpen, Belgium: https://www.researchgate.net/profile/Dimitri_Renmans2/publication/311654929_Analysing_Performance-Based_Financing_through_the_Lenses_of_the_Principal-Agent_Theory/links/5852b65608aef7d030a512f5.pdf

Roessler, B. (2011). Meaningful work: Arguments from autonomy*. Journal of Political Philosophy, *20*(1), 71–93. https://doi.org/10.1111/j.1467-9760.2011.00408.x

Rosso, B. D., Dekas, K. H., & Wrzesniewski, A. (2010). On the meaning of work: A theoretical integration and review. *Research in Organizational Behavior*, *30*(2), 91–127. https://doi.org/10.1016/j.riob.2010.09.001

Salge, C., & Polani, D. (2017). Empowerment as replacement for the three laws of robotics. *Frontiers in Robotics and AI*, *4*(25). Retrieved from https://doi.org/10.3389/frobt.2017.00025

Schmidt, W., & McKnight, C. (2015). *Inequality for all: The challenge of unequal opportunity in American schools*. Teachers College Press.

Schneider, D. (2017). New report highlights dangers of hacked factory robots. Retrieved from https://spectrum.ieee.org/automaton/robotics/industrial-robots/report-dangers-of-hacked-factory-robots

Smids, J., Nyholm, S., & Berkers, H. (2019). Robots in the workplace: A threat to—or opportunity for—Meaningful work? *Philosophy & Technology*, *5*(17). https://doi.org/10.1007/s13347-019-00377-4

Taylor, A., Lee, H. R., Kubota, A., & Riek, L. D. (2019). Coordinating clinical teams. *Proceedings of the ACM on Human-Computer Interaction*, *3*, 1–30. https://doi.org/10.1145/3359323

Thompson, F. (2018). Agency theory in organizations. In A. Farazmand (Ed.), *Global encyclopedia of public administration, public policy, and governance* (pp. 262–266). Springer International Publishing. Retrieved from https://doi.org/10.1007/978-3-319-20928-9_88

Wallen, J. (2005). Robotic scanner preserves the pages of history. *Industrial Robot: An International Journal*, *32*(3), 2–32. https://doi.org/10.1108/ir.2005.04932cab.002

Wisskirchen, G., Biacabe, B. T., Bormann, U., Muntz, A., Niehaus, G., Soler, G. J., & von Brauchitsch, B. (2017). Artificial intelligence and robotics and their impact on the workplace. *IBA Global Employment Institute*, *11*(5), 49–67. Retrieved from https://www.researchgate.net/profile/Mohamed_Mourad_Lafifi/post/World_population_and_jobs_market_where_are_we_going/attachment/5c13b4a03843b006754b5976/AS%3A703736632586245%401544795296667/download/AI-and-Robotics-IBA-GEI-April-2017.pdf

Zardkoohi, A., Harrison, J. S., & Josefy, M. A. (2017). Conflict and confluence: The multidimensionality of opportunism in principal–Agent relationships. *Journal of Business Ethics*, *146*(2), 405–417. Retrieved from https://doi.org/10.1007/s10551-015-2887-7

Research Cases

2.13 Facial Recognition
Legal, Ethical, and Cultural Concerns

Yuchen Lin, Zhehen Zheng, and Yang Zhou
University of Canterbury, Christchurch, New Zealand

INTRODUCTION

With the rapid development of modern technology, facial recognition technology has been trending in different fields such as smart security, banking, marketing, retail, and healthcare (Sajjad et al., 2019; Sennaar, 2019). For example, law enforcement agencies are employing facial recognition technology as smart security to uncover criminals and protect innocents (Sajjad et al., 2019), and banking is adopting machine learning algorithms to identify fraud detection, minimize the requirement for traditional passwords, and increase the capability to distinguish between a photograph and a human face. In the marketing field, facial recognition technology can potentially improve customer loyalty by installing a facial recognition camera at the business entrance and sending customized promotions to different customers based on their social media's "likes" (Wright, 2019). Retail stores are implementing facial recognition technology to identify known shoplifters to prevent loss and reduce chargebacks (Efron, 2019) In the healthcare industry, facial recognition technology and machine learning are combined with a computer to increase the accuracy of tracking patient medication consumption and assist in the pain management program (Sajjad et al., 2019).

CASE INTRODUCTION

The American technology company "Clearview AI" published its facial technology software, which is also called "Clearview AI," in 2017, by its CEO, Hoan Ton-That (O'Sullivan, 2020). Hoan Ton-That is a techie from Australia. His most significant hit is that he invented an iPhone application that enables people to put Donald Trump's unique yellow hair on various photos (Hill, 2020). Regarding Hoan Ton-That's latest product, the facial recognition software "Clearview AI," the company claimed that it spent merely two years accessing and collecting more than three billion photos and images that were posted publicly on the Internet, including various social media platforms, such as Facebook, Twitter, and Instagram (O'Sullivan, 2020). The company downloaded all the photos and images from the Internet and stored them as data in their database. Additionally, the company captured the faces of primary

DOI: 10.1201/9781003264415-30

individuals and other peoples' faces in the background of the photos. After generating the massive dataset, the company sells its software database to its customers, such as law enforcement agencies. Afterward, the company sets up an API of the facial recognition software for customers to match the unknown faces to the faces in the database (O'Sullivan, 2020). The significant value that distinguishes Clearview AI from other facial recognition technology vendors is that Clearview AI can provide 99% accurate results by enhancing its algorithms among different races (O'Sullivan, 2020). Due to its massive database, Clearview AI reduced the geographical limitations that frequently happened in law enforcement agencies' traditional facial recognition technology.

Target Customer

According to Hoan Ton-That, Clearview AI targets this software to law enforcement agencies, such as police departments (O'Sullivan, 2020). Moreover, Hoan Ton-That claimed that there are more than 600 law enforcement agencies in the US and Canada currently using this software (Hill, 2020; O'Sullivan, 2020). The business goal of Clearview AI is to assist law enforcement agencies in stopping criminals and protecting innocents (O'Sullivan, 2020). For example, the New Jersey police department used Clearview AI as a tool in a child predator incident investigation (Musil, 2020; O'Sullivan, 2020). Clearview AI assisted the police in identifying the man before he showed up for what he believed was a meeting with a minor (Marr, 2019).

Moreover, Clearview AI helped the police investigate the man's background to identify if he was likely to be armed (O'Sullivan, 2020). The company had licensed the Clearview AI software to a handful of private companies for security objectives (Hill, 2020). For example, Hoan Ton-That stated that some banks are using Clearview AI as a tool in fraud investigations, but none of these banks are willing to speak out (O'Sullivan, 2020).

Existing Problems

Thus far, there are many existing controversies of Clearview AI in the public. First, although the company claimed that Clearview AI software could increase effectiveness and efficiency for law enforcement agencies, the truth is that after reaching a successful investigation with the assistance of Clearview AI, the state of New Jersey and its Attorney General have banned 21 counties from using Clearview AI due to security concerns (Connor, 2020; O'Sullivan, 2020). Second, various social media platforms, such as YouTube, Facebook, and Instagram, had sent Clearview AI cease-and-desist letters to state that Clearview AI had violated their regulations and policies by accessing and gathering users' photos and images from their platforms (O'Sullivan, 2020). In reply, however, Hoan Ton-That refuted that the photos and images in their databases were all publicly posted and shared online (O'Sullivan, 2020). He is ready to defend his company in court in the future.

A third existing problem is that the company claims its target is to sell Clearview AI software to law enforcement agencies, but a previous data breach of Clearview AI's customer lists revealed that Clearview AI also sold its software to commercial and non-governmental agencies (Muhlberg, 2020). A fourth troublesome scenario is the justified usage of facial recognition technology by law enforcement agencies may lead to misidentification, lack of regulations and legislation on monitoring the use of facial recognition technology, abuse by law enforcement agencies that may harm or violate human basic rights and increase privacy issues (Shiri, 2019).

Finally, Dubai cybersecurity firm CSO, Mossab Hussein, has found Clearview AI's repository which contains Clearview AI's source code (Whittaker, 2020). Even though the repository has been protected by passwords, there is still a misconfigured setting which enables the public to register and log in to the code-storing system (Whittaker, 2020). Therefore, the lapse exposed Clearview AI's internal files, applications, and source code (Whittaker, 2020). Moreover, Mossab Hussein stated that Clearview AI's Slack token was also exposed which enables anyone to reach the company's private messages and examine sensitive communication without any password (Whittaker, 2020). After the data breach incident, three things were exposed: Clearview AI's customer list, the frequency amount of customers searching faces in the database, and the account number that had been created by the customer was published to the public (Whittaker, 2020). This incident further exposed the truth that counter to Clearview AI's claim to sell facial recognition technology only to law enforcement agencies, their customer list contained some private business companies such as NBA and Walmart (Whittaker, 2020).

LITERATURE REVIEW

Facial Recognition Technology in Law Enforcement Agencies

The major difference between facial recognition technology and other biometrics systems is that facial recognition technology can facilitate surveillance capability (Woodward Jr, Horn, Gatune, & Thomas, 2003). Law enforcement agencies have been adopting the facial recognition system to identify and locate specific criminals, terrorists, and lost children since the early 2000s (Davis, Lander, Evans, & Jansari, 2016; Garvie & Frankle, 2016; Woodward Jr et al., 2003). Generally, law enforcement agencies collect and compare the faces with the clear mugshots of existing criminals and suspects with closed-circuit television (CCTV) and police surveillance cameras (Davis et al., 2016; Garvie & Frankle, 2016; Woodward Jr et al., 2003). In the early stages, these police surveillance systems were not that accurate and reliable (Garvie & Frankle, 2016). With the maturity of technology over the years, law enforcement agencies are now able to gain a more complex and comprehensive surveillance system (Garvie & Frankle, 2016). For example, Florida sheriff departments have been allocated smartphones or technical devices with facial recognition systems (Garvie & Frankle, 2016). The Florida facial recognition system database contains all driver's license images so that the police can capture the criminal, identify the driver who refused to present his license, and verify the suspects in real time (Garvie & Frankle, 2016). However, these governmental facial recognition technologies have an algorithm accuracy rate of potential concern across different races due to the poor photo quality (Davis et al., 2016; Garvie & Frankle, 2016). Whenever there is a weakness, there is an opportunity for error. Clearview AI's CEO Hoan

Ton-That would argue that the existence of Clearview AI is based on providing a more effective and efficient alternative to facial recognition technology, connecting with a larger database, and building up more advanced algorithms for more accurate results (O'Sullivan, 2020).

Advantages of Facial Recognition Technology

There are some benefits in terms of utilizing the facial recognition technology. First, video-based facial recognition technology facilitates contact-free surveillance (Woodward Jr et al., 2003). Public faces can be captured from various channels in real-time, such as the CCTV in the elevator, in the business area, and street. For example, the police desire to search a suspect or criminal face to identify whether the person is the right man to arrest. They can conduct the facial identification process without physical contact and minimize the criminal's suspicion of being captured (Woodward Jr et al., 2003). Second, facial recognition technology can be integrated with other biometric identification technology to enhance information security (Sim, Zhang, Janakiraman, & Kumar, 2007; Xiao & Yang, 2010). There are already some access control technologies which have been adopted in companies to secure information security (Sim et al., 2007; Xiao & Yang, 2010). However, those may not be sufficient and effective for high information security (Sim et al., 2007; Xiao & Yang, 2010). Adopting facial recognition technology with a combination of other biometrics, such as fingerprints, can assist companies in securing their resources by processing continuous verification (Sim et al., 2007; Xiao & Yang, 2010). For example, companies can also adopt facial recognition technology to verify the users periodically through facial recognition software on their devices (Xiao & Yang, 2010). Once the faces are not successfully verified, the system will lock the monitor of the device to prevent further actions (Xiao & Yang, 2010). The combination of facial recognition technology and biometric technologies adds an extra layer to the company's information security.

Disadvantages of Facial Recognition Technology

Privacy Risk

With the development of technology, the world is full of facial recognition systems and closed-circuit video cameras, which gives the state and law enforcement agencies huge power to keep everything under constant surveillance and monitor their populations (Lemboye, 2019; Leong, 2019). From the perspective of the governed and consumers, facial recognition technology erodes privacy. The misuse of the facial recognition system by the government may put innocent people on the watching lists. Compromising the privacy of these innocent people due to the doubt without cause is unfair. At the same time, businesses would use facial recognition to illegally discriminate against their customers based on their private data.

The rapid development of information technology gives social media companies a chance to collect and store a lot of personal data and facial images of the consumer. With these data, a sufficient database is built to support the implementation of facial recognition, which may raise privacy risks. This risk has emerged because privacy security and the collection of personal data are not under the protection of any federal law (Brinckerhoff, 2018).

With an unconnected security framework and the lack of federal laws to manage the storage and usage of personal data, the privacy of social media users is especially vulnerable and in danger of data misuse. In this way, self-regulation is not enough to control the operation of social media companies or prevent them from taking advantage of innocent consumers of social media platforms and applications (Brinckerhoff, 2018).

Most Individuals would not read the policies of the social media platform before authorization, so, they do not know how their facial images and personal data would be used by the platform. Privacy concerns are raised by the usage of FRTs on these social media platforms and social applications and the lack of a protective method to deal with the privacy thread (Lemboye, 2019).

Uncontrolled Conditions

Machines are not always right and may encounter problems when they conduct facial recognition based on surveillance. In some specific situations, the accuracy rate of the automated system of facial recognition function is quite low, such as comparatively high fidelity and long distances. In terms of the basic level, searching for a face from an electronic image is already a difficult technical problem. The target object should be perfectly captured under a tightly organized situation. For example, each target object should look directly into the camera and stay in a specific area of the image to accurately identify the face in the image. Therefore, even though the accuracy rate of some facial recognition software is quite high, the technology is not an exact tool to be generated for important situations (Woodward Jr et al., 2003).

Uncontrolled environmental conditions will influence the accuracy of facial recognition technology, such as the color of the people's skin, lighting, and camera angle. The recognition ability of the system would also be dramatically influenced by other parameters (Woodward Jr et al., 2003). In addition, tests show that facial recognition technology can be led by disguises and have some limitations on operation in terms of changing lighting, the angle of shooting, and variables on naturally aging faces ("Flaw In Facial Recognition System Could Delay Deployment," 2003).

Risk of Unethical Behavior

Without letting the public know, Clearview AI has made this vast FacePrint database available to private companies, police departments, government, and wealthy individuals, allowing them to use facial recognition technology to secretly track and target anyone they want. Countries like the US, UK, and Australia are now using facial recognition technologies and operating them in different public spaces, like cafes, airports, and shopping centers, the underlying goal for using this technology is to track people's movement. The use of Clearview AI in public spaces violates the fundamental right to privacy, and some proponents say that when people are in public spaces, they do not have a strong expectation of privacy. According to Helen Nissenbaum, who argued that even when people are in public, the expectation of privacy is diminished, but they still have a reasonable expectation of privacy.

One premise of Nissenbaum's argument for privacy is when citing empirical data, many people are worried when they become aware that their personal information is being gathered without their permission, even when they are in public. Many people claim that they do have a confident expectation of privacy when they are in public, so it is unethical for Clearview AI to monitor everyone without their consent.

Also, another issue with Clearview AI is that in this case, Clearview AI collected over three billion images or photos that were publicly posted on the Internet over two years. The social media sites like Facebook, Twitter, and Instagram were the targeted platform that Clearview AI used to collect people's photos (O'Sullivan, 2020). Clearview AI downloaded all the photos and stored them in the database, and after gathering a massive amount of data, the company sold the data to other organizations. This is also considered unethical because Clearview AI did this without letting the people involved know. Even after downloading personal photos from social media sites, Clearview AI did not ask for permission from those sites but sold them to other organizations. The problem is that the understanding of all users on social media sites is that their photos posted on social media sites are only used on their page, they did not know that Clearview AI collects their data for profit. Another potential risk not knowing what the purchased data is used for, maybe other organizations bought the data for a malicious purpose, for example, to launch phishing attacks. Also, Clearview AI does not guarantee that data which has been sold to other organizations will be protected, a data breach can happen in another organization. As a result, some social media websites, including Facebook, Twitter, and YouTube, sent cease and desist letters to Clearview AI. According to YouTube's terms of service, collecting data from YouTube is forbidden, and Facebook also claimed that they would not allow Clearview AI to collect information from Facebook because the software company is violating Facebook's policies (Ng & Musil, 2020).

Possible Future Development of Clearview AI

In February this year, Clearview AI expanded to 26 countries, including Belgium, Finland, France, Italy, and Switzerland. Clearview AI also engaged with the law enforcement agencies and police forces in these countries. However, there is a chance that Clearview AI is illegal in Europe. According to the European Data Protection Board, the services provided by Clearview AI might breach the EU data protection regime (Shead, 2020). Meanwhile, in the US, over 600 law enforcement agencies implemented Clearview AI in the last year, andthey had used Clearview AI to help trace criminals, victims of crimes, murder, etc. (Hill, 2020). The possibility of weaponizing this technology is enormous and disturbing. Using Clearview's technology, criminals can easily find out where people on the street live or work. Foreign adversaries can quickly gather information about targeted individuals for blackmail. From the cases above, we can see that Clearview AI or facial recognition technology is not welcomed everywhere; this technology is breaching some states' laws. So, in the future, if Clearview AI wants to expand its business, it must amend its policies or technology to suit the particular laws. Otherwise, the scope of their business will be limited. The use of facial recognition can be extensive, for example, hotels can use this technology to help customers to check in automatically. Additionally, in restaurants and Cafés, customers' information is stored in the database, from recognizing their faces, the restaurants or cafés will be able to know what the customers like and what they usually order. In the financial sector, facial recognition technology can be useful for banks to identify people. In China, facial recognition technology is adopted in supermarkets when people check out, they do not need to take out their wallet, they just need to face the camera and the payment will be made.

Historical and Social Context

In a historical context, authorities have been using facial recognition technology for a long time. Authorities and privacy bodies remain far apart on the appropriate usage of facial recognition technology. For example, the Centre for Privacy & Technology at Georgetown University law school issued a lawsuit against the New York Police Department, because NYPD does not govern its use of facial recognition software rightly ("Privacy bodies across the US challenge authorities on facial recognition," 2017). However, the software's success rate has been low because of the not-big-enough database and geographical restrictions. Facial recognition technology incorrectly targets 81% of innocent people as suspects, according to a report from the University of Essex (Martin, 2019).

In a social context, information explosion and technological development give Clearview AI the chance to create a dataset of over three billion face images and develop a facial recognition technology with 99% accuracy. We currently live in a society where people love to share personal information and photographs on the Internet. The Internet can expose more of our private information to the public than any technology in history. Meanwhile, the cost of image storage decreases due to the exponential increase in data processing and storage capabilities. The information explosion provides opportunities for new technologies, as well as increases the risk of information security. As personal information has become a kind of network resource, the boundary between public and private information blurs. As it happened in the case, Clearview AI continues to use the images that people have already deleted from the Internet. Defining whether these images are public information or private information is difficult. What's more, due to the competitive incentive, enterprises may take aggressive action to increase their advantage, such as illegally obtaining information from personal computers or emails.

Web of Relationships

After placing Clearview AI on its web of relationships, one can find some related risk factors in its profit model. In the web of relationships, the downstream customers of Clearview AI are big institutions such as law enforcement agencies, the justice department, and banks. As a tiny start-up company, Clearview AI is in a weak position when dealing with its downstream customers. To sell its service to its strong potential customers, Clearview AI needs to offer a 30-day free trial version before its potential customers decide whether to purchase the service. Meanwhile, the primary suppliers of images and photos for Clearview AI are weak individuals who unknowingly provided free resources to Clearview AI through social media platforms. The following paragraphs elaborate on the risk factors in the web of relationships of Clearview AI.

After providing the technical service to clients, supervising the use of this technology is hard and the information may be used inappropriately. Because of the 30-day free trial version, some individual clients may use this technology without rules and limitations. For example, users only need a law enforcement email address to sign up for the service, and some New York police department officers are loading it onto their devices to search for anyone they want, such as their friends and family (Lynch, 2020). Furthermore, some officers appear to have been using Clearview AI after the agency's official facial recognition department refused the technology. Without enough training and management of software usage, the use of Clearview AI is totally out of control in most police departments. Simultaneously, Clearview AI seems to have used its technology to identify a journalist working on a story about its product (O'Sullivan, 2020).

These actions also raise ethical issues. Ethics is a set of behavioral standards either shared by a group or community or held individually (Whitman & Mattford, 2019). Many professional disciplines have explicit rules governing the ethical behavior of members; however, IT security does not have binding codes of ethics. Therefore, there are troublesome ethical issues in law enforcement agencies and Clearview AI. In addition, the enforcement of ethics and policy is not as strong as the law. The main difference between laws and ethics is that laws are executed by a governing authority while ethics are socially acceptable behaviors (Whitman & Mattford, 2019).

To evaluate whether the behavior of Clearview AI is ethical or not ethical in general is hard. When analyzing the same scenario from different perspectives of business ethics theories, the results are very different. From the perspective of the social contract theory, this is not an ethical choice for Clearview AI to use individual's images for its profit purpose. Based on social contract theory, businesses are obligated to operate for the welfare of all society (Whitman & Mattford, 2019). The facial recognition technology is the erosion of the privacy of citizens and the violation of human rights and thereby against the welfare of all society. From the perspective of stockholder theory, this is an ethical choice for Clearview AI to maximize the interests of shareholders through its advanced technology. Stockholder theory believes that companies should operate for the benefit of the stockholders, and companies are structured as contracts among their stockholders (Whitman & Mattford, 2019). From a third perspective called stakeholder theory, the service is ethical because law enforcement agencies enhance their efficiency and effectiveness through cooperation with Clearview AI. Stakeholder theory believes that managers should balance the needs of various parties who would be interested in the operation of the company, rather than focus on the benefit of the stockholder (Whitman & Mattford, 2019).

When collecting information from suppliers, Clearview AI violates social media platforms' policies by using images shown on these platforms for profit and retaining access to some information that is no longer public. This is a serious abuse of data and of its source of the images and photos. Policy is a series of organizational guidelines that dictate behavior within the organization. The policy protects the confidentiality and availability of information when it is housed and transferred. A company can have multiple policies that make up its overall security policy. For example, YouTube's policies of service explicitly prohibit data collection that can be used to identify individuals. This policy is used to protect the privacy of individual users.

Therefore, Clearview AI secretly collects personal image data without any consent and misuses the data in violation of the policies of social media platforms. In this situation, for YouTube to defend its rights with the help of this policy is not easy. Policies function like laws, and they must be established with the same care as laws; policymakers must ensure that the policies are appropriate and fairly applied to each stakeholder. However, policies are more flexible than laws. Moreover, there is no excuse for ignoring the law while ignoring policy is a viable defense. With social media platforms' challenge of multilingual translations, distributing their policies to all individuals who are expected to comply with them or be understood by all stakeholders is hard. That is why the cease-and-desist letters from social media companies don't faze Ton-That, the founder of Clearview AI.

Today, technology enables the easy violation of privacy more than ever. Clearview AI is famous because it works better than other big players in the facial recognition market, such as Facebook and Google. Clearview AI catches lots of free photos from Internet platforms, specifically images of individuals' faces. Then, Clearview AI uses its own technology to label these photos for facial recognition function. However, the facial recognition technology of Clearview AI is a radical erosion of privacy, which would end people's ability to walk down the street anonymously. In the context of information security, privacy is the right of individuals to protect their information from unauthorized access. Like security for an organization, privacy for individuals also requires a trade-off with openness. Each person decides for himself whether there is value in giving up their privacy. In this case, images collected, used, and stored by social media platforms are to be used only for the purposes stated by the social media company at the time data is collected. Furthermore, privacy means that the images will be used only in the way approved by the individuals who provided them, such as showing their lifestyle on social media platforms. Hence, for Clearview AI to use these images for facial recognition purposes is an invasion of personal privacy.

Information Gathered

In these times, people live a very active social media lifestyle due to the maturity of technology and the rise of different mobile devices. At their fingertip, people gradually fall to sharing their selfies, life video logs, and group photos online. Clearview AI seizes upon these trends to generate a more comprehensive and multiple-dimension database of faces. Clearview AI mined various social media platforms and downloaded the users' photos, which include selfies, group photos, and any photos of the users who are tagged in their account. After collecting various user images, Clearview AI stores the images in its databases. Since the photos contain users' facial features from different angles, Clearview AI takes advantage of this and builds up an innovative facial recognition algorithm. Moreover, Clearview AI builds up a user interface of the facial recognition software, which enables the users to search for any desired face on a computer or tablet device. Therefore, Clearview AI aggregates all the image data and facial recognition software to sell it as a commodity to law enforcement agencies. Clearview AI enables its customers to search a particular individual's facial image in their database by taking a photo of the individual in advance. After inputting the individual facial images into the system, the system will conduct a background calculation and comparison among the faces in the database to figure out the accurate person. The major difference between Clearview AI and previous governmental facial recognition technology is that Clearview AI's algorithm tends to be more accurate in returning the results. Since there are many laws and regulations that restrict and govern the facial images used in law enforcement agencies' facial recognition systems, law officers can only use facial photos from limited sources such as driver licenses and identification licenses. Hence, the limited sources of facial images affect the result of facial recognition and cause misidentification in different races. Therefore, Clearview AI invented a more comprehensive facial recognition technology for law enforcement agencies to identify suspects and criminals more accurately and effectively.

Authors' Personal Relation with the Case

In the last few years, facial recognition technology has been implemented across different realms of society, mostly for public safety. Facial recognition technology is also being introduced to school environments to address issues like school security and student emotion detection (Andrejevic & Selwyn, 2019). Just like everyone else who goes to public places, students such as this case study's authors are very close to this case. Without them knowing, camera surveillance and the Clearview technology capture and trace people's movement. The student authors' role in this case study will be the same as everyone else which is part of the data, with their posted pictures captured from online social platforms and their images caught by cameras when out in public. The students' data is very likely already in someone's or some organization's database which Clearview AI collected without students' consent. Another concern is that studies prove any picture would be stored in Clearview AI's database even when the user deletes the photo, which means there is no going back once Clearview AI collects one's data. If someone wants to delete their data from Clearview AI's database, they must prove their identity by providing a personal ID issued by their government as well as headshots. Even when Clearview AI deletes the data, that which they sold to their clients is tough to track back. Therefore, once one's personal data is sold and dispersed in unknown databases, deleting all that personal data is impossible.

EXPECTATION AND ACTUAL CASE

Creators of facial recognition technology expect that it should be very welcomed by the government and police departments. However, after reviewing some troubling cases of misidentity, that is not the truth, even though over 600 US law enforcement agencies have been using Clearview AI technology. Evidence proves that this technology is helpful in tracing criminals (O'Sullivan, 2020), but there are some places within the US like New Jersey and part of Europe which have banned this technology. Some giant tech firms like IBM and Amazon announced they were getting out of the facial recognition business. Amazon said it intended to stop selling its Rekognition platform to police for a year while the tech giant made the platform more reliable and adhered to regulations on facial recognition technology to govern its ethical use (Shead, 2020). If agents use facial recognition technology in the right way, it will bring more benefits than harm. At the same time, the safety of the data is also a concern to many people.

The face recognition system is not foolproof and unbreakable. Attackers use loopholes to invade the system and directly modify the face information in the device to "cheat" the face recognition system with any face and open the door. On October 24, 2017, at the Shanghai GeekPwn2017 International Security Geek Competition, a hacker used the vulnerability of the face recognition device itself to successfully pass through the access control system with any face in just two and a half minutes.

Nowadays, face recognition has slowly become a new security authentication method and penetrates people's lives. In airports, customs can use facial recognition technology to replace manual recognition and companies use facial recognition technology to check-in employees and visitors. The uses of face recognition technology are getting broader and broader.

Nevertheless, while enjoying the convenience brought to us by high technology, hidden dangers must be considered. For example, in the "face payment" process, the face data may be attacked by hackers, viruses, etc., during the transmission process. Besides, in the process of analyzing face information on the back-end server of the system, the analysis result can also be tampered with and stolen, which may lead to payment errors or theft of money by criminals. With the increasing artificial intelligence technology, more and more artificial intelligence technologies are used in people's work and life, such as the intelligent lending system of online finance; and more and more artificial intelligence products have penetrated people's lives, such as intelligence Speakers and home robots. Many of these artificial intelligence applications require data collection, analysis, and processing to be realized.

Nevertheless, everything has two sides. The more data in an artificial intelligence system, the higher the data threat, because the possibility of data leakage, alteration, loss, etc., will increase. While data helps artificial intelligence, it also threatens artificial intelligence. How to protect data and improve artificial intelligence security has become an essential part of the development of AI.

According to Buzzfeed, the US Customs and Border Protection (CBP) recently announced that its subcontractor had privately and without authorization from federal agencies transferred the border patrol database that identifies passengers and license plate images. In the process, the subcontractor's network was the victim of a "malicious network attack" by hackers, which led to the disclosure of photos of inbound and outbound passengers in the database, explicitly involving images of vehicles passing through lanes at the entry for the past months. CBP added that its system had not been affected. In the announcement, CBP claimed that it did not identify any image data on the dark web or the Internet. In other words, no one discovered evidence that the leaked data was spread on the Internet. Besides, CBP did not explicitly mention the name of the subcontractor, the number of affected people, and the composition of the affected group (US citizens or non-citizens). However,

CBP did mention the name of the manufacturer Perceptics in the documents submitted to the Washington Post (Alba, 2019). So, when people's data has been collected, there is the potential risk that the data might be leaked, either by a hacker's attack or mistakes made by the people who are managing the data. In relation to this case study, the downside of facial recognition technology is that hackers steal the data collected and then use the images and photos in inappropriate ways, causing damage to the owner of the picture. The safety of data is critical.

Other than this, the accuracy of facial recognition is also a concern. In this case, Clearview AI claimed that the facial recognition technology they developed has 99% accuracy and it keeps the high accuracy no matter what color of the people, which is higher than any other competitors. However, when the observation target is captured while in a non-cooperative state, the accuracy of face recognition technology will be significantly reduced. The technical errors in face recognition spring from camera angles, lighting, etc., will all have an impact on the exactness of the face recognition system, but for the collection of face images in non-cooperative circumstances, the occlusion problem is severe. Particularly in the actual observation context, the observation target often wears accessories such as sunglasses and hats, which result in incompletely captured face images. This poor image has a huge negative impact on recognition and even causes the face recognition algorithm to be invalid. Nevertheless, these poor images are used in large-scale applications. In the environment, how to maintain or improve accuracy is also a fundamental issue at present.

Clearview's face recognition technology has rapidly risen to the forefront of national research. What is particularly prominent is the increasing concern from activists and politicians over how face recognition technology involves the violation of privacy and human rights and the effect of adopting the technology too quickly based on its false or misleading claims.

Amazon Rekognition, a cloud-based facial recognition product, has also received similar criticism for selling its technology to law enforcement. However, some scholars and activists have repeatedly expressed concern that when used to identify darker-skinned female individuals, this technology has pronounced flaws. ACLU used Clearview to match photos of members of Congress with a result of 28 congress members being incorrectly identified as other people who were already in jail. The results also show that false match from different genders and races is different. Nearly 40% of the wrong match was found to be of people of color while other studies found that the actual accuracy of facial recognition technology is lower than what providers claimed. Incorrect matching can lead to innocent people being wrongly arrested. Furthermore, if Clearview fulfills its reported ambition to provide software to many governing bodies, including authoritarian regimes with a long record of human rights abuses, the risks could become more significant.

There are some governments investing more resources in facial recognition technology, especially the US and China, which are the market leaders. The US government has started to step up airport security by using facial recognition systems to identify and register passengers. Several US states allow law enforcement to search their databases – which include details of driving licenses and photographs of identity cards. China has already started several facial intelligence projects, while other countries are still in the planning stages. The technology is being used all over the world, and many benefits are being derived from it. In India, banks are using facial recognition technology to prevent fraud at ATMs.

The technology will continue to develop over the next few years and will generate huge revenues. Surveillance and security are significant industries that are strongly influenced by technology. Schools, universities, and even medical institutions also plan to implement facial recognition technology in their offices for better management. The sophistication used in facial technology is also entering the robotics industry.

MULTIPLE INTERPRETATIONS

There are four perspectives among those participating in the case. From the law enforcement agency's perspective, they bought the facial recognition software from Clearview AI legally. Because they need this technology to help them solve heinous crimes for the benefit of the public, they have decided to adopt the technology even though it violates individuals' privacy. With the help of Clearview AI, they can work more efficiently and effectively. The Indiana State Police only spent 20 minutes to solve a case with the application of Clearview AI's technology. Here is the case: one man shot another man in a park and a passerby recorded the gunman with a phone, thus the police got an image of the criminal's face. However, the criminal did not own a driver's license and had never been arrested, so he was not in government databases. Then, the police tried to recognize the image in Clearview's app. The criminal appeared in a video captured by Clearview AI from a social media platform and his name was in the caption of the video. Therefore, the criminal was arrested and charged immediately. Without the facial recognition technology and the huge database of Clearview AI, identifying the criminal in such a short period is tough. This success made the Indiana State Police become Clearview's first paying customer (Hill, 2020). In this case, Indiana law enforcement agencies placed a priority on fighting crime and that has come at the expense of privacy. So, they have good reasons to use Clearview AI. In the future, law enforcement agencies will keep exploiting new technologies like Clearview AI and use them as their secret weapon. Central Intelligence Agency and its secret agents have come to a crisis because of the development of facial recognition technology. Because of the advanced biometrics and AI technology, for agents to hide their identities when they are going on a mission is getting increasingly challenging (Crane, 2019).

From the individual's perspective, they become unwitting information providers and therefore unhappy about the erosion of their privacy. For Clearview AI to use public citizens' personal information to make profit is unfair. As an individual user of a social media platform, people think their images are inaccessible to the public after deleting or adjusting the privacy settings on the platform. Their images should be confidential and be protected from disclosure or exposure to unauthorized individuals or systems. However, in this case, once their images are public, they will be stored forever in the database of Clearview AI. Therefore, it is accessible to their customers. Clearview AI will profit from these free images without the consent of the image provider (O'Sullivan, 2020).

It might be not appropriate for private companies to use facial recognition technology and that is why Clearview AI tries to sell this technology to law enforcement agencies. Probably someday in the future, the law enforcement agencies would use their images against them. Asking a person to show his identification without a reason is illegal in most states, while facial recognition technology could find a person's identity without permission, be sold, then possibly be used in a wrong or even dangerous way. That is why people should take action to protect their privacy, as once stated by a privacy researcher, "Humanity is protected through the protection of privacy" (Wingreen, 2020).

"Security is not about how well you know your enemy, but how well your enemy knows you. If your enemy knows nothing about you, then there's nothing for him to attack" (Wingreen, 2020). If connected to the internet, any device or data is not secure, no matter what the quality of your security framework or the security policy. Sometimes old-school security is better. If people really want to protect their privacy, they should never post their personal images on the Internet.

From the perspective of Clearview AI, they believe that they use public information to help law enforcement agencies. At the same time, they comply with laws in different regions. The policies of social media companies do not apply to Clearview AI because they are not government-legislated guidelines or laws. Policies are more flexible than laws and ignoring policy is a viable defense. For social media platforms, distributing their policies to all individuals who are expected to comply with them or be understood by all stakeholders is hard, which creates a policy loophole. Obviously, Clearview AI has benefited from this policy loophole.

In terms of technology, Clearview AI revisited the value of public facial images through facial recognition technology. A thing only has value if someone believes it is valuable or uses it to create value. Collecting publicly available facial images from social media platforms is cheap and easy for Clearview AI because nobody thinks these images are valuable. Using their algorithm, Clearview AI adds value to the database and turns it into a valuable thing.

From the social media companies' perspective, their policies are violated, and they need to protect their information security by defending their policies. Obviously, the behavior of those downloading and storing pictures is against most of the major social media platforms' policies (O'Sullivan, 2020). If these policies do not work, collecting and combining public personal information from several different sources is easy. This kind of behavior is known as information aggregation. It could result in databases that might be exploited in a way the original data holder has not granted or even knows about (Whitman & Mattford, 2019). Most people believe that their family and friends know them the best, yet the information aggregator may know everything about them. Information aggregators could also make profit through selling the information to marketing departments for advertisement or law enforcement agencies for surveillance purposes. Thus, social media platforms think the profit model of Clearview AI is questionable as it engages in Information aggregation. However, social media companies face severe problems of information security governance. Because the companies have virtually no control over what happens to images after these images are downloaded from their platforms, this puts a negative spin on their reputation. Therefore, individuals might be reluctant to post pictures on a social media company's platform anymore.

At the same time, there is a huge difference between private and public organizations depending on the agency theory. Agency theory focuses on the distribution of risk in the organization. In an ideal situation, decision-makers should get the rewards of right decisions and tolerate the bad consequences of their wrong decisions. Private organizations have lots of internal disciplines to manage customers' privacy to satisfy their customers. If they fail to protect the privacy of the customer, they will face loss of reputation and financial loss, which are bad results of wrong decisions. However, for public organizations, they do not need to bear any consequence of wrong decisions regarding privacy issues. So, the erosion of privacy brought by Clearview AI is a big crisis for social media platforms, while good news for law enforcement agencies.

LESSONS LEARNED

There are some different perspectives that arise from this case. First, the data breach of customer lists in Clearview AI posed a serious security issue. A hacker accessed Clearview AI's complete customer list without authorization, and this is the first time Clearview AI faced data breach problems. After the data breach happened, Clearview AI sent out notifications to its customers (Muhlberg, 2020). The notification from Clearview AI ensured the security of customers' servers and that there was no compromise of Clearview AI's system. Moreover, Clearview AI stated that it has fixed the vulnerability part which led to the data breach and will work continuously to enhance the security of its system (Muhlberg, 2020). However, as the hacker is able to access Clearview AI's client lists, he has access to its other data in the system simultaneously. Clearview AI needs to enhance its confidentiality of computer security to take better control of the risk of external attacks and secure its information (Whitman & Mattford, 2019).

Second, there are some ethical issues in Clearview AI. The employees of law enforcement agencies intend to use the Clearview AI software for personal use without authorization, which is an unethical behavior (Whitman & Mattford, 2019). For example, some law enforcement employees used Clearview AI software to search facial images of friends, family members or celebrities without authorization (Mac & McDonald, 2020). An even worse scenario is if these employees use Clearview AI for malicious or illegal purposes; such actions may damage both Clearview AI and law enforcement agencies' reputations and increase company liabilities (Whitman & Mattford, 2019). To avoid unethical behaviors, such as ignorance of ethical actions, accidental incidents, and other intents, Clearview AI should provide training on up-to-date InfoSec topics since key studies show that education could level up the ethical perceptions in small populations (Whitman & Mattford, 2019). With training, education, and awareness of proper ethics, Clearview AI could create informed, skilled, and low-risk employees and systems users (Whitman & Mattford, 2019).

Third, social media platforms, such as Facebook, Instagram, and Twitter, failed to set up effective policies to protect their users' privacy and rights while uploading their images to the platform. Moreover, they provided the pipes for Clearview AI to download these publicly posted photos and images as free resources without any fees or official contracts. Clearview AI had violated the users' privacy and broke the social media platforms' policy while Clearview AI stated that they only gathered publicly posted photos for justifiable purposes without breaking the law. In this case, the policies set by various social media platforms are not effective enough to protect their users' privacy and rights. To cease the actions of Clearview AI effectively, these social media platforms should think of adjusting or adding more useful policies. These would ban not only Clearview AI but also other external vendors from continuously gaining increasing amounts of data from the platforms to protect their users' privacy and rights. Since deterrence, such as policy, is only effective if Clearview AI is fearful of penalty, the possibility of being caught and for the penalties to be administered must be made high (Whitman & Mattford, 2019). Otherwise, those social media platform policies will do no harm to Clearview AI.

Fourth, facial recognition technology has a lot of gray areas and debate in regard to civilians' privacy, basic liberties, and human rights (Whitman & Mattford, 2019). Seemingly there is no existing global legislation to monitor the usage of facial recognition technology. In the real-world situation, every state and every county has its own legislation regarding facial recognition technology. For example, Texas, Illinois, and California have set up different laws which target the use of biometric information. However, since the contents of each law are different, publishing a comprehensive law to monitor and assess the usage of facial recognition technology is hard. Moreover, another problem is that the law is effective only when there is someone enforcing it (Whitman & Mattford, 2019). Hence, as facial recognition technology is an increasing trend, how to protect civilians' privacy and monitor the performance of facial recognition technology to prevent illegal actions is increasingly crucial.

Fifth, every individual and company is to take responsibility for protecting their own information security in regard to data confidentiality, integrity, and availability (Whitman & Mattford, 2019). Based on this concept, there are four different players in this case. The first is Clearview AI, which should be responsible for its information security to ensure the confidentiality of customer data. For example, Clearview AI did not successfully guard or evaluate the system risk which provided the hackers a pipeline to gain access to the system and steal the sensitive data. This may decrease customer trust and affect Clearview AI's reputation. Hence, Clearview AI should take the responsibility to evaluate their system risk, assess the system vulnerability, and set up a relative incident response and disaster plan in advance to reduce the damage of any threat and sustain business continuity.

The second player are various social media platforms; they should be accountable for the users' information security. As stated in the previous paragraph, the social media platform policies do not stop Clearview AI from downloading users' images online. Moreover, the cease-and-desist letter cannot prevent Clearview AI's actions. Therefore, to protect the users' image privacy and maintain customer trust, those social media platforms should be aware of their responsibility for users' information security. Signing a business contract with Clearview AI to regulate Clearview AI's actions might be helpful. However, there will undoubtedly be a time-consuming negotiation process between social media platforms and Clearview AI to reach an agreement.

For the third party in this case, law enforcement agencies have the responsibility to take control of employees' behavior through better management methods to prevent or minimize unethical behavior. For example, previous content has mentioned the unethical behavior of law enforcement agency employees intentionally using the system without authorization to search for friends' and family members' facial images. The law enforcement agencies should take this situation into consideration in advance. There are some actions that the law enforcement agencies can take such as setting up precise operation standards to regulate the business processes, establishing effective system-specific security policies, providing SETA programs to educate and test the employees' understanding of the policies, and allocating security personnel to take control of protecting information security. These steps can generate more well-behaved system end users by increasing the awareness of policies and directing them to comply with the policies. Afterward, the law enforcement agencies can minimize unethical behavior among employees.

The fourth player, social media users are in charge of their personal information security. Even though people have the right to pose any personal photo online, they should be aware of protecting their personal information at the same time. There is always a trade-off between posting images online and the sacrifice of privacy. Once the photos are shared with the public, there is no privacy since every single user can access and view the photo. Hence, before uploading a personal photo, the users should consider whether this action is more valuable than placing their personal information at risk. Once the person has decided to expose his personal information online, he/she needs to bear the consequences that there might be some privacy risks. In contrast, if the person thinks that privacy matters more to him/her, not posting any images online is the most straightforward method to maintain his information security.

Today in this technologically advanced world, information can be easily collected and stored from various channels. Thanks to the development of technology and the rise of mobile services, people tend to be more socially connected to each other. By clicking on a single button, personal information can be shared and transferred between various devices and networks. As life is becoming more convenient than before, people prefer to forget the importance of securing their own privacy and information. Hence, every individual should take responsibility to determine the value between the privacy vs risk trade-off, and every corporation should also be responsible for controlling their system and information security (Whitman & Mattford, 2019).

DISCUSSION QUESTIONS

1. How do we protect each other's privacy while posting images online?
2. Do you agree with Clearview AI's point of view that their Information aggregation is legal since the information gathered is all publicly shared? Why?
3. What is your perspective of law enforcement agencies using facial recognition technology as assistance in matching suspects?
4. What are the other possible unethical behaviors that may occur in Clearview AI?

REFERENCES

Alba, D. (2019, June 11). Photos of travelers coming in and out of the US have been hacked and stolen. Retrieved August 10, 2020, from https://www.buzzfeednews.com/article/daveyalba/the-us-governments-database-of-traveler-photos-has-been

Andrejevic, M., & Selwyn, N. (2019, November 5). Facial recognition technology in schools: Critical questions and concerns. Retrieved August 09, 2020, from https://www.tandfonline.com/doi/full/10.1080/17439884.2020.1686014

Brinckerhoff, R. (2018). Social network or social nightmare: How california courts can prevent facebook's frightening foray into facial recognition technology from haunting consumer privacy rights forever. *Federal communications law journal*, *70*(1), 105. Retrieved from http://www.fclj.org/wp-content/uploads/2018/04/70.1-Brinckerhoff.pdf

Connor, T. (2020, January 24). NJ Bans cops from using clearview AI, facial-recognition app that scrapes social media. Retrieved from https://www.thedailybeast.com/nj-bans-cops-from-using-clearview-ai-facial-recognition-app-that-scrapes-social-media

Crane, E. (2019). Will the digital age kill off spying?. Retrieved from https://www.dailymail.co.uk/news/article-7837767/CIA-faces-crisis-intelligence-gathering-digital-footprints.html

Davis, J. P., Lander, K., Evans, R., & Jansari, A. (2016). Investigating Predictors of Superior Face Recognition Ability in Police Super-recognisers. *Applied Cognitive Psychology*, *30*(6), 827–840. doi:10.1002/acp.3260

Efron, J. (2019, November 26). Facial Recognition to Reshape the Retail Industry in 2020. Retrieved from https://shuftipro.com/blogs/facial-recognition-to-reshape-the-retail-industry-in-2020

Garvie, C., & Frankle, J. (2016). Facial-recognition software might have a racial bias problem. *The Atlantic*, *7*.

Hill, K. (2020). The secretive company that might end privacy as we know it. Retrieved from https://www.nytimes.com/2020/01/18/technology/clearview-privacy-facial-recognition.html

Lemboye, M. (2019). Facial Recognition and Mass Media Use: Effects of Technology on Perceptions of Privacy. Ph.D. dissertation, University of Arkansas at Little Rock.

Leong, B. (2019). Facial recognition and the future of privacy: I always feel like … somebody's watching me. *Bulletin of the Atomic Scientists*, *75*(3), 109–115. doi:10.1080/00963402.2019.1604886

Lynch, J. (2020). Clearview AI—Yet another example of why we need a ban on law enforcement use of face recognition now. Retrieved from https://www.eff.org/zh-hans/deeplinks/2020/01/clearview-ai-yet-another-example-why-we-need-ban-law-enforcement-use-face

Mac, R. H. C., & McDonald, L. (2020, January 28). Clearview AI once told cops to "Run Wild" with its facial recognition tool. It's now facing legal challenges. Retrieved from https://www.buzzfeednews.com/article/ryanmac/clearview-ai-cops-run-wild-facial-recognition-lawsuits

Marr, B. (2019, August 19). Facial Recognition Technology: Here Are The Important Pros And Cons. Retrieved from https://www.forbes.com/sites/bernardmarr/2019/08/19/facial-recognition-technology-here-are-the-important-pros-and-cons/#658de59514d1

Martin, A. M. R. (2019). 81% of 'suspects' flagged by Met's police facial recognition technology innocent, independent report says. Retrieved from https://news.sky.com/story/met-polices-facial-recognition-tech-has-81-error-rate-independent-report-says-11755941

Muhlberg, B. (2020, March 11). Clearview AI data breach exposes facial recognition firm's client list. Retrieved from https://www.cpomagazine.com/cyber-security/clearview-ai-data-breach-exposes-facial-recognition-firms-client-list/

Musil, S. (2020, June 10). Clearview AI still backs facial recognition, despite competitors' concerns. Retrieved from https://www.cnet.com/news/clearview-ai-still-backs-facial-recognition-despite-competitors-concerns/

Ng, A., & Musil, S. (2020, February 6). Clearview AI hit with cease-and-desist from Google, Facebook over facial recognition collection. Retrieved August 09, 2020, from https://www.cnet.com/news/clearview-ai-hit-with-cease-and-desist-from-google-over-facial-recognition-collection/

O'Sullivan, D. (2020). This man says he's stockpiling billions of our photos. Retrieved from https://edition.cnn.com/2020/02/10/tech/clearview-ai-ceo-hoan-ton-that/index.html

Privacy bodies across the US challenge authorities on facial recognition. (2017). *Biometric Technology Today*, *2017*(6), 3,12–13,12. doi:10.1016/S0969-4765(17)30109-1

Sajjad, M., Nasir, M., Ullah, F. U. M., Muhammad, K., Sangaiah, A. K., & Baik, S. W. (2019). Raspberry Pi assisted facial expression recognition framework for smart security in law-enforcement services. *Information Sciences*, *479*, 416–431. doi:10.1016/j.ins.2018.07.027

Sennaar, K. (2019, November 22). Facial recognition applications – Security, retail, and beyond. Retrieved from https://emerj.com/ai-sector-overviews/facial-recognition-applications/

Shead, S. (2020, June 12). Facial recognition tech developed by Clearview AI could be illegal in Europe, privacy group says. Retrieved August 9, 2020, from https://www.cnbc.com/2020/06/11/clearview-ai-facial-recognition-europe.html

Shiri, G. (2019, December 10). How to avoid a dystopian future of facial recognition in law enforcement. Retrieved from https://www.vox.com/recode/2019/12/10/20996085/ai-facial-recognition-police-law-enforcement-regulation

Sim, T., Zhang, S., Janakiraman, R., & Kumar, S. (2007). Continuous verification using multimodal biometrics. *IEEE Transactions on Pattern Analysis and Machine Intelligence*, *29*(4), 687–700. doi:10.1109/TPAMI.2007.1010

Whitman, M. E., & Mattford, H. J. (2019). *Management of information security* (6th ed.). Cengage Learning.

Whittaker, Z. (2020, April 4). Security lapse exposed Clearview AI source code. Retrieved from https://techcrunch.com/2020/04/16/clearview-source-code-lapse/

Wingreen, S. C. (2020). Quoted from a lecture on, “Risk Management”, in the course, “Information Technology Security and Risk Management”, delivered in the Fall 2020 semester at the University of Canterbury.

Woodward Jr, J. D., Horn, C., Gatune, J., & Thomas, A. (2003). *Biometrics: A look at facial recognition*. Retrieved from https://www.rand.org/content/dam/rand/pubs/documented_briefings/2005/DB396.pdf

Wright, E. (2019). The future of facial recognition is not fully known: Developing privacy and security regulatory mechanisms for facial recognition in the retail sector. *Fordham Intellectual Property, Media & Entertainment Law Journal*, *29*(2), 611. Retrieved from https://ir.lawnet.fordham.edu/iplj/vol29/iss2/6/

Xiao, Q., & Yang, X.-D. (2010). Facial recognition in uncontrolled conditions for information security. *EURASIP Journal on Advances in Signal Processing*, *2010*, 1–9.

2.14 Aadhaar *The National Identity System of India*

Anjali Krishnan
University of Canterbury, Christchurch, New Zealand

Shameer Thottoli
Acclaim Group Limited, Auckland, New Zealand

INTRODUCTION

In all parts of the world and in recent years, digital transformation from paper-based legacy systems has been on the rise in every field ranging from agriculture, education, and healthcare to governments. A game-changer for many, digital transformation benefited a lot of people, and many organizations yielded better operational efficiency and profit. To realize the potential benefits of moving to a more advanced technology was the main objective behind these transformations. This paradigm shift in the way everyone worked was to achieve more efficiency and effectiveness than the old paper-based systems could ever offer. The creation of a digital identity in any form is one of the important aspects of digital transformation. A digital identity system aims to get greater access to public services or other services offered commercially. Most of these digital identity systems have biometric data included in their data about each person apart from personal data like name, address, and nationality. Although there are beneficial factors in having a digital biometric identity system, there is also a threat or risk associated with it. There are instances where a digital biometric identity system was abused to oppress, exclude, or discriminate [1].

India, a country of 1.366 billion people [2], currently has one of the largest digital biometric systems in the world called Aadhaar. In the year 2009, the government in power created Aadhaar to avoid fraudulence, black markets, and tiresome bureaucracy load. To further explain, India has a large part of its population below the poverty line and the government of India is trying to provide them services in the form of subsidies and other public welfare schemes. The Public Distribution System in India is responsible for distributing some of these services, especially food to poor people. However, there are a lot of intermediaries that riddle this system for personal gains and thus hamper the distribution. In response to the former, the government of India created Aadhaar to systematize the procedure to collect data and to make it

DOI: 10.1201/9781003264415-31

easier to distribute government services to the public, especially to the people who are below the poverty line.

Aadhaar assigns a 12-digit unique number to all residents in India and the numeric identifier is like a social security number (SSN) in the USA. The procedure to receive an Aadhaar Card involves providing personal data like name, date of birth, and address along with additional biometric data like fingerprints, retina scans, and face photos. Founded by the government of India as per the Aadhaar Act of 2016 under the Ministry of Electronics and Information Technology, the Unique Identification Authority of India (UIDAI) is the regulatory authority that manages Aadhaar-related services. The vision of UIDAI [3] is to "To empower residents of India with a unique identity and a digital platform to authenticate anytime, anywhere." The mission includes delivering good governance, providing subsidies to the deserved, creating "policy, process and system" for collecting biometric and demographic data, collaboration, innovation, availability, proportionality, and restoration of technology infrastructure, sustainability, and providing valuable insights. The values of UIDAI are integrity, commitment to building a nation, holistic approach, best service, continuous learning, quality improvement, innovation, transparency, and free access. The headquarters of UIDAI is in New Delhi and has eight Regional Offices (RO) across the country. The chief executive officer is Pankaj Kumar. As mentioned on the UIDAI website, under the Aadhaar dashboard there are a total of 1,259,684,005 enrolments which accounts for 92.2 percent of the total population in India.

The background of Aadhaar can be traced back to 2009 when the then-ruling party United Progressive Alliance first conceived the idea [4]. UIDAI was formed as an extended agency to the Planning Commission of India which is also a government-funded policymaking agency. The venture was headed by Nandan Nilekani, a founding member of tech giant Infosys. The main problem for him to solve was to find a solution that could simplify the administrative or rule-bound way of availing government schemes in India. Making matters plainer and requiring a lot of tax money to distribute government benefits to the beneficiaries was exceedingly difficult. The amount of paperwork, time, cost, and effort to get a government subsidy was so humongous that many poor people were unable to find it easy to do the same. Hence, the benefits offered by the government were not reaching the deserving people. To put an end to this agony, the government introduced Aadhaar and with just one Aadhaar number any government scheme can be availed.

Published by Tech Crunch, the source article for the case study in this research case is [5] titled "Indian state government leaks thousands of Aadhaar numbers." The article talks about reports on the security lapse that resulted in the leakage of 100,000 Aadhaar numbers. A website that was used to record the attendance of civil servants of an Indian state named Jharkhand was exposed without any password with data since 2014 thereby allowing anyone to access the names, designations, and partial contact numbers of about 166,000 public servants. The photo of each record was saved with the corresponding Aadhaar number which anyone can access from the website. Though the data leak is not linked to UIDAI's central database, it is still connected to an Indian state government authority that was supposed to protect the data. The article then claims that Aadhaar numbers are not actually a secret number but are like SSNs in the US. Citizens can use their Aadhaar card to avail government services like subsidies, welfare, and voting. People even use their Aadhaar number to book an Uber, open a Facebook account or a bank account, get a mobile SIM card, or even to do online shopping. Aadhaar numbers getting leaked has facilitated the illegal trade of citizen data in the black market and dark web.

Regarding the Aadhaar leak, the team of Tech Crunch requested the help of Baptiste Robert, a French security expert who goes by the online name Elliot Alderson. He has previous experience with Aadhaar data leakage issues. He showed Tech Crunch that, a mere 100 lines of Python code is required to completely download all the Aadhaar numbers and other citizen data from the whole website. To check if the Aadhaar numbers they were able to collect were valid, they crosschecked some of the data on the UIDAI official site. To their surprise, all were positive matches which indicates all the leaked data are personal data of an Indian citizen. With these findings, the Tech Crunch team sent an email to UIDAI and the Jharkhand government. However, there was no response from either of the parties to Tech Crunch, but UIDAI quickly pulled the website offline.

The government and UIDAI frequently claim that the Aadhaar database is leak-proof and there is no way anyone can get access to citizen biometric and demographic data. Though this case study only exposes a very small percentage of citizen data leaked, it still accounts for violation of privacy. UIDAI is hesitant to accept these security incidents and constantly claims that its system is impenetrable. The article accuses UIDAI of portraying a long history of rejecting any news about data breaches and claiming that news is fake. Cases were filed against some of the reporters who brought the data breaches to light. The leakage of Aadhaar numbers is not as sensitive as a direct breach of biometric data but by using the Aadhaar number identity theft and other fraudulent activities can happen. Some people think the Aadhaar system provides a way to surveil citizens for the government.

In light of the Aadhaar data leak case study, the problem statement of focus is, "Risks involved in sharing demographic data and biometric data with government." To define the problem statement, we need to first investigate the significant difference in potential risks while sharing our own data with a government when compared with a private organization. The above-mentioned problem statement emphasizes the risks involved with sharing biometric data or demographic data with the government and this case study presents a good example of how data leaks are occurring from government-controlled agencies. The Aadhaar data leaks prove that security and protection offered to citizen data which is handled by the government is not fault proof and data breaches do happen to government databases. Hence, this case study analyzes the risks involved in sharing personal data which contains information like biometrics to the government.

To address the issue presented in the problem statement, this study prescribes a few frameworks that might be helpful. The Aadhaar data leak affects the government and the citizens, and addressing the risks involved in data sharing between a citizen and the government should include the empowerment of both parties involved. This study presents two framework models, one for each of the parties, to analyze and potentially address the problem statement. The first one is a technical framework for government-to-government data sharing (G2G) called SecureGov [6] which will address the risks associated with data sharing between government agencies. As most of the data leaks happen when Aadhaar details are shared between government agencies [7], SecureGov will provide a technical framework model that the Indian government can consider for implementation to avoid data breaches. The fraudulent activities in many accounts and identity thefts where the biometric data of a citizen is misused occur when an Aadhaar leak happens. To safeguard citizen data, the Indian government can implement the second, as well as a framework for data protection and privacy like the General Data Protection Regulation (GDPR) as created and used by the European Union [1]. Adequate data protection and privacy regulations are required to ensure accountability by UIDAI and government agencies.

LITERATURE REVIEW

This section of the research case study reviews three sources of current literature that directly or indirectly address the problem statement, "Risks involved in sharing demographic data and biometric data with government."

Choi, Chun, Kim, and Keromytis (2013)

Choi et al. begin by briefly explaining what open governments or paperless governments are and how they allow governments to share information to create a system with collaboration and encouraging innovation. This system creates improved services for citizens and the whole society, eventually leading to a transformed government which can be called an e-government or smart government. This paper then describes several types of information sharing in this context. The first type is G2C sharing which is the data sharing done by the government to the citizens. For instance, information about new services, laws, etc., is shared by the government with citizens. This type of data sharing can be termed G2C sharing. The second type is C2G sharing which indicates data being shared by citizens to the government. For example, citizens can raise their voices against any new policy created by the government, or citizens can share their feedback on an existing government service. This type of information sharing can be classified as C2G sharing. The third type of data sharing is G2G sharing which indicates the information sharing between government agencies, including citizen-related data. Most of the government services are to be done with a combined effort of different government agencies or departments. Hence, citizen data is mostly shared between various departments of a government to provide services to the society.

In G2G sharing, the key factor is to ensure there is a secure transaction of citizen data between different government departments. Any data breach that can occur to sensitive citizen-related data will tarnish the reputation of the government. This paper presents a security framework for safe and secure data sharing between government agencies called G2G. Choi et al. call their novel security framework Government Data Sharing or SecureGov. This framework ensures safe and secure sharing of data with no privacy breaches or data leaks and is created based on Service Oriented Architecture with layers of security with Public Key Infrastructure (PKI), technology for access control, and another technology that prevents forgery. The authors aim to address any kind of data leaks or data alterations during the transit of data between sender and receiver, unauthorized access to external parties, and modifications or circulation of citizen-related data in illegal ways. The authors gave the new security framework SecureGov to South Korea to be implemented in their Public Information Sharing Centre (PISC).

The first of the four main components of the proposed security framework SecureGov is PKI which produces the public key certificates for service requestor identity authentication. The second component is the Digital Rights Management (DRM), a server component which evaluates the roles and verifies the rights permitted for the requestor and creates the document rights container or e-Container. The third component is the Digital Rights Management Packager which does the encryption of documents at document source organization or the service provider. Last, End User Access Control (E-DRM) is a client component at the requestor end that controls the received data manipulation and ensures the rights. The combination of all the four security systems summarizes the new security framework called SecureGov and the overall system named Government Information Sharing System (GISS).

The paper concludes by stating the major challenges that lie ahead. Most citizens want government services to be available through mobile and hence a security framework that can be implemented in mobile technology must be formulated for secure data sharing. Another area of concern is to what extent a developing country like India can invest in SecureGov to ensure safe data sharing between government agencies. When Aadhaar data leaks are taken into context, a security framework like SecureGov is a security framework model that can be imitated to avoid future data leaks especially when Aadhaar data is shared between government departments.

Dixon (2017)

Dixon first explains the role of biometrics in digital identity systems by stating examples of various instances where biometrics like fingerprints had a key role in determining identity in most criminal cases. The author also highlights the usage of biometrics in the healthcare sector to uniquely identify each patient. Using deep learning and neural networks, the error rates in biometrics are constantly decreasing. Next, the author evaluates the digital biometric identity system like Aadhaar and weighs the pros and cons of the sudden digital transformation of the process of availing government benefits in India. The author states that there was no foundation of any legal framework before the Aadhaar system was introduced in India. There are also no comprehensive information protection and privacy laws that explain how Aadhaar can be used.

The author then points to the robust information protection laws that guide the European Union, namely the General Data Protection Regulation (GDPR). The GDPR includes the protection and safety of biometric data. The author says GDPR is a very complex and lengthy law that contains a sophisticated, inclusive approach to data privacy and includes any new technology that can impact human anatomy. The law also ensures any outside entity that wants to use EU citizen data must comply with GDPR, and this leads to a notable change in behavior toward data protection and privacy worldwide. The paper also discussed the important principles that govern biometric data under GDPR. The author also gives us an idea about the Sectoral Framework implemented in the US to address data protection and privacy laws which is not as wholesome and adequate as EU's GDPR.

Under the Discussions section in Dixon's paper, the author shares his opinion that all three jurisdictions explained in this paper are entirely different from each other. The regulations pertaining to data protection and privacy laws especially applicable to biometric data of India, the EU, and the USA are compared to each other. According to the author, there are no laws or regulations that provide protection for biometric data contained in Aadhaar while the EU has an "omnibus data protection and privacy law" that has specific sections for biometric data protection and is comprehensive. Concerning the USA, the author claims there are no overall data protection laws that unify all states in America and biometric data protection laws are different for each state. The author terms these differences as a "sectoral framework" because there is no regulation in the USA that is applicable to everyone with US citizenship.

The author also highlights the importance of consent in biometric data collection. Now that India has made Aadhaar mandatory, citizens who need government benefits must submit their biometric data. The author calls this Coerced Consent because people who need government benefits have no other choice but to share their biometric data for survival. The author reiterates that mandatory biometrics use proposals in India must be addressed urgently.

The author then shares that parties must strictly adhere to four concepts irrespective of socio-economic and cultural boundaries that must be a foundation for digital biometric identity. The first concept is Do No Harm, which indicates avoiding any kind of harm caused by hostile hands or misuse of digital biometric identity and requires continuous oversight, foresight, and deep evaluation. The second one is Policy before Technology which highlights the importance of having proper legislation in place before the actual deployment of a digital biometric identity system. The third concept discussed is Ethical Data Use Guidelines which indicates the significance of charting the guidelines of data usage in an ethical way by combining opinions from all stakeholders. The fourth concept is Privacy by Design which means having no room for faults while designing a digital biometric identity system. When a system is created that uses a strong identification feature like an iris scan, the author states that the designer of the system needs to take full responsibility to ensure privacy is a mandatory factor and that the system can never fail. This privacy-mandated design of a system must be in addition to the supportive regulations for data protection and privacy. The author is of the opinion that India's Aadhaar digital biometric identity system must go beyond the four principles stated above and should take inputs from the OECD Fair Information Practices, EU's GDPR, ID4D Principles on Development, India's Group of Experts report, and Do No Harm principle. All these principles are basic foundations for legislative guidance while creating a digital biometric identity system. In conclusion, the author emphasizes the fact that no matter what the socioeconomic or cultural difference is, any country that implements a digital biometric identity system like Aadhaar must adhere to the Do No Harm principle and rigorously scrutinize its system for the risk of surveillance, discrimination, and exclusion.

AlKalbani, Deng, and Kam (2015)

AlKalbani et al. focus on the significance of information security in e-government development. Information security compliance must be considered widely to make sure information security is maintained in public organizations. The authors develop and validate a conceptual model for compliance with information security built upon organizational security culture and the moderating role of social pressure.

Organizational security culture is the belief of every single employee in an organization on the merits of complying with information security policies and standards. Employers can instill this culture of information security in employees by naturally imbibing info-security values in their daily work lives. The three main factors of organizational security culture are management commitment, accountability, and awareness about information security.

AlKalbani et al. develop three hypotheses that all the three major factors mentioned above have a positive influence on information security compliance. Social pressure is nothing but the protection of socially desirable needs that must be appropriately addressed by government services. In noting social pressure significance, the authors create three other hypotheses stating that it positively moderates the relationship between the three major organizational security culture factors and information security compliance. The first three hypotheses try to find if management commitment, accountability, or awareness about information security is positively influencing information security compliance by a government providing its services digitally like Aadhaar. The last three hypotheses are related to social pressure and its role as a moderator in the relationship between the three factors mentioned above and information security compliance.

After conducting the research, the results obtained indicate all three factors of organizational security culture, which includes management commitment, accountability, and awareness about information security have a great positive influence or impact on information security compliance. Hence, the results support the first three hypotheses. Social pressure is a moderator only for the relationship between accountability and information security compliance. In the case of management commitment and information security awareness, the interaction effects of social pressure are negligible. These insights can be applied to Aadhaar by making sure all government employees handling Aadhaar data follow organizational security culture and thereby increasing information security compliance. In addition, social pressure is a major moderating factor that can ensure accountability for all government employees and the pursuit of information security compliance.

ANALYSIS

The Need for Citizen's Personal Information

Documents or information provided to the government by people is the new radical state's essential slave. Collecting truthful information from the people to give welfare payments of government aid is necessary for the government, and Information collection from them is an extremely long, complicated, and unmanageable task. The government controls almost every aspect of people's lives, particularly, though not limited to, our job practices; the health of our workplaces and their availability to people with disabilities; the quality and efficacy of the food, medicines, and pesticides that we purchase; the nature of the aircraft, automobiles, and trains that we drive; and financial institutions such as banks.

The government cannot accomplish its legislative mandate until those exposed to the legislation supply accurate and appropriate information which will help authorities to track enforcement practices of the controlled agencies. Similarly, Aadhaar aims to establish a system for identifying people to supply utilities, incentives, and other benefits for the country's citizens. Digitalization needs indices or special IDs and usually local IDs such as ration numbers or work card numbers are being used in social protection programs. Standardizing the procedures of holding digital records across geographic regions and product lines and connecting local IDs with Aadhaar's unique national identities ensure that the various digital database tables are effectively integrated into one. While digital information will also be dispersed globally, real-time data connectivity can be given to approved national and regional authorities for evaluation, reporting, and research and planning applications using the Aadhaar IDs as keys. This way one's Aadhaar number can be used as a common index for all the utilities.

Moreover, the Aadhaar program will also offer the requisite momentum to standardize and digitize other areas, most of which are long past due. The Aadhaar IDs can conveniently be utilized to establish local IDs for scanning new lines of business. Most notably Aadhaar will promote the relation of local IDs to independent sectors such as population data, schooling, medical and immunization data, information concerning birth and death, registry of property, taxation, accounts, mortgages and debts, police and compliance monitoring, control of disasters, protection and surveillance, and others. Along with facilitating the effective planning, distribution, tracking, and assessment of services in each area, Aadhaar is also expected to use modern data analytics technologies to identify large-scale interaction between user information, which can promote better implementation of government policy interventions, early detections, and alert systems for anomalies. For example, being able to compare literacy rates, household income, and diet through the larger community can be incredibly insightful; or to compare the spreading of infection in relation to citizens' wealth and education is valuable. More broadly, modern data analytics technologies can facilitate quantitative research, observational studies, automated detection of latent subjects, and relationships among variables across numerous financial realms.

A large recall of several of the most well-known and influential interests of the government will help us to understand why the use of such personal information taken from the people by the government is a prevalent and critical issue.

DATA LEAKS BY GOVERNMENT VERSUS DATA LEAKS BY PRIVATE ORGANIZATION

Aadhaar details have been leaked several times exposing the personal data of millions of Indians since the inception of this biometric identity program. Aadhaar is now an essential document required to make use of basic government benefits like ration, phone connection, and healthcare. However, Aadhaar data breaches are happening either through technical error due to weak technical architecture or human error. Many cybersecurity warriors have complained and warned UIDAI as well as the government in this regard. But both the government and UIDAI are rebuffing all security breach incidents as fake news. According to a tweet by UIDAI, they claim the Aadhaar database is impenetrable. They openly ridicule people who raise their voices against the Aadhaar system's weak data protection. Various reports citing responses from UIDAI and the government regarding Aadhaar data leaks demonstrate that there is no accountability from any regulatory body.

When comparing this scenario with a data breach happening with a private organization, we can find a stark difference in handling such cases. Most private organizations have risk management policies and data protection regulations formulated by directorial board members which they follow in their work culture, and steps are taken to imbibe the same data protection values in their employees. Hence, when a security breach incident occurs at a private organization, most of the time the top management releases a statement in which they take due responsibility and hold any guilty parties or system failures accountable for what happened. Soon after this, the victims of the security incident are compensated, and new reforms are introduced to make data protection regulations more efficient. If the breach is due to a technical error, most companies review their technical security framework and implement necessary changes. If the security incident is because of human error, strict actions are taken against the person who erred. In the case of India's Aadhaar data leak cases, there was not a single instance where UIDAI has acknowledged and taken accountability. As soon as a security incident is reported, the government is quick to take down the website like what happened in the case study published by "Tech Crunch." The way in which the Indian government and the UIDAI have responded shows a lack of accountability.

In addition, when a data breach happens with a private organization, the number of affected people will be limited to their customers or their employees. Most countries have legal provisions where the victim can sue the private company and get compensated. In case of a data leak happening at the government end, the victims are citizens who have no other choice but to abide by the ruling government. If there are legal provisions to file a case against the government or the regulatory body, then there is a chance to get justice from the judicial court. However, as stated in the article by "Business Standard" [8], currently there is no provision in the law that allows a citizen to file a case against the UIDAI for Aadhaar data breaches. Many legal experts are urging the ruling government to empower citizens by making an amendment that allows citizens who are victims of Aadhaar data leaks due to inefficiency from UIDAI to file a case.

PROBABLE FORMS OF PRIVACY VIOLATIONS IN AADHAAR

Three ways a person's privacy can be violated in a system like Aadhaar are listed and defined below.

Correlation of identification across domains: This can be achieved using its internationally accessible Aadhaar numbers in Authentication User Agencies (AUAs) and various fields of operation. It will proceed without an agreement to recognition.

Verification without permission using Aadhaar data: The use of biometric authentication to unlawfully identify individuals may be prohibited. These breaches involve recognizing persons through improper matching of fingerprints, retinal scanners, or facial images held in the Aadhaar database, including the use of demographic profiles to classify people without their permission.

Illegal surveillance of people: People can be registered or recorded, using verification and identity documents and traces, in a database or in one or more AUA databases, without sufficient authorization or legal authorization. Usually, these documents contain details concerning the exact location, period and history of authorization, and usage facilities.

The second and third threats above are even more probable if the intruder cooperates with an individual having access to different Aadhaar system components.

AADHAAR DATA LEAKS: TECHNICAL OR HUMAN ERROR?

The case study discussed in this research case is about a government website without any username or password protection that leaked the personal data of some of its employees with their Aadhaar number saved as the filename of their photos. There was another incident reported by "Tribune India" [9] in which one of their reporters was able to get a login ID and password to the Aadhaar database by paying a mere amount of 500 Rupees through a WhatsApp group. The reporter also got the "software" that facilitated the printing of an Aadhaar card for any Indian citizen with an additional payment of 300 Rupees. Further investigation by the reporter has revealed that access to the Aadhaar database was possible via operators hired by the Ministry of Electronics and Information Technology. Both data breaches that happened with Aadhaar indicate that both human error and technical error are reasons for Aadhaar data leaks. A technically weak government website with zero to little protection has caused Aadhaar data to be leaked. Similarly, operators working under the Ministry of Electronics and Information Technology with access to UIDAI data are also responsible for leaking Aadhaar data. Hence, to completely ensure Aadhaar data protection, UIDAI and the Indian government should make sure their technical infrastructure is competitive with current cybersecurity challenges as well as instilling information security compliance among their employees.

Social Corruption

The Aadhaar program would be the largest and most far-reaching breach of personal freedom to privacy that this world has seen, especially because almost 1.4 billion of the world's population is concerned. Aadhaar was once marketed to the Indian public as a help to tackle inefficiency and the misuse of money funded with public finances. Aadhaar is now clearly just a "big data collection exercise" that enables private corporations to reach and exploit the highly confidential details of Indian people in whatever way they desire from inside Indian government frameworks.

Several recent reports and opinions shed light on the many anomalies in India in relation to the biometric data collection system. Although farfetched, much of the gathered evidence will reveal the outcomes and trends that are already obviously portrayed in the popular "Black Mirror" TV show. The contemplation of how such scenarios can be exploited in a society with many illiterate people is psychologically and morally disturbing.

Validation of Identification Against Authentication

Aadhaar is a venture for national identification, but the implicit distinction between identity verification and authorization is not well established by itself, and this contributes to misunderstandings in policy making and implementation. Before analyzing further, distinguishing between the two definitions is important.

A security principle (a person or a device) must have two separate pieces of information, "identification and authentication," according to traditional conceptions of electronic authorization when demanding access to a device. While identification gives an explanation to "who are you?," the authorization is a hurdle-response mechanism which, usually using an access code, offers "evidence of identity assertion." Login IDs, email addresses, ATMs, or smart cards are typical examples of identification. But PINs, passwords (including OTPs), and confidential encryption techniques are some common credentials for authorization.

Identification may be deemed open information, but it is important to have an authorization code in secret. A password must be available to the consumer only. In fact, authorization must be a deliberate mechanism involving active user involvement, but not strictly for person identification. For example, a bank might require an identification check before opening a bank account and, at that point, no secret such as a code is normally needed, but a customer may verify a PIN for payments such as cash withdrawals. No publicly available details will be used as authorization keys. In the Aadhaar scenario, the government has given the identification but not the authorization to citizens for their identity.

RIGHT TO PRIVACY VERSUS RIGHT TO LIFE

The primary intention of the Aadhaar biometric identity system was for seamless flow of government benefits to the intended and deserving people without any hindrance from intermediaries and to avoid cumbersome paperwork. Aadhaar was a means to make people who live by receiving government benefits much simpler and easier. However, the data breaches which happened with Aadhaar indicate that a serious privacy violation has happened to citizens of India. Citizens of India have provided their personal data as well as biometric data to get an Aadhaar card issued in their name. The only voice of dissent in the Indian Supreme Court's Aadhaar verdict of 2018 was that of Justice D.Y. Chandrachud, and he penned a 480-page opinion on Aadhaar. In his note [10], he raises concerns about how technology (in this case Aadhaar which is a digital biometric identity system) is taking over the role of the state. He voiced that "technology confronts the future of freedom itself." He continued his note by raising three important queries in the light of the Aadhaar project,

"Are there competing interests between human rights and 'welfare furthering technology' in democratic societies?" Chadrachud doubts that and continues, "Can the same technology which was supposed to bring growth opportunities cause violation of fundamental human freedom?"

> "If the answer to the first [question above] is in the affirmative, how should the balance be struck between these competing interests?"

The above-stated questions, from our perspectives, point to the question, should people forfeit their personal and biometric data to get access to basic human needs like food, shelter, and healthcare? Chadrachud also says in his dissent note that the world's largest biometric and demographic database must be examined carefully to verify its compliance with fundamental human rights which include the right to privacy and the right to life.

Integration of AI (Artificial Intelligence) and Aadhaar

In the article by Mardav Jain [4], he shared his views on the impact of AI research and how it is connected to Aadhaar. India is only behind China and the US in terms of advancements in artificial intelligence. AI has a significant role in the development of a country and can impact nearly all industries of a country including education, healthcare, manufacturing, and agriculture. In a report published by the National Institution for Transforming India, the same expert group that conceived the idea of Aadhaar, AI has the potential to add $1 trillion to India's economy if steps are taken to integrate AI in all fields.

Aadhaar database when integrated into AI will enable it with both demographic and biometric data of a billion people. In addition, there are many entities like a SIM cards, landline connections, Facebook accounts, bank accounts, food delivery apps, or movie booking apps to which an Aadhaar number is linked. AI can have access to not just the personal and biometric data of citizens of India but also to every other entity to which the Aadhaar number is linked.

With details of all Indian citizens in a central database and an AI which is integrated into it along with the numerous other entities that Aadhaar is linked to, the Indian government can easily track even the daily activities of any Indian citizen. Any malicious act can be tracked, and the suspect can easily be identified because there are fewer loopholes to avoid technological footmarks in the current fourth industrial revolutionized world. Integration of AI and Aadhaar will be mostly used in the policing front as authorities can check a citizen's activities, their money transactions, and even their location details with the help of AI-Aadhaar integration and find any patterns that can be marked as suspicious. This application will be particularly useful in reducing crime and terrorism activities but also might lead to a fully autocratic surveillance state.

In the Indian state of Punjab, police are already using AI within the Punjab Artificial Intelligence System (PAIS) which is a face recognition algorithm that is widely used to identify and catch criminals. The police officers of Punjab are already looking for ways to integrate their AI system with Aadhaar to make the system more accurate and efficient in identifying criminals. Another government-sponsored program called Crime and Criminal Tracking Network & Systems (CCTNS) holds the biometric data of all convicted people in India and system users are planning ways to integrate their system with AI for the effective and efficient criminal biometric database. They are also planning to get DNA details of criminals or convicted people of India to be added along with Aadhaar. All these are prime examples of a failed democratic society turning into an autocratic state with mass surveillance that denies basic human rights.

Supervision and Discrimination

The government of India is moving and ensuring that even more data are gathered by making the Aadhaar compulsory and finding a wider rollout for this savagery and censorship of the modern age. The government would simultaneously disregard consequential steps to breach privacy every day, all over Indian government departments for all time to come. The drive for the "key Identification" program is now underway that links all government facilities and any citizen's personal information. An Aadhaar number is now linked to all savings accounts, tax information, driving licenses, welfare assistance, and every other part of the life of an individual. The Aadhaar number will soon become the ultimate instrument for successful policy supervision. So, this system is going to be an effective instrument for ethnic cleansing in the unknown future. This notion may have been deemed ridiculous if relatable and tangible evidence had not proven the contrary. Disability critics have long argued that the Aadhaar System discriminates by neglecting disabled people's interests. In addition, the aged and elderly tend to face significant difficulties in enrolment with the consistent denial of numerous vital facilities.

By linking every dimension of a human being's life to this one number, the centralized bureaucratic structure of government can conveniently and effectively keep track of the activities of each person residing in India. Through this tracking program, if confidential information from people comes into mistaken hands, the consequence could be potentially disastrous and cause individuals and communities harm.

Significantly, the requirements of the Aadhaar Act might in the future allow the usage of DNA as "biometric identifiers." Despite today's multiple developments in genomes, a DNA-enabled Aadhaar card would create the opportunity to segregate, separate, or exclude people under any repressive future rule depending on their ethnicity and race. Even the boundaries of one's creativity might determine how to divert these instruments from original pieces to be used by someone in India to achieve twisted results, despite the centuries-old tribal divergences and racial divides. The above declaration must be valid. Every Indian is of the same population. Nonetheless, a new study has found that there are various racial and genetic differences among Indians.

These allegations may seem dark and extreme, but all the materials already exist to make these forecasts a reality. The rise of the ruthless dictator who has a burden of hate toward a type, belief, religion, or ethnicity of individuals, is all that is necessary to drive this defeatist future into our present. Under the given racial and ethnic divisions in India and the changing political environment across the world, one cannot completely exclude the possibility of an individual with a range of phobias, strong intolerances, and the power to act openly on hatred-filled thoughts to take over a region or even the nation.

History shows us plainly the crimes various dictators like Adolf Hitler committed with their agenda and tools available; some of the former might have seemed benign at first but would later become heinous and destructive. The consequences of using the Aadhaar system as a surgically accurate technique for massive tracking are clear, particularly when one is concerned about how the Aadhaar system is structured and is now being placed as necessary on a predominantly unconscious populace. An even worse example has happened in the Republic of Rwanda, where national ID was used to trigger massacres, which was believed to have killed 800,000 people in 100 days.

The Aadhaar card for a person can easily be a one-stop means for collecting and monitoring all information about any entity. If a person uses the identification system linked with her or his Aadhaar number, notifications about each operation and all the activities will be sent to various organizations. The disruptive social consequences of the Aadhaar system would be devastating when this instrument is used for sinister reasons, considering the degree to which people engage in and are tracked across any number of the following: government initiatives, utilities, or other more public practices such as railway, land ownership, and selling, as well as purchasing and banking utilities. The development and production of such an ever-growing weapon of a bad nature does not provide guarantees or protect the people and has a high price on taxpayers. From this perspective, Aadhaar does not seem to serve any human or social function. In India, there is a sluggish development in consciousness of the potential ways in which Aadhaar can be abused. Sadly, a lot of people do not realize the consequences of breaching their personal privacy, or of tracking. In the meantime, political debate on this issue has absolutely faded away as the government and big business forces have moved the issue of the Aadhaar system simultaneously from all political platforms with intense vigor.

Cultural Context: An Arrogant Threat to Privacy

Due to their cultural and social circumstances, the private space is an abstract term to be grasped by most Indians. No phrase remains for "privacy," notably in most languages of India. In a recent lingua franca article on privacy, perforce, terms that mimic and even cover elements of "secrecy" are pervasive in India. It is a necessity: since privacy does not correspond to semantic anonymity, no one can linguistically synonymize it. The definition of privacy is primarily extra-terrestrial for several Indians in the way it is actively known in the West. They do not really attempt to find out what privacy can mean, and how loss of privacy might be dangerous for them both financially and culturally in a global setting.

Historically in India, citizens had huge families and stayed together. Family concepts became a pattern where parents lived under one roof with several kids, their in-laws, and grandkids. One explanation for this intergenerational life relates first to the economy, whereby a large family was or is unable to buy or rent a bigger house. Consequently, most families value the close relationship between representatives. With such an atmosphere where visitors and relatives can come and private conversations are held, sometimes one's own living room is the only space available for such an activity. Newly married couples have also realized that sharing a room with family members is necessary.

To understand this, there are 4.91 citizens per Indian householder on average, compared to an average of 2.64 individuals per US home. A sealed system in India has not only been used to inculcate but has also served to solidify some unfamiliarity or perceptiveness toward a more global conceptualization of private life and the wider social standards of privacy around the globe. The aboriginal cultural structures have also helped many of those in India to visibly express an utter disrespect to and for the "requirement to value the space of the human." To put it another way, India's collective mindset has conditioned the people to agree that people do not have privacy. This is not surprising because sociologists have established that the opinions of the people regarding "privacy" usually constitute a direct result of those conservative values and cultural norms that they carry and consciously display in everyday life. Clearly speaking, the inner battle of India to advance into the new world in every aspect of society is made more difficult by deeply ingrained cultural values and customs, particularly in the domain of economics. If India wants to be a powerful economic member of the international community, there must be a harmonization of individual opinions, societal perceptions, and national structures for the preservation of privacy, security, and reintegration processes for Indian citizens despite the divided democratic and socio-economic structure of Indian states and their corresponding institutional systems. Such pillars are extremely critical and essential to allow Indians as well to thrive as global customers.

LESSONS LEARNED

Aadhaar is a well-meaning government program that has gone wrong. What the government conceived as an idea with the good intention to decrease the cumbersome bureaucracy and intermediaries is now becoming a threat to data privacy and personal data protection of all Indian citizens which is an infringement of basic human rights.

The government has made Aadhaar mandatory to receive food, shelter, and healthcare benefits. Even to get admission to a government-run school, authorities require Aadhaar as proof of identity. Food, education, shelter, and healthcare are basic human rights irrespective of any country and socioeconomic-cultural differences. UIDAI and the Indian government have made it compulsory for everyone to use it as proof of identity for all Indian residents. This resulted in Aadhaar having the power over an Indian citizen's access to basic government services, mobility, and occupation, and even has a significant impact on their identity as a full Indian citizen. Many cases report that the Aadhaar biometric digital identity system has led to exclusion and discrimination affecting the most vulnerable – people who are unable to enroll in this program or people who have lost their biometric identity due to aging or accidents. Although these issues are not linked to IT security and risk management, biometric identity cannot be made as the only proof of identity or for verification in a digital identity system.

The myriad of security issues pertaining to Aadhaar has made the biometric data of most Indian citizens vulnerable to a variety of identity thefts and other abuses both internally and externally. As discussed earlier, the Analysis part of this research case reports on the potential risks of mass surveillance on Indian citizens under the Aadhaar system which is an infringement of constitutional rights as well as basic human rights. The integration of Aadhaar and AI where biometric and demographic data in Aadhaar is fed to state-controlled machine learning programs is also a major concern that experts say will result in the denial of basic human rights and profiling of Indian citizens. There is no doubt that the Indian government is finding more ways to expand policing and surveillance activities over their citizens even if doing them means going against democratic values.

If considering Aadhaar as an IT project in an organization, there are a few points one can learn from security breach incidents which happened with Aadhaar. [11, 12]

Do not Ignore Fundamentals

Being passionate about studying cutting-edge protection techniques is fine, but do not forget the fundamentals. Although not everyone is technologically inclined, a bit of preparation will go a fair way toward alerting the staff to cyberthreats and data breaches. To improve awareness, have regular meetings with consumers and solicit input about the compliance steps the business is implementing to prevent risks that might contribute to data loss.

Additional safety bases should be taken such as establishing clear password policies, periodically updating fixes, allowing authentication of two variables, and tracking and enforcing access controls. Goofing in any of these important facets of defense may have far-reaching implications.

Engage Top Management

Leaving all the burden to the users is not appropriate, and even IT still requires assistance from top management. High-level management must take security very seriously and actively promote the implementation of existing safety guidelines and regulations. A top-down strategy would allow staff to incorporate security measures into their routine activities.

Good interaction with top management is crucial when it is time to upgrade cybersecurity technologies. Maintaining the connection now will make any increase in communication and utilization simpler in the future to apply for cybersecurity investments.

Improve Security at the Endpoint

In an age of remote gadgets, connecting to the network with protection at endpoints is much more critical now than before. Not only can attackers reach the network via conventional endpoints, such as PCs, but they can also use unpretentious tools, such as the copy machine. Output devices such as printers are part of that system and almost always do not require access authorization, although they can preserve information of sensitive business records. This lack of authorization leaves preserved sensitive data vulnerable and exposed to threats if not adequately covered. Every minor mistake here could be a possible vulnerable link in the overall defense of one's network.

Investing in machines that upon purchase have built-in security features is better, such as printers that instantly auto-heal from threats by activating a reset. This feature protects these kinds of endpoints. The government can also partner up with a supplier of Device as a Service (DaaS) to rent gadgets rather than purchase them. With a trustworthy provider willing to track government machines and upgrade them with the latest security updates and with much minimal hassle from the government side, the government can make their system more secure.

Access Reviews Are Key

The largest disclosures connected with Aadhaar have emerged from insiders up to this level. Restricting access to confidential information on a "Necessity to Know Only" condition is therefore essential. As stated by an Indian media, the sale of Aadhaar website accessibility was found by a village development organization, whose function had become redundant after the government canceled their licenses to register fresh applicants for Aadhaar. Breaches such as this further reinforce the necessity of regular upgrading of access privileges in accordance with the progression of the internal responsibilities of network members.

Compliance and Safety Education of Associated Officials

The Aadhaar breaches also emphasize the significance of educating authorities on effective information management. This will include developing a comprehensive enforcement framework for the protection of data and guaranteeing that the parties involved are informed on the framework.

Learn from Hackers

The Aadhaar program's survival relies on UIDAI's rapid reaction toward recent infringements, thereby improving its protections to deter potential accidents. Data protection is always an ongoing war which means the construction of an infallible program is unlikely. Nonetheless, the likelihood of compromises can be reduced by setting up a program for constant tracking and training, to defend against a changing security risk environment.

Apart from the above points discussed, there are a few other points recommended by Marzene Fuller who is the chief security officer at Signal Fx [13]. These points can be included in best practices for better IT security.

First, security requires accountability at the management level, cooperation from the full management committee, and sufficient staff numbers and budgets. The following procedures will be introduced after these technical criteria are fulfilled to resolve the underlying causes of data loss types mentioned above.

- Effective design lifecycle where architecture and safety work seamlessly toward the same goal
- Risk analysis to recognize main properties, hazards, attack particles, and potential mitigations
- An architectural review board including examination of security architectures against negotiated security criteria and best practices
- Use of OWASP stable coding processes and correct checklists

- The study of static and dynamic code
- Daily pen-tests by an unbiased foreign entity

The concerns regarding the protection of Aadhaar should be an insight or wake-up call for IT administrators everywhere. A robust data management strategy is expected to improve network security and prevent breaches. Now is the time to take these steps in every organization, so that they can secure the most important assets and information of their firm.

DISCUSSION QUESTIONS

1. What are the differences between when a private organization leaks sensitive data and when a government does?
2. Is the "Right to Privacy" a democratic value or a basic human right? Is the Aadhaar biometric identity system a violation of the "Right to Privacy"?
3. Is Aadhaar making Indian citizens trade their "Right to Privacy" for survival?
4. Is Aadhaar making it easier for the government to hold unjust power to surveil its citizens?

REFERENCES

[1] AlKalbani, A., Deng, H., & Kam, B. (2015, July). Organisational security culture and information security compliance for e-government development: The moderating effect of social pressure. In *Pacific Asia Conference on Information Systems (PACIS)* (p. 65).
[2] Choi, J., Ae Chun, S., Kim, D. H., & Keromytis, A. (2013, June). SecureGov: Secure data sharing for government services. In *Proceedings of the 14th Annual International Conference on Digital Government Research* (pp. 127–135).
[3] Dixon, P. (2017). A Failure to "Do No Harm"–India's Aadhaar biometric ID program and its inability to protect privacy in relation to measures in Europe and the US. *Health and Technology*, *7*(4), 539–567.
[4] https://data.worldbank.org/indicator/SP.POP.TOTL?locations=IN
[5] https://uidai.gov.in/ml/about-uidai-ml/unique-identification-authority-of-india-ml/vision-mission-ml.html
[6] https://jsis.washington.edu/news/the-aadhaar-card-cybersecurity-issues-with-indias-biometric-experiment/
[7] https://techcrunch.com/2019/01/31/aadhaar-data-leak/
[8] https://www.business-standard.com/article/economy-policy/aadhaar-verdict-can-you-sue-uidai-and-its-licensed-entities-for-offence-118092800250_1.html
[9] https://www.tribuneindia.com/news/archive/nation/rs-500-10-minutes-and-you-have-access-to-billion-aadhaar-details-523361
[10] https://theprint.in/india/governance/our-privacy-cant-be-stripped-for-right-to-life-what-lone-dissenting-judge-said-on-aadhaar-act/125378/
[11] https://www8.hp.com/in/en/tektonika/index.php/2018/05/17/data-security-lessons-from-the-aadhaar-security-breach/
[12] https://procivis.ch/2018/02/13/lessons-from-the-worlds-largest-e-identity-program-indias-aadhaar/
[13] https://www.darkreading.com/abtv/vulnerability/lessons-learned-from-2018-security-breaches/a/d-id/749478

Section 3

Process Cases

Teaching Cases

3.1 NHS

COVID-19 Research Ethics and Governance

Roydon Dominic Correya
MYOB, Christchurch, New Zealand

Tushar Ashok Sagathia
ASB Bank, Auckland, New Zealand

CASE SUMMARY

In the wake of the current COVID-19 pandemic caused by the Novel Coronavirus SARS-CoV-2, many countries have adopted various strategies to control the spread of this virus and to efficiently manage the economic and humanitarian crisis caused by this virus. The pandemic has forced many countries to take drastic measures to keep their populations safe, ranging from public lockdowns, mass testing, spreading public awareness of hygienic practices such as the use of face masks and sanitizers, and practicing social distancing to keep people safe. Governments around the world have also used technology to a large extent to tackle this new threat to their countries. This is an alarming situation and people need to be more cautious about their rights and liberties in times of crisis like this. The possibility of the governments taking advantage of this situation to carry out otherwise considered unethical activities is great. According to Kolstoe (2020), the confidential medical data of people in the United Kingdom (UK) is being used by researchers without their consent if they had contacted the National Health Service (NHS) for any COVID-19-related health concerns. The data being used for research may include the medical records of the patients and any other data which the healthcare providers may have collected concerning the patients' care.

This data is being used by researchers in universities, government, and private companies to learn more about the way the virus spreads and is transmitted and to develop a potential vaccine or find the best treatment options for infected patients. This seems like a necessary action required in times of a crisis such as the current pandemic but the data being shared for research is medical data which otherwise is treated as confidential. Many people want to keep their medical information private and do not wish to share it with others.

DOI: 10.1201/9781003264415-34

The two main laws that deal with the privacy of healthcare data in the UK are the Europe-wide General Data Protection Regulation (GDPR) and UK-specific Common Law Duty of Confidentiality (CLDC). In this pandemic, the second UK-specific law has been relaxed to a certain extent and the authority of an important review body has been suspended. The UK Data Protection Act 2018 turns the principles of GDPR into law. It also provides details of how personal data should be handled and safeguarded to prevent misuse. This requires the consent of the individual and it's also legal under GDPR to access personal data without consent if an alternative legal basis is declared. The additional UK-specific CLDC requires the consent of the individuals for their data to be used for research purposes and is based on a 'no surprises' philosophy where a patient should not be surprised by how their data is being used. The law also recognizes that under special circumstances confidential patient information can be shared without consent with the approval of the Secretary of State for Health and Social Care under a section of the Health Act 2006. The approval is based on advice from a special committee called the Confidentiality Advisory Group (CAG). This committee reviews the applications submitted by researchers to access patient data without consent, the committee doesn't itself approve applications but provides detailed advice that normally must be followed for the approval of applications by the Secretary of State for Health and Social Care.

Simon Kolstoe, a research committee chair of the CAG which reviews the research applications, mentions concerns relating to the access of patient information to researchers. The authority of the CAG has been replaced by informal advice which does not necessarily need to be followed to review research applications for use of patient information without consent. Consequently, many researchers are 'covidizing' their research projects to access confidential patient information. This data is being shared even if the individual has requested to use the National data opt-out service in the past, which mirrors the GDPR's 'right to erasure' also known as the 'right to be forgotten'.

CASE ANALYSIS

The NHS

The National Health Service (NHS) is the publicly funded healthcare system in the United Kingdom (UK). The NHS has been funded by general taxation since after the Second World War to provide free healthcare to the people residing in the UK. Until 2013, the NHS in England was accountable to the political body of the country (UK Government) but was passed to NHS England, an executive non-departmental public body (NDPB) of the Department of Health and Social Care (Wikipedia). The new NDPB status of the NHS makes it a quasi-autonomous non-governmental organization (quango) which gives it autonomy over the execution of policies and distances itself from the government. The idea of this publicly funded system would be appealing to many when healthcare in competitive markets is extremely expensive such as in the United States of America, but this system can have its flaws. Giving it autonomy and making it less accountable by distancing itself from the government can have some disastrous consequences. As discussed in the case summary, Kolstoe (2020) discusses how confidential patient information is being accessed without consent by researchers bypassing the proper review of the CAG.

Considering the current crisis, when there is not much data available about the behavior of the virus this might seem necessary, whereas Zwitter and Gstrein (2020) warn us about the very strong temptation of doing 'whatever is necessary' in this pandemic. COVID-19 requires comprehensive measures but that does not mean anything goes. The authors also mention the importance of the contextual nature of data, that the same datasets can be sensitive in different contexts. They also emphasize the need for appropriate governance frameworks to make sure this data is being stored, analyzed, and shared legitimately and responsibly. In turn, the opposite of this can be seen in the case of the NHS where instead of adopting new and better frameworks suited for times of crisis like these, the existing frameworks are being weakened.

Privacy Laws in the UK

The Data Protection Act 2018 (PDA) in the UK turns the principles of the European Union's GDPR into law, providing details of how personal data should be handled and safeguarded to prevent misuse. The PDA includes a list of possibilities for the lawful basis of storing and processing an individual's personal information which includes consent, contract, legal obligation, vital interests, public tasks, and legitimate interests (Kolstoe, 2020). Hence, requires the consent of the individual to access any personal information.

The existing framework of the UK-specific CLDC required the CAG committee to review the researchers' access to patient information, which has now been changed and is not legally required to be followed. This weakens the system and concerns are raised over the recent 'covidizing' of research projects and rapid construction of projects made recently and prioritized because of COVID-19. This is against the 'no surprises' philosophy on which this law is based, where an individual should not be surprised by the way in which their personal information is being used without their consent. The loopholes in the regulatory frameworks are exploited and everything is disguised under the utilitarian mindset of working for the 'greater good'.

Ethics in a Pandemic

A Utilitarian mindset of doing things for the 'greater good' in this pandemic can prove fatal to people's liberties and freedom. Zwitter and Gstrein (2020) suggest that any data is contextual and can be sensitive in different situations; one should not forget that the unconsented access to data given here is confidential patient information. In any context, medical records and accompanying patient profiles are considered sensitive data and can be harmful if misused. The authors also discuss the sharing of location data by the government which then developed applications for tracing infected patients and alerting other people. This sharing of location data was enforced in the People's Republic of China, and in other countries such as India, the people willingly participated in this endeavor. The whole act of giving away sensitive data is a major threat to the liberties and freedom of individuals in a nation and can weaken a democratic system.

The use of AI surveillance during the pandemic has also been observed which affects the privacy of an individual. The AI-based technologies have much to offer in combating COVID-19 but also pose a risk to privacy (Shachar et al., 2020). The authors stress the fact that in times of a health emergency such as the ongoing pandemic the authorities and private actors must not only comply with the existing regulatory frameworks but also consider the broader privacy questions of the individuals in society. Privacy violations are unlikely to be remedied in the aftermath of a pandemic. This can lead to a totalitarian state which enforces its rule using AI and technologies, wherein the state monitors people, enforces lockdowns or restrictions for the prevention of disease, and or isolates possible carriers of disease in this or a future pandemic event. This kind of totalitarian rule also leads to inequality as seen in the case of Jew Ho v. Williamson, where a federal court struck down a quarantine because it was racially motivated. People whose liberty is restricted whether by AI, technologies, or policies are entitled to judicial reviews, and when governments enforce quarantines they must meet individuals' basic requirements of healthcare, food, and sanitation (Parmet & Sinha, 2020).

Existing Loopholes

The loopholes in the CLDC and the Health Act 2006 seem to be already exploited for gaining unconsented access to patient information as discussed in the above sections, but this looks like a common theme or practice for avoiding data privacy laws. Gutmann and Warner (2019) address this in their paper, 'Fight to be Forgotten' wherein the authors discuss how privacy is threatened by decommissioned storage devices as the data in the devices can be retrieved by simple forensic analysis tools.

The foundational concept behind this paper is the way data is stored on physical devices, in the form of binary numbers (0's and 1's). Those who mine for personal data can scan the physical memory of the device by using forensic tools and can reconstruct the deleted data, which is a breach of privacy of the user's data without their knowledge. The authors mainly identified the inconsistencies between the language used by two popular operating systems in the terms 'delete' and 'erase', which do not have the same meanings and can mislead the users to believe the data is erased when it is only deleted. Deleted data can be recovered by using software tools whereas one can follow practices such as overwriting the physical memory of the device with scrap data to make sure the original data has been truly erased from the physical memory of the device.

On a similar note of data privacy and the GDPR, Shintre et al. (2019) discuss in their paper titled, 'Making Machine Learning Forget' how machine learning models generally overfit data while training and are indirect stores of personal data of individuals. This data can be retrieved with inference attacks on these indirect stores of data, which can compromise the privacy of an individual. The authors suggest that the GDPR Article 17 (Right to be Forgotten) should extend to data stored in machine learning models or such indirect sources.

In conclusion, there are loopholes in the current system which can be exploited to compromise data privacy which in turn compromises the liberties and freedom in a democratic society. The NHS case study summary and analysis is an example of privacy breach; governments must improve the current system and laws whereby authorities and private players accept more ethical responsibility for the storage and processing of data using the regulatory frameworks.

LESSONS LEARNED FOR ETHICS AND GOVERNANCE

This NHS case study teaches important ethics and governance lessons for exploring how the current systems and laws are lacking and what measures can be taken to improve these current regulatory frameworks.

Shintre et al., (2019) show that data can be extracted from indirect data sources such as machine learning models. Hence the definition or the scope of the regulations should be broadened to make sure all instances of data and devices/software containing data are included under these regulatory frameworks, leaving less chance for finding loopholes or risking the privacy of user data. A similar measure of broadening the scope of the laws can be applied to the issue addressed by Gutmann and Warner (2019) concerning data which can be extracted from decommissioned devices using simple forensic analysis tools. The authors suggest that the laws and regulations can mandate an ethical duty of operating system manufacturers to use the right terminology or create awareness about how best practices can be used to make sure the devices can be disposed of securely.

An additional lesson learned is that information systems can be designed to enhance privacy as Bincoletto (2019) explains by creating an Electronic Health Record System (EHR) based on the regulations so that the risk is mitigated in the initial stages. Another lesson would be to speed up the adoption of existing emerging technologies such as 'Blockchain' as it can provide us with the features of security, transparency, and immutability – all highly desirable in health information systems at this time. ZHANG et al. (2018) designed a blockchain-based Precision HealthCare system which shows how confidential patient information such as Genetic Data can be stored and processed securely. This shows us that technological viability exists today but there is no mass awareness or adoption of this by the authorities or private players. These are emerging technologies and can be considered still a work in progress, but the fundamental adoption of these and working from there should be necessary for protecting the liberties of individuals in today's world.

Zwitter and Gstrein (2020) also suggest taking guidance from humanitarian organizations such as The International Committee of the Red Cross and The Dutch Red Cross as they have developed guides for data legislation where legal frameworks fail. Their research also suggests referring to institutions such as the International Covenant on Civil and Political Rights, the European Convention on Human Rights, and the legal and institutional framework designed by the Council of Europe, as these Constitutions and human rights are designed by keeping crises like the current pandemic in mind. Shachar et al. (2020) suggest that the data used can be deidentified by using proper mechanisms or using anonymized data. Another group of researchers Qadir et al. (2016) discuss the use of big data and analytics in the situation of a crisis and define how a data exhaust is formed like a vehicle exhaust pipe which emits traces of our online digital activity. These suggestions can be effectively introduced in the current regulatory framework and in future scenarios.

DISCUSSION QUESTIONS

1. Is it okay to share all personal information for the greater good?
2. Can governments change policies that decide the liberties of individuals under any circumstances?
3. Would you be comfortable with sharing your personal medical data?
4. Can we guard our data from being misused by the governments and how?
5. Is what is legal and what is ethical? Is there a difference? Who decides?
6. How important is medical data to you? How important is confidentiality?

REFERENCES

Bincoletto, G. (2019). A data protection by design model for privacy management in electronic health records. In *Privacy Technologies and Policy* (pp. 161–181). https://doi.org/10.1007/978-3-030-21752-5_11

Gutmann, A., & Warner, M. (2019). Fight to Be Forgotten: Exploring the efficacy of data erasure in popular operating systems. In *Privacy Technologies and Policy* (pp. 45–58). https://doi.org/10.1007/978-3-030-21752-5_4

Kolstoe, S. (2020). Coronavirus: researchers no longer need consent to access your medical records. The Conversation. https://theconversation.com/coronavirus-researchers-no-longer-need-consent-to-access-your-medical-records-138567

Parmet, W. E., & Sinha, M. S. (2020). Covid-19 — The law and limits of quarantine. *The New England Journal of Medicine*, *382*(15), e28.

Qadir, J., Ali, A., Rasool, R.U., Zwitter, A., Sathiaseelan, A., & Crowcroft, J. (2016). Crisis analytics: Big data-driven crisis response.

Shachar, C., Gerke, S., & Adashi, E. Y. (2020). AI surveillance during pandemics: Ethical Implementation Imperatives. https://pubmed.ncbi.nlm.nih.gov/32596887/

Shintre, S., Roundy, K. A., & Dhaliwal, J. (2019). Making machine learning forget. In *Privacy Technologies and Policy* (pp. 72–83). https://doi.org/10.1007/978-3-030-21752-5_6 Wikipedia. National Health Service. https://en.wikipedia.org/wiki/National_Health_Service

Zhang, X., Sharma, R., & Wingreen, S. (2018). Block-chaining in Precision HealthCare: A design research approach. https://aisel.aisnet.org/pacis2018/253/

Zwitter, A., & Gstrein, O. J. (2020). Big data, privacy and COVID-19 – Learning from humanitarian expertise in data protection. *Journal of International Humanitarian Action*, 5(1). https://doi.org/10.1186/s41018-020-00072-6

3.2 Marriot Data Breach
A Case Study Analysis

Areeb Amanulla
Kainga Ora – Homes and Communities, New Zealand

Saadik Niyaz
Waste Management NZ, Ltd, New Zealand

CASE SUMMARY

Marriott International is one of the main leisure and tourism corporations around the globe, functioning in over 6500 sites across the world presently. Marriott International controls many of the globally renewed hotel chains such as "Sheraton Hotels and Resorts", "The Ritz Carlton", and "Westin Hotels and Resorts". The leisure and tourism giant makes over $20 billion in annual revenue.

In 2018, Marriott International made a public announcement stating that there had been unauthorized access to their system. This system had sensitive information of approximately 500 million customers. As a response to this breach, Marriott International hired an expert consultancy team to identify the approach of the breach and what information access the hackers gained. After their investigations had commenced, the consultancy team identified that individuals had unauthorized access as early as 2014, which was four years before the present breach was identified. The investigation also identified that the hackers had gained access to various passport details and credit card information, email addresses, passwords, departure, and arrival dates of customers.

Apart from the loss of secure data, the immediate worry for Marriott International was to identify the reason the breach went undetected for four years until they had finally identified it. As one of the leisure and tourism giants in the world, they should have had the best technology and toughest security mechanism in place. Furthermore, the investigation revealed that Marriott International was the victim of an attack made through an SQL injection (cf. Ruben, 2018; Sucuri, 2018). After deep analysis, this common web hacking technique injection was probably why these data breaches went undetected for such a long period of time.

Additionally, the investigation also identified that this particular data breach had only taken place for the chain of hotels at "Starwood" which was a hotel and resort chain that was just acquired in 2014 by Marriot International. This was one major acquisition out of the many merges that formed the world's largest hotel chain which

DOI: 10.1201/9781003264415-35

exists at present. Moreover, the analysis identified a certain pattern which gave the management answers to certain important questions to which they were trying to make sense, for example, two months after the merger with Starwood Hotels and Resorts, the guest reservation database of Starwood had a weakness which provided a pathway for unauthorized access. These early breaches were overlooked by the newly merged team at Marriott International.

Marriot and its investigative team recognized that the technology used by "Starwood" was the main cause for this breach, as their reservation system did not have sufficient security features, and their system also had outdated versions of Windows Server on their personal computer systems. Furthermore, Starwood had multiple databases – the reservation database, the guest member database, and the data warehouse. Also, Marriott International allowed remote access to their systems through the open internet.

These major security failures of Marriot International and Starwood's information systems allowed the hackers to access the database for over four years through SQL injections while being undetected. The attackers accessed the servers remotely and then installed a remote access trojan. The remote access trojan helped cover the tracks of the attackers by encrypting and deleting stolen data.

CRITICAL ANALYSIS

Timeline of the Breach

During the in-depth investigation, quite a lot of insights were revealed including a series of events that occurred prior to the discovery of the final data breach of 2018. Summarized below are the major phases discovered in this data breach.

- **July 2014**: The attackers penetrate the system of Starwood Hotels and Resorts.
- **September 23, 2014**: The acquisition of Starwood Hotels and Resorts is officially completed by Marriott International.
- **September 7, 2018**: A contractor named Accenture who maintains and manages the databases of Starwood Hotels and Resorts for Marriott International notices an unusual SQL query, which is then alerted.
- **September 8, 2018**: Accenture officially notifies the breach to Marriott International.
- **September 10, 2018**: Marriott International hires an expert external investigation team of consultants that can help identify the scope of the data breach.
- **September 17, 2018**: A Remote Access Trojan (RAT) which was installed in the system of Starwood Hotels and Resorts is identified by the investigators.
- **October 2018**: During the investigation, the investigators identified that a tool commonly used to hack passwords of computers known as "Mimikatz" was in the Starwood system.
- **November 2018**: Further revelations through the investigation demonstrated that the activity of unauthorized hackers has been taking place since July 2014.

- **November 13, 2018**: Moreover, the team of consultants identified that two encrypted, compressed document files were previously removed permanently from the system by the attackers.
- **November 19, 2018**: The discovered files are decrypted and identified to be files that were also part of the data breach.
- **November 26 and 26, 2018**: Investigators revealed that two files from the guest reservation database of Starwood Hotels and Resorts were copied between 2015–2016. These files could not be recovered.
- **November 29, 2018**: Marriott International informs the Federal Bureau of Investigation, information security regulators, credit card companies, and other related authorities about the data breach. They issued a public statement explaining the details of the data breach.

Technology Used Which Led to the Breach

The focus of the case analysis shifts to the technology used by Starwood Hotels and Resorts and Marriott International which led to the data breach. First, the systems of Starwood Hotels and Resorts had an out-of-date version of Windows Server. A Windows Server is defined as a system operational server that provides a processor the ability to take upon various network roles such as domain control, file server, and web server. It can also be utilized as a distinct platform to initiate server applications such as SQL servers (Techsoup, 2019). The issues related to an out-of-date Windows Server affecting entities are summarized below.

Increased Security Vulnerabilities

After a Windows Server has exceeded its restitution date it does not get any client support or updates from its facility provider. A non-updated Windows Server is exposed to the incapability of counter-attacking the newest forms of cyber attacks produced by the distribution of ransomware. This permits hackers to simply evade even the toughest firewalls and anti-virus software. Therefore, as a result, a non-updated Windows Server permits hackers the capacity to get into the system and monitor events, access applications, and duplicate data (Roos, 2019).

Noncompliance with Regulations

The Privacy Act of 1988 entails entities that collect, use, and store consumer data to meet firm data standards to ensure consumer privacy and integrity are not conceded. To certify compliance all entities must ensure that they accumulate data in a protected platform. Failing to upgrade the data servers could result in a breach of the guidelines as the platform is no longer protected (Messageops, 2019).

Higher Operating Costs

Even though entities are reluctant to upgrade software as a method to bring down costs, this is not constantly the circumstance with Windows Server. As a result of out-of-date software, the systems crash much more often than they would normally, which as a result needs more attention from the IT support team. Moreover, if the IT support team is subcontracted, the cost will be even more than usual, as IT support entities charge higher markups to uphold outdated software purely because it needs more attention (Roos, 2019).

Furthermore, investigators understood that the attackers used an SQL injection technique to hack Marriot International's system. SQL injection, a very common web hacking technique (Imperva, 2019; Ruben, 2018; Sucuri, 2018), is a method used by attackers where a malicious code is applied in SQL statements to obtain access to a database.

LESSONS LEARNED FROM THE MARRIOT INTERNATIONAL DATA BREACH

The data breach that occurred at Marriott International gave them an unwanted record, of being the organization which had the world's second-largest data breach in history (behind the Yahoo data breach). Among the numerous, a few key learnings that one can take out of this historical breach are summarized below.

Cyber Insurance

The investigation at Marriott International took place over a time period of three months from its inception to the end. A major part of this analysis had to be devoted to the effort of analyzing a data log of over four years. However, once the analysis was accomplished, Marriott International had to incur a cost of $71 million for the consultancy company that carried out the investigation. Even though it was an enormous sum of payment, Marriott International had to only bear a cost of $1 million because they were able to get a cyber-insurance entitlement of $71 million.

A key experience to be learned from this event is the importance of having cyber insurance, which many entities ignore at present. Insured as they were, Marriott International only had to pay a small percentage of the total costs incurred for the investigations carried out. This clearly highlights the value of having cyber-security insurance. The top management of Marriott International who spoke to the media concerning the breach praised the importance of cyber-security insurance but also stated that acquiring it will likely be inaccessible in the future.

Data and Network Security

Inappropriate data management practices at Starwood and Marriott International were a reason for the data breach to take place. The breached data contained information on 5.25 million passport records as well as credit card data for all the customers of Starwood Hotels and Resorts. The data was very sensitive and should not have been stored unless it was secured via encryption. However, subsequently Marriott International publicized that going forward, it plans to store passport and credit card data in an encrypted form. They further specified that sensitive data will be stored at the location of each hotel or resort to safeguard and minimize the impact of future breaches (cf. Johnston, 2019).

Further review revealed that technical resolutions caused an installation of 70,000 end-point defenses on the network of Starwood Hotels and Resorts. Moreover, this also resulted in a fast-tracked arrangement of an additional 200,000 end-point protection devices on all company networks. Marriott International also introduced IP whitelisting for all databases that were breached through the employment of network segmentation which will safeguard sensitive data.

Ideally, these resolute actions taken by the management of Marriott International should have already been introduced prior to such an attack. This highlights the importance of Network segmentation; this can be used for compliance with PCI-DSS, which is the standard that oversees companies collecting, using, and storing payment card information. Marriot International's experience in this case underlines the importance of complying with such standards.

DISCUSSION QUESTIONS

1. Should Marriott International have gotten the help of an expert consultancy firm to perform a safety check on the Starwood systems before the merger went through?
2. How was Marriott International's response to the data breach sufficient, and how did their actions going forward ensure a secure or not secure enough system?

REFERENCES

Imperva. (2019). Structured Query Language (SQL) Injection. Imperva. Retrieved from https://www.imperva.com/learn/application-security/sql-injection-sqli/

Johnston, R. (2019). The Pros and Cons of a Single Database for a multi-site WMS implementation. Sight. Retrieved from https://www.go4sight.com/blog/pros-and-cons-single-database-multi-site-wms-implementation

Messageops. (2019). If youre using outdated servers what does that mean for your security. Message Ops. Retrieved from https://messageops.com/if-youre-using-outdated-servers-what-does-that-mean-for-your-security/

Roos, A. (2019). Is outdated Microsoft software putting your business at risk. Empower IT. Retrieved form https://www.empowerit.com.au/blog/software-applications/outdated-microsoft-software/

Ruben, P. (2018). How to prevent SQL injection attacks. Esecurity planet. Retrieved from https://www.esecurityplanet.com/threats/how-to-prevent-sql-injection-attacks.html

Sucuri. (2018). What is an SQL injection attack. Sucuri. Retrieved from https://sucuri.net/guides/what-is-sql-injection/

Techsoup. (2019). Microsoft windows server overview. Techsoup New Zealand. Retrieved from https://www.techsoup.net.nz/help/microsoft-windows-server-overview

3.3 ChoicePoint

Huixin Zhang and Zuejie Tai
University of Canterbury, Christchurch, New Zealand

CASE SUMMARY

Headquartered in Atlanta, ChoicePoint is a data economy company which focuses on providing sales information and data services to the insurance industry, enterprises, government agencies, and direct sellers. The company claims to be not only the main source of information for the US insurance industry but also can identify, store, analyze, and transmit relevant data. Since the company is not legally regarded as a financial services company, it is not subject to relevant data laws (Culnan & Williams, 2009).

To expand the company's scale and increase its sales, ChoicePoint obtained data from various third parties, which include public records, insurance claims data, and credit reports. Almost all the information of American adults is collected and stored by ChoicePoint. This data includes more than 19 billion pieces of information. Even though the company is not subject to legal restrictions, ChoicePoint, having such a large amount of private information, should use advanced technology to protect data. Interestingly, the theft of ChoicePoint's data discussed in this case is not the result of computer hacking but thieves' purposeful misrepresentation of business license and identity to steal data.

Located in Hollywood California, Kinko's copy center has a business license and can apply for personal data from ChoicePoint. Olatunji Oluwatosin, a Nigerian who claimed to be the collection agent of Kinko's copy center, is one of such thieves referenced above. ChoicePoint's business policy requires customers to upload a copy of their business license, but the company usually does not verify the identity of the applicant to ensure its legitimacy. Olatunji Oluwatosin applied to ChoicePoint for permission to download personal reports. After submitting a fake business license, he downloaded personal reports of individuals at $15 each. In October 2004, Oluwatosin was arrested and charged with identity theft. He admitted that he had obtained the private data of more than 1,000 people.

It was not until 2005 that ChoicePoint was forced by Senate Bill 1386 and public opinion to notify the victims. Initially, ChoicePoint only notified victims in California and deliberately concealed the breach from victims in other states. To protect the rights of citizens, the attorneys general of 18 other states also required ChoicePoint to make public disclosures. So far, 145,000 victims have received notification from ChoicePoint that some of their personal information may have been stolen. The

DOI: 10.1201/9781003264415-36

notice signed by Derek Smith, the CEO of ChoicePoint, only described in detail the process by which the company discovered that the data had been breached and did not provide victims with potentially stolen information (Culnan & Williams, 2009).

Before 2005, the company had discovered that other thieves used previously stolen personal data to create a company and opened at least 50 private accounts. These personal data include name, address, social security number, and credit report. However, ChoicePoint not only failed to disclose it privately to the government and law enforcement officials, but it also did not disclose the news to the public. ChoicePoint was rightly subject to a class action lawsuit, even though ChoicePoint publicly apologized on its website and changed the company's operating strategy. This is not the first time ChoicePoint's review policy has failed. In 2000, identity thieves used forged documents to impersonate legitimate customers, but ChoicePoint did not respond to this.

This series of private information leakage incidents resulted in ChoicePoint being ordered to compensate the victims 15 million US dollars and 15 million fines. The Federal Trade Commission (FTC) requires the company to conduct an independent assessment every two years and provide a copy for review. In addition, ChoicePoint is restricted in the types of information it sells and the types of customers. This chapter presents a critical analysis of ChoicePoint and its major leaks as well as lessons for the company in terms of two topics: breach laws, and the ethics and morals of information.

CRITICAL ANALYSIS

Breach Laws

Since 2001, cases of data theft due to security breaches have been increasing. In addition to the customer's name and phone number, the stolen data also included sensitive information such as driver's license number, consumption records, bank account numbers, and balances. One example of a security breach is from 2001, when the cashier at Bank One provided the thief with account information and social security numbers for at least 250 people. The bank fired the cashier but did not notify the victims until the first victim appeared. The same thing happened in 2005. As a major provider of legal and consumer research information, LexisNexis lost nearly 30,000 pieces of private data. More and more security experts realize that businesses need to implement higher levels of data security.

In 2003, California Senate Bill 1386 (S.B. 1386) required companies to immediately notify California residents when it was found that the security of personal data was violated. The bill applies to any business that has personal records of California residents, even if the business is not in the state. Companies that fail to comply with the law may receive civil lawsuits, including class actions. The federal government's policy encourages industry organizations to conduct self-regulation, which includes companies' responsibility to publicly disclose the arrival of internal data security issues to the government and law enforcement officials, not secretly as these companies may prefer.

The legislative intent of S.B. 1386 is to protect the security of consumer privacy information. Companies or institutions that store consumer private information must immediately disclose all violations of information security to victims to protect consumers' property safety. Victims have the right to file a civil lawsuit against companies and institutions that violate this clause and demand compensation for their losses. The bill is aimed at all businesses that store the private information of California residents, which includes small, medium, and large companies regardless of the location of the company. In addition, the bill also includes many positive steps that companies can take to avoid serious non-compliance consequences (Tinseth & Richlin, 2005).

In this case, after ChoicePoint learned that other thieves illegally used stolen information to create fake companies and obtain benefits, the company should have complied with the requirements of S.B. 1386 by immediately notifying the victims of California residents. However, the company did not notify but rather concealed this fact for several months. This violates the relevant provisions of the S.B. 1386 Act.

To ensure the fairness, accuracy, and privacy of personal information contained in credit reporting agency documents, the Fair Credit Reporting Act (FCRA) was passed in 1970. FCRA is an act that supervises credit reporting agencies and forces them to ensure that the private information stored and sold is a fair and accurate summary of consumers' credit history (Debt.org, 2012). Companies and institutions collect personal private information for a variety of purposes, such as deciding whether to provide loans or sell insurance to consumers. Therefore, the Act also stipulates that companies not sell consumers' personal private information for disallowed purposes and fails to maintain reasonable procedures to restrict the provision of consumers' personal private information for permitted purposes stipulated by the law (Feldman, 1973).

In addition, FCRA also restricts which organizations or individuals are allowed to view credit data and under what circumstances. For example, the employer can request the interviewer's credit report but must obtain the interviewer's permission (Kagan, 2020).

In this case, ChoicePoint should have reviewed the identity information of the recipients of personal privacy information through the normal business review process. The company too often ignores the step of censoring identity, so it provides opportunities for thieves to profit. This behavior violated FCRA, so ChoicePoint was sued and fined.

Ethics and Morals of Information

Ethics is a set of behavior standards shared or held individually by groups or communities. Most professional organizations and institutions should have codes of conduct or ethics related to their businesses, such as doctors, lawyers, and information professionals. Both individuals and organizations should have a belief in judging right from wrong, this is morals.

There are two aspects of ethics, business ethics and information ethics. Business ethics describe the relationship between business and ethics, and information ethics more specifically describes the related ethics responsibilities of information collection and use. According to Mason (1986), there are four key areas to measure business ethics and information ethics. First, privacy refers to the power an individual has, which includes what personal information is disclosed under what circumstances. Second, the organization's policies and standards can protect personal information from leaking. Third, clear property ownership is necessary, as in who owns the data and how to use the private data. Access authority is the fourth information technology to which attention needs to be paid. This technology protects personal privacy data by examining and restricting the identity of the visitor.

Similarly, moral responsibility has two aspects, which includes the morals of individuals and organizations. This means that the organization and its employees are responsible for protecting the security of sensitive data. The result of non-compliance with morals is not only a huge loss for the organization but also for individuals who are expelled or prosecuted for fraudulent practices. For example, a cashier of the First Bank personally violated the morals of the bank staff and sold customer information to thieves. This breach compromised the bank's reputation, and the First Bank disciplined the cashier.

In the United States, the federal government and state governments have established rules and procedures for how to operate businesses. Companies that do not follow these rules often face huge fines. Although larger organizations sometimes believe that the fines paid for violating the guidelines are less than the benefits, the costs of continuing violations of the guidelines are enormous.

In this case, several behaviors of ChoicePoint violated the ethical standards to which the organization should abide. First, ChoicePoint is not bound by the relevant data laws, because the company is not legally considered a financial services company. Therefore, ChoicePoint should formulate ethics and moral standards in line with its business processes to regulate the behavior of itself and its employees.

Second, although reviewing the identity of applicants is a business process of ChoicePoint, the company usually does not verify the identity of applicants to ensure their legitimacy. This means that identity thieves can easily steal copies of other people's business licenses to obtain personal privacy data, which ChoicePoint does not know and cannot stop.

Third, ChoicePoint stores the personal private information of almost all American adults, which has more than 1.9 billion pieces of information. As a company with such a large amount of private information, it should use advanced technology to protect customers' private data, which is the responsibility of ChoicePoint in the ethical category. However, due to the company's imprecise review process, the private information of 145,000 people was leaked.

Fourth, when it was discovered that the previously stolen data was illegally used by other thieves, it was ChoicePoint's moral responsibility to publicly disclose and notify all victims. Initially, S.B. 1386 forced the company to only notify the victims in California, which violated the most basic ethical responsibility within the company. If the attorney generals of other states had not strongly demanded that ChoicePoint make a public statement, tens of thousands of victims would have been kept in the dark.

LESSONS LEARNED

Breach Laws

Protecting consumers' personal privacy information is becoming more and more important because identity theft can reveal the victim's personal privacy information and bring economic losses to the victim. According to Kahn and Liñares-Zegarra (2015), identity theft can have a negative impact on credit cards, stored value cards, bank account payments, and online payments. Regarding the use of payments, these researchers found that certain types of identity theft had a significant negative impact on the use of cash, money orders, and credit cards. In addition, the authors proved that the above findings are reliable in different types of transactions (Kahn & Liñares-Zegarra, 2015).

Compliance with relevant laws and regulations is very important for companies or institutions that store personal information of the public. These laws are not limited to national laws but also include the state level. In particular, FCRA and FTC update and improve relevant laws and regulations every year, which not only protect the security of the public's personal information but also regulate the code of conduct and business processes of related organizations and enterprises, such as reviewing the identity of the person purchasing the information. Violation of laws and regulations will not only lose the trust of customers but also cause companies and institutions to pay huge costs, such as brand and reputation loss and fines (New Zealand Government, 2020).

Ethics and Morals of Information

Both companies and individuals need to fulfill ethics and moral responsibilities because unethical behavior may have far-reaching consequences, including sensitive data leakage, loss of profit, and psychological damage (Lindebaum et al., 2017).

Companies can reduce the risk of unethical behavior by formulating a series of ethics and employee codes of conduct. In addition, a lack of business plans and untrained employees will lead to lower work efficiency. Enterprises can conduct supervision from two aspects: setting business goals and training employees. On the one hand, setting realistic goals for employees can prevent employees from engaging in unethical behaviors to achieve unattainable business goals. On the other hand, proper training of all employees can prevent employees from cutting corners and making excuses for not completing work as required (Zeiger, 2019).

DISCUSSION QUESTIONS

1. How does the organization formulate a code of ethics suitable for itself?
2. What are the reasons that cause the individual or organization to violate the ethics code?
3. What are the challenges in the process of solving the leakage of private information?

REFERENCES

Culnan, M. J., & Williams, C. C. (2009). How ethics can enhance organizational privacy: lessons from the choicepoint and TJX data breaches. *Mis Quarterly*, 673–687.

Debt.org. (2012). *Fair Credit Reporting Act*. Retrieved 15 July, 2020, from https://www.debt.org/credit/your-consumer-rights/fair-credit-reporting-act/

Feldman, S. (1973). The fair credit reporting act-from the regulators vantage point. *Santa Clara Lawyer*, *14*, 459.

Kagan, J. (2020). *Fair Credit Reporting Act (FCRA)*. Retrieved 15 July, 2020, from https://www.investopedia.com/terms/f/fair-credit-reporting-act-fcra.asp

Kahn, C. M., & Liñares-Zegarra, J. M. (2015). Identity theft and consumer payment choice: Does security really matter? *Journal of Financial Services Research*, *50*(1), 121–159. https://doi.org/10.1007/s10693-015-0218-x

Lindebaum, D., Geddes, D., & Gabriel, Y. (2017). Moral emotions and ethics in organisations: Introduction to the special issue. *Journal of Business Ethics*, *141*(4), 645–656.

New Zealand Government. (2020). *Protecting customer and employee information*. Retrieved 10 July, 2020, from https://www.business.govt.nz/risks-and-operations/it-risk-and-avoiding-scams/protecting-customer-and-employee-information/

Tinseth, A., & Richlin, D. (2005). SB1386: One Year Later. *Information Systems Security*, *13*(6), 43–50. https://doi.org/10.1201/1086/44954.13.6.20050101/86220.7

Zeiger, S. (2019). *Effects of a Lack of Ethics on a Business Environment*. Retrieved 15 July, 2020, from https://smallbusiness.chron.com/effects-lack-ethics-business-environment-23332.html

3.4 TJX

Huixin Zhang and Xuejie Tai
University of Canterbury, Christchurch, New Zealand

CASE SUMMARY

Founded in 1976, TJX is one of the largest apparel and home fashion retailers in the US budget apparel market. In 2006, the company became a leader in the US inexpensive clothing market. In 2019, TJX was ranked 85 in the 2019 Fortune 500 company listings. As of the end of 2019, the company had annual sales of $42 billion, with more than 4,500 stores and four e-commerce sites in nine different countries and regions. Since its establishment, TJX has formed a chain store with brands from many countries and regions, such as TJ Maxx and Homegoods in the United States (TJX, 2020a, 2020b).

The value proposition of TJX is to provide customers with great value on a global scale. To this end, they provide high-quality products, brand names, and designer products. The prices of these products are 20–70% lower than the prices in department stores or specialty stores. The company's business model is very flexible because its inventory changes rapidly with market trends and customer preferences. A flexible business model can not only help TJX expand and contract product categories but also help it better understand current fashion and price trends (TJX, 2020a).

With the development of the Internet, TJ Maxx and Homegoods under TJX launched e-commerce websites in 2004. For large retailers like TJX, the most significant advantage of e-commerce is low financial costs. Other advantages include international sales, personalized online experience, gain of easy access to customer data, and processing a high number of orders. For example, most consumers feel uncomfortable when providing email addresses to physical retail stores (Ferreira, 2019). TJX can obtain customers' personal information through registration in e-commerce, which can provide customers with better services and after-sales services. e-commerce needs the support of a strong IT system. The IT system of TJX plays a vital role in its value chain to establish the relationship between the company and its suppliers and customers.

The e-commerce site mentioned above was closed after a year because TJX discovered a security breach and intrusion in the IT system located in Framingham at the end of 2006. The system processes and stores information related to the customer's bank card and the returned goods, but there is no receipt. General Dynamics Corporation and IBM were hired as security consultants for internal investigations. Although TJX notified law enforcement officials and financial institutions such as banks and debit card companies of the intrusion, only two months later did the company publicly disclose seven security vulnerabilities in its computer system. These vulnerabilities

 DOI: 10.1201/9781003264415-37

included a lack of regular audits, lack of processing logs, and lack of data encryption during transmission as well as inadequate wireless network security. The data breach affected the security of customer privacy data in many countries and regions.

Beginning in 2005, thieves destroyed TJX's computer network and stole more than 45 million consumer payment information and customers' personal privacy information within 18 months. TJX's inability to implement reasonable information security procedures is also one of the reasons for security vulnerabilities, including not storing and transmitting sensitive information in a civilized form, not encrypting stored sensitive information, and not restricting wireless access to its IT system network.

By allowing the breach of security as described above, TJX violated the Federal Trade Commission (FTC) Act and Payment Card Industry (PCI) security standards, and there is a huge fine of up to $156 million. In addition, the FTC requires TJX to implement a comprehensive information security plan, conduct an independent evaluation every two years for up to 20 years, and submit a copy to the FTC. Consumers whose private information was leaked, and related banks and credit card companies have been affected to varying degrees. In this chapter, there are critical analysis of TJX's major leaks and lessons for the company in terms of three topics, which are laws, information privacy and ethics, and risk management.

CRITICAL ANALYSIS

Breach Laws and the FTC Act

Information security is not only a priority factor for the FTC, but it is also essential to all companies in the United States. The FTC Act prohibits unfair or deceptive behavior by companies or institutions that store sensitive consumer information. If the company can reasonably avoid the damage caused by the violation, the failure to implement reasonable safety measures may result in unfair trade transactions. Companies or institutions may also be prosecuted for false privacy policies (Federal Trade Commission, 2020).

In this case, as a leader in the US inexpensive clothing market, TJX collects and stores sensitive information of more than 100 million consumers. FTC Chairman Deborah Platt Majoras believes that companies that collect sensitive consumer information are responsible for ensuring the security of this information. According to the complaint of FTC, TJX failed to take reasonable security measures to prevent unauthorized visitors from downloading and using consumer personal information on the information system network. This breach resulted in not only the theft of tens of millions of credit and debit card sensitive data of TJX consumers but also the loss of personal information of approximately 455,000 consumers who had returned goods. The thieves used these sensitive data to illegally obtain tens of millions of dollars from the bank (Federal Trade Commission, 2008).

Payment Card Industry Data Security Standards (PCI-DSS)

In recent years, identity theft (including credit card data leakage) has been the biggest threat to computer network security. Large-scale security breaches involving payment card data have caused consumers to pay attention to security issues and have spawned litigation to clean up legal liability for damages related to these violations. In the United States, there is a lack of government supervision of the security of personally private data. So far, fines have replaced legal prohibitions because they provide a basis for security. Within the payment card industry (PCI), PCI-DSS is an effective model to provide additional protection for consumer data. In addition, it also poses many challenges in PCI. For example, how to protect the effectiveness of consumers (Morse & Raval, 2008).

PCI-DSS clearly stipulates that the organization will not store sensitive consumer data, such as card-validation code (CVC) numbers and personal identification numbers (PIN) after receiving payment authorization. However, TJX's customer records contain the consumer's CVC and PIN. PCI requires organizations to conduct annual self-assessment questionnaires and scan them by approved vendors. Since the scanning process did not include checking the unencrypted data in the organization's database, TJX's failure to encrypt data violated PCI-DSS. In addition, according to the Privacy Act and PCI standards, e-commerce sites can only store the last four digits of consumer credit cards. The organization needs to delete the old credit card information in the database manually. Automatic network scanning cannot find related software, so TJX retains consumers' old private information, which also violates PCI-DSS.

INFORMATION PRIVACY AND ETHICS

Information privacy is a multidimensional concept that depends on context, which changes with people's life experiences. Hence, it is also an information issue that needs to be taken seriously. Due to defects in the control of information by consumers, they are at a disadvantage when dealing with enterprises. The United States has adopted a sectoral approach to managing data privacy rather than complying with fair information practices. Culnan and Williams (2009) believe that incorporating ethical reasoning into privacy procedures can enable organizations to value their moral obligations and take reasonable precautions to protect consumer information privacy. Privacy issues can be divided into two types, which are information reuse and unauthorized access to personal information. Both types of activities can damage the private data of individuals and organizations. Especially, unauthorized access will not only violate confidentiality and economic loss but also lose the trust of customers (Culnan & Williams, 2009).

The vulnerability of moral responsibilities means that vulnerable groups suffer from a lack of information and control. Collecting large amounts of personal information is usually the root cause of large-scale privacy breaches. For TJX, its consumers are vulnerable. Today, most "transactions" between consumers and businesses are "second transactions", which means that consumers provide personal information in exchange for something valuable, such as home delivery and personalized services. Consumers are not aware of the misuse of credit card data caused by TJX's loose security measures, and they have no ability to stop it. Consumers believed that TJX could protect their private information after the transaction, but TJX failed to do so. At a minimum, TJX's failure to take reasonable measures to avoid harm to consumers violated its moral responsibility.

Risk Management

Risk management is a program for companies to manage risks, which includes a process from risk identification, risk analysis, and risk evaluation to risk treatment (Whitman & Mattord, 2018). These threats or risks that companies encounter could originate from various sources. According to Kaplan and Mikes (2012), risks are separated into three categories. The first category is preventable risks which are internal risks. Employees and managers might bring risks to their companies by their unauthorized and illegal actions. Companies should try to eliminate these risks because they will not get any benefits from these illegal actions. Companies can take active preventions which involve monitoring processes and guiding employees' behaviors. The second risk category is strategy risks which differ from preventable risks. Some enterprises accept some risks for their business strategies since they can gain great benefits although they might face high risks. The third category is external risks that happen from outside the company (Kaplan & Mikes, 2012). For example, risks could come from natural disasters and cyberattacks. TJX suffered a security breach from external intrusion. In terms of different sources, companies should take different actions to address these risks.

Risk management is important for organizations. Initially, organizations can save plenty of money through implementing a risk management plan, because it can help companies minimize negative impacts and prevent potential threats or risks. In addition, a more secure environment for staff and customers can be established due to the risk management plan. Moreover, a risk management plan enables companies to reduce legal liability and increase the stability of companies' operations. Without a risk management plan, companies might lose their business directions.

There are several processes that companies can follow to conduct risk management. In terms of risk management preparation, organizations need to establish contexts which include external and internal contexts. The external context involves the business environment, threat environment, and support environment, such as governments and professional associations. The internal context includes organizations' governance structure, security infrastructure, and culture (Whitman & Mattord, 2018).

After establishing the context, identifying risks is necessary, which includes the identification of information assets and threats to the assets. Information assets contain people, data, software, hardware, and networking components. Identification of information assets allows companies to know what exactly is to be protected. Threats should be identified in the process of identification since companies should use different methods to address various threats. For example, espionage, human error, information extortion, software attacks, and technical hardware or software failures are involved in threats to information security. After organizations identify threats they face, they should list these threats and prioritize them so that they can determine which one should be solved for its severity. TJX could find many vulnerabilities that they had, such as insufficient network security, inadequate physical security, and lack of firewalls when they identified their information assets and threats.

Risk analysis is the third process of risk management. The objectives of risk analysis are to figure out how these threats occur and to assess the negative impacts of these threats. Risk analysis helps companies further understand each risk

(Whitman & Mattord, 2018). For instance, TJX had a lack of physical security because they were negligent to monitor and secure in-store assets. In addition, since TJX did not have a strategy for risk mitigation, the management could not find its vulnerabilities immediately and mitigate potential risks. Risk evaluation is also significant since it can assist companies in determining whether these assets are exposed to risk for cyberattacks and data breaches. The final process is risk treatment which is the process of choosing methods to avoid and minimize risks. In terms of the TJX case, TJX should update its network security protocol to protect its physical assets. The implementation of firewalls is also essential for TJX to control access to the system. Additionally, security audits should be performed on a regular basis.

Therefore, TJX can make effective risk management according to its processes and can build its internal and external contexts. In fact, TJX has considered its external contexts, such as its competitors from Ross and Overstock, and threats from its substitute products and new entrants. However, TJX's internal governance, security system, and culture are less considered. There were no incentives for employees to report security problems. Due to the less consideration of internal contexts, a variety of vulnerabilities in TJX's networks and systems were embodied after the computer intrusion. TJX had lack of in-store physical security, firewalls, data encryption, regular audits, and processing logs, and the company also violated PCI standards. Security professionals at TJX should analyze these risks and explore the best ways to avoid and eliminate these risks.

LESSONS LEARNED

Breach Laws

Hiring a security consultant and strengthening the security of its system is one of TJX's responsibilities. TJX needs to solve the problematic insecure system urgently to become PCI-compatible through five measures. First, maintaining an information security policy is the basis for compliance with PCI standards. Second, establishing and maintaining a secure network environment can effectively reduce the risk of intrusion. Third, upgrading the computer system to fix vulnerabilities can protect the security of customers' private data. Implementing strong access control measures is the fourth necessary operation. Finally, TJX needs to monitor and test the network regularly to ensure the security of sensitive data. In addition, recruiting new IT security positions is one of the company's internal control methods.

Information Privacy and Ethics

In terms of information privacy and ethics, enterprises need to design a code of ethics to regularize employees' behaviors so that employees are more likely to consider companies' ethical rules and customer rights when they attempt to conduct some activities for their own benefits (Manifest, 2018). Organizations can also reward ethical behaviors to encourage ethics in the workplace. An incentive mechanism is important in the workplace since it can stimulate employees to follow a professional code of ethics. Ethics is the priority when organizations conduct recruitment. Additionally, leaders in organizations could be trained to recognize and appraise employees with good behaviors. Nomination systems can also be created to recognize and comment on employees' ethical behavior.

Risk Management

For risk management, organizations not only consider their external environment but also consider internal contexts. Some organizations focus on their competitors, substitute products, and any threats from their competitors and new entrants, but ignore their employees, business processes, IT networks, and systems. Although many attacks originate from outside companies, inadequate security infrastructure enables malicious actors to have chances to conduct attacks.

In terms of the internal context, companies can re-establish it from the following aspects. First, companies can start from the perspective of personnel. Providing staff training is necessary so that they can follow the company's rules and effectively respond to any security incidents. Internal incentive mechanisms can be constructed since incentives or rewards can encourage employees to report any threats or risks that they discover. Second, companies should seriously focus on IT governance which ensures business activities and oversee any potential risks. Strategies for risk mitigation and management should be developed. Some companies are confident in their risk management, but actually, there are many loopholes in risk management. Companies should make detailed strategies which describe personnel responsibilities and actions on how to minimize and avoid similar threats and attacks. IT networks and systems need to be considered as well. Companies can update their network security protocol if they have weak network security protocols. Various detection, monitoring, and encryption tools can be installed since they help companies monitor and detect any threats from the internal or external and increase the security of sensitive information.

Companies can conduct risk identification after establishing the internal and external contexts. Some organizations do not encourage their employees to identify risks because they regard those activities related to risk management as not helpful and troublesome, but risks could be only addressed effectively after risk identification. In addition to risk identification, mitigation plans for identified risks can be determined through understanding users' needs, seeking experts, and recognizing opportunities arising from risks.

Although authors have discussed some measures to help these companies which are affected by different attacks solve threats or risks, there are still several problems to be discussed.

DISCUSSION QUESTIONS

1. How should companies ensure that every affected customer is protected by breach laws in different states?
2. How do companies make sure that every staff follows a professional code of ethics even though companies formulate ethics rules?
3. What should companies do to ensure the effectiveness of risk mitigation strategies?

REFERENCES

Culnan, M. J., & Williams, C. C. (2009). How ethics can enhance organizational privacy: Lessons from the choicepoint and TJX data breaches. *Mis Quarterly*, 673–687.

Federal Trade Commission. (2008). *Agency Announces Settlement of Separate Actions Against Retailer TJX, and Data Brokers Reed Elsevier and Seisint for Failing to Provide Adequate Security for Consumers Data*. Retrieved 12 July, 2020, from https://www.ftc.gov/news-events/press-releases/2008/03/agency-announces-settlement-separate-actions-against-retailer-tjx

Federal Trade Commission. (2020). *What We Do*. Retrieved 10 July, 2020, from https://www.ftc.gov/about-ftc/what-we-do

Ferreira, N. M. (2019). *20 Ecommerce Advantages and Disadvantages You Need to Know*. Retrieved 8 July, 2020, from https://www.oberlo.com/blog/20-ecommerce-advantages-and-disadvantages

Kaplan, R., & Mikes, A. (2012, June). *Managing risks: a new framework*. Retrieved from Havard Business Review: https://hbr.org/2012/06/managing-risks-a-new-framework

Manifest, T. (2018). *How can companies encourage ethics in the workplace?* Retrieved from Medium: https://medium.com/@the_manifest/how-can-companies-encourage-ethics-in-the-workplace-5c36318e98a4

Morse, E. A., & Raval, V. (2008). PCI DSS: Payment card industry data security standards in context. *Computer Law and Security Review: The International Journal of Technology and Practice*, *24*(6), 540–554. https://doi.org/10.1016/j.clsr.2008.07.001

TJX. (2020a). *A Global Off-Price Retailer*. Retrieved 5 July, 2020, from https://www.tjx.com/company

TJX. (2020b). *OUR HISTORY*. Retrieved 4 July, 2020, from https://www.tjx.com/company/history

Whitman, M., & Mattord, H. (2018). *Management of information security* (6th ed.). Cengage.

3.5 Case Study: A Data Breach on Flipboard

Joe Mani Kuttikatt
Z Energy NZ, New Zealand

Rion Jovial Luis
TussockLine, New Plymouth, New Zealand

INTRODUCTION

Flipboard is a news and social networking aggregator app established in July 2010, in California. Flipboard has been designed to connect different users' favorite news outlets with their social content to give a deep look at everything from political concerns and technological developments to inspiration for travel. Flipboard, an excellent news-reading application that consists of articles, has the mission to "Advance conversation through important and influential stories that not only keep people informed but also inspire them to engage, learn and lead". Flipboard was initially developed exclusively for the Apple iPad and later was launched for the iPhone and iTouch in the year 2012. Flipboard application developers finally launched for Android phones beginning with the Samsung S3. The application got a positive response when launched and was named as a killer app for iPad. The applications userbase has grown drastically from a whopping 100 million to 145 million users and the company Flipboard, hugely popular and third amongst its competitors, is estimated to be valued at around 800 million. The operating systems that Flipboard is available in are Android, IOS, and Windows. The company has about 19 investors, JP Morgan Chase & Co being the lead investor.

SUMMARY OF CASE STORYLINE

The platform was repeatedly hacked twice between 2 June 2018 and 23 March 2019 and 21 to 22 April 2019, for an enormous period of 9 months. The company Flatboard stated that not all accounts were hacked but did not disclose the exact figures of how many users were the victims of this breach, thereby compromising personal information. Flipboard terms this incident as unauthorized access to their database. When the Engineering team started an investigation, they found that an unauthorized person had accessed and possibly attained duplicates of certain databases which contained information regarding the users. The users'

DOI: 10.1201/9781003264415-38

account information that was accessed included the individual's name, Flipboard Username, cryptographically protected password, and email ID.

Flipboard was using a hashing mechanism called salt hashing to protect user passwords. The Flipboard users could log in through third-party accounts such as Google+ or any other social media account. Flipboard's team did not find any evidence which proved that the hackers got access to any of the third-party information. As a precaution, the team decided to replace or delete all digital tokens. The case also particularly mentions that this incident did not involve any social security numbers, government IDs, bank accounts, credit card details, or any other kind of financial details being leaked. On 23 April 2019, the engineering team recognized the illegal activity that took place from 21 to 22 April 2019 and further examined the distrustful activity that had arisen on 23 March 2019. The Flipboard team had to make an official announcement to their users about the security breach that occurred and inform users what data had been compromised. The team also had to inform users that the data compromised also consisted of passwords meaning users had to reset their passwords even though the passwords were cryptographically protected. These steps were taken by the users as additional precautions.

The company also had to educate the users regarding their official emails so that users would not entertain emails from unknown sources who claim to be Flipboard. This knowledge was important for the users as it would help them prevent being victims of an email phishing attack. They also informed that the official emails sent by Flipboard would not consist of any attachments or links that would redirect users to other pages or websites. As an additional security measure, Flipboard set a prompt to change the password if users had signed in to a new account.

Flipboard also had to inform and engage with law enforcement authorities and cybersecurity agencies. Contracted to investigate the incident, the company Blackfog Inc. had to get insights into the happenings, strengthen the security of the application and systems at Flipboard, and enhance security measures to prevent future breaches and attacks.

The CEO and founder of Blackfog Inc. stated that the incident was widely concerning as the unauthorized hacker or attacker who had access to the database had entered the database for a long period, could have accessed the database for more than 9 months, and could make copies of the database (Musil, 28 May 2019).

CRITICAL ANALYSIS OF THE CASE

The case reflects on the data breach that took place at Flipboard twice from 2 June 2018 to 23 March 2019 and 21 to 22 April 2019, wherein the hacker had access to the database for more than 9 months. The user data and information compromised were usernames, Flipboard usernames, cryptographically protected passwords, and email addresses. Flipboard had always secured the passwords of users cryptographically using salt hashing. The advantage of hashing passwords is that unique salt is used for every password in the blend and the password is never stored in plain text. The use of these hashing algorithms makes it very difficult for hackers to crack the passwords, as this can only be performed with the help of some vital computing resources. On 14 March 2012, Flipboard started using additional measures to protect the passwords by hashing the passwords with a computing function called bcrypt. Bcrypt is a password hashing method that uses a blocking mechanism against a cypher cryptographical algorithm and other security features as well as an additional round computing algorithm to provide additional protection for password cracking.

The users who would like to connect to Flipboard through other third-party applications or social media accounts do this with the help of digital tokens. In this case, there is no real evidence that the unauthorized person had access to these accounts but as an additional security measure, the company had to replace or delete all digital tokens that were connected to the user application.

The term "Data breach" is an incident related to security where confidential, safeguarded, or sensitive data are copied, observed, stolen, transferred, or used by unauthorized hackers which may cause a rising risk to the company or society (Jackson et al., 2019). A data breach can have a huge adverse impact on an organization and its consumers and can also impact the organization's goodwill and reputation in the market to a great extent.

These incidents were identified by Flipboard's engineering team when they came across unauthorized and suspicious activities that took place on 21–22 April 2019 and further investigated suspicious activity that took place on 23 March 2019. The team then immediately informed the higher authorities who then quickly took help from law enforcement authorities and got in touch with an external cybersafety and security firm called Blackfog Inc.

As a consequential safety measure the company had to reset all the user passwords even though they were cryptographically protected, and they also made it mandatory for users who signed into their existing accounts using new devices to create a new password. As another important additional precaution, Flipboard disconnected digital tokens that were previously being used to connect the third party and deleted or replaced all tokens with collaboration with their partners where it was appropriate.

As an additional way to secure its users and to prevent some likely incident from happening in the future, Flipboard implemented enriched security measures and are determined to look for additional ways to toughen the security of the firewall and systems.

Flipboard also had to give knowledge to users about what information they store and what is not stored. They also informed the users that this incident did not involve any sensitive personal information like government-issued IDs, Social Security numbers, driver's license numbers, payment cards, bank accounts, or any other information related to the finances of the user.

As a safeguard, they recommended users to change any similar password they have used for other different accounts which are related to their Flipboard account password. They also stated that the user should change their password regularly and should not use similar passwords for different online accounts.

LESSONS LEARNED

The above case helps us learn a lot from the mistakes made by Flipboard. The company is a huge organization that generates high revenue and is one of the top companies with regard to its category. They also have many investors who believe the company can be one of the best by continuing to compete with its competitors. Having lots of articles and news related to different kinds of topics makes Flipboard's website and system strongly vulnerable to different kinds of attacks and the company's mistakes increase their vulnerability to malicious intrusion.

In a large organization like Flipboard, there needs to be a separate department altogether to consider Cybersecurity. Delegating a unit to look after Cybersecurity enables the company to monitor the unusual activities there constantly and continuously. We also learned that a company needs to have a security audit every quarter.

The case clearly points out that Flipboard was dependent on the engineering team to investigate and monitor the security of the company. All of its employees need to be educated regarding the different ways a security breach can take place by giving them strict instructions not to click on random links and banners.

Flipboard can also use a threat detection mechanism which will help them analyze and stay safe from such kinds of attacks. By researching and referencing common attacks that are happening around the world, companies such as Flipboard should conduct incident analysis and document the same as this could help the employees and management learn more about keeping their work environment and users' data safe.

Remote monitoring is another thing that seems to be lacking in the case of Flipboard. Remote monitoring provides an organization with an IT team to continuously monitor and be aware of all incidents that take place in the system. Putting your system logs on to monitor internet activity can help administrators understand if there is any unauthorized access (Hojeige, 2019). A cost-efficient option for big MNCs is to collaborate and work with a good Managed IT, service provider (MSP) as this will help the staff working in the company to manage and perform other managerial functions better.

The number of data breaches affecting the US alone, from 2005 to 2017 increased tenfold from 157 attacks to 1,632. These types of attacks have become very common in the years 2017–2020, making it necessary for organizations to have a dedicated security team to keep the work environment and user data safe. Taking precautions before the risk arises is widely important for any big or small organization.

DISCUSSION QUESTIONS

1. What are the benefits of hashing passwords in the system?
2. What are the steps to be taken by a parent company to protect their client data effectively after the security breach?
3. What are other ways in which organizations can track unauthorized access and log files other than commonly used manual intervention?
4. Why is it important to conduct security auditing on systems, networks, and servers on a regular basis or at least quarterly?
5. How should companies frame policies to make sure employees don't violate security regulations passed by the organizations?

REFERENCES

Hojeige, G. (2019). The How-To: Protect Your Business From A Data Breach.

Jackson, S., Vanteeva, N., & Fearon, C. (2019). An investigation of the impact of data breach severity on the readability of mandatory data breach notification letters: Evidence from U.S. Firms. *Journal of the Association for Information Science and Technology*, *70*(11), 1277–1289. https://doi.org/10.1002/asi.24188

Musil, S. (2019, 28 May). Flipboard database hacks exposed users' account information. CNET. https://www.cnet.com/news/privacy/flipboard-database-hacks-exposed-users-account-information/

3.6 What Happens in Vegas
Breach of Data Confidentiality

Satish Prabhu
Capgemini, Christchurch, New Zealand

Abhishek Kumar
University of Canterbury, Christchurch, New Zealand

SUMMARY

MGM Resorts International is a chain of resorts in the US located in the cities of Las Vegas, Mississippi, Detroit, New Jersey, and Maryland. In July 2019, guests' data from MGM Resorts of Las Vegas were leaked into the public domain as well as on a hacking forum. More than 10.6 million guests' information, including that of Twitter CEO and Justin Bieber, had been exposed. A technology news website, ZDNet had verified and informed that the leaked data includes home addresses, emails, phone numbers, and date of birth. However, these data didn't involve any financial information or payment card or password data. The spokesperson of the hotel stated that the leaked data were old and impacted guests were notified in accordance with applicable state laws. But the issue of concern was – when and how the data security was breached. The hotel spokesperson said that in the summer of 2018, an unauthorized access to a cloud server had been discovered with a loss of a limited amount of information. They had hired two cybersecurity forensic firms to do the internal investigation of a breach of data security from the server. Following this fiasco, hotel management also took responsibility to strictly protect guest information and decided to enhance and strengthen the security network to prevent any further data security breach.

CRITICAL ANALYSIS

Modern societies are increasingly dependent on newer information and communication technologies which provide both opportunities and challenges to improve the quality of life of the people and the communities where they live. The large volume of transactions, complex reporting requirements, and international communication demand newer technologies in this industry. These can improve almost all areas such as sales, guest services, reservations, food service catering, food and beverage management, maintenance, and hospitality accounting. Therefore, to prevent the loss

 DOI: 10.1201/9781003264415-39

of data, all organizations observe and monitor their computer network for security threats from various sources, such as espionage, computer-assisted fraud, sabotage, hacking, vandalism, phishing, system failures, flood, and fire. To maintain public trust and to prevent hackers from attacking the computer system, most of the hospitality industry tries not to reveal the instances of data breaches and cyber-attacks on the computer network.

Each level of resort and hotel management (e.g., user level, property level) encompasses a computer network and a server system. The property level involves a local area network where the front office, reservation, accounting, payroll, human resources, restaurant management, and other systems are located. At the customer or user level, people browse the hotel website to enquire about the hotel or guest house and to confirm their reservations in the hotels. Such interaction in a computer network environment increases the susceptibility to the security threat for user data privacy and security. The unencrypted information transmitting on Internet Protocol/Transmission Control Protocol (IP/TCP) can be stolen by unauthorized people. The information security issues as experienced by most of the employees in hotels who work on computer systems (e.g., accountants, front office managers) are not IT experts and have only some or no familiarity with computer systems. As new advances in IT systems emerge such as social media, cloud computing, and big data, integrating into hospitality industries' systems, cybersecurity and privacy are becoming an increasing issue.

Like other organizations, the IT system in the hotel and resort industry consists of both hardware and software. The software includes the point-of-sale system (POS), property management system (PMS), accounting system, and call accounting system (CAS). The hardware in a hotel or resort industry includes POS terminals, front-desk computers, back-office computers, routers, firewalls, printers, switches, and network cables. The computer systems, printers, and POS terminals are connected to switches and routers with network cables that facilitate communication between the devices. Routers also bond hotel networks with the internet and other networks. The attached firewall prevents the hotel network from an external network. Whereas, Local Area Network (LAN) comprises the devices within the premises of the hotel. All these information systems (IS) are meant to maximize revenue generation and better customer service. But they also ensure the users minimize the damage caused by security threats providing three main services, i.e., availability, integrity, and confidentiality are implemented.

Using the IS system, the hotel industry may have various privacy and security threats which include inside attackers for network access, viruses, spoofing, laptop theft, unauthorized outside attacks, unauthorized insider access, phishing attacks, ransomware attacks, and denial of service (DoS) attacks. With the advancement of the IT system, the nature of cybersecurity attacks is also advancing to breach the privacy and confidentiality of the users' data. Hacking or cracking is the most common crime in the hotel industry where the computer systems or servers are connected by an unauthorized network to obtain confidential information of users. However, there are other security attacks and privacy breaches possible.

Resorts and hotel industry have a large amount of sensitive data, especially personal and financial information of guests which makes them much more susceptible to cyber-attacks. They have a low level of cybersecurity and become more vulnerable to cybersecurity breaches. They even don't have security audit cycles and this issue may put the guests in a situation of a high risk of privacy and security breaches. Therefore, to ensure the longevity of the hotel and resort industry, securing all the sensitive data of guests is critical. If the guests feel that their personal and financial data are not secure to share or they are a risk to be exposed then, they would hesitate to or possibly refuse to share such information with the hotel industry. Thus, to secure sensitive data, management needs to take proactive steps to prevent security and privacy breaches.

In this quickly developing threat landscape, the hotel or resort industry must secure its data center against the potential threat. Las Vegas is one of the busiest tourist locations; millions of tourists come to this city every year. Therefore, hotel management should be concerned about a cyber-attack that could jeopardize the sensitive personal data of guests and tourists. After the Las Vegas incident, MGM Resorts submitted to investigation which revealed several mistakes in data security and confidentiality. Their limited security system was insufficient to protect the personal information of guests from the wide range of probable cyber-attacks. MGM Las Vegas had failed to use the most available security system, i.e., firewall. They might have stored credit card details in the cleartext. They also failed to implement required information security measures before connecting local computer networks to corporate-level networks. Addressing any security vulnerabilities on cloud servers might not be considered. Sometimes hotel management uses the default username and passwords for access to the servers. Employees were not trained to use complex user IDs and passwords. MGM Resorts had not taken any security measures to scrutinize unauthorized computer accesses. They never conducted security investigations.

Despite the several factors of data security and privacy breaches, hotels are also not wholly innocent and without fault. They can take several proactive preventive measures for these cyber-attacks. To take action against cyber risks, implementing different security tools that are critical and necessary for the hospitality industry is required. Mostly, this industry doesn't have enough knowledge or sometimes, no knowledge of data security, which is their biggest drawback. Therefore, cyber attackers and hackers will have more chances to take advantage of these shortcomings, making hotel industries easy targets for them to gain. Also, hackers are not only able to exploit the computer system of the hotels and guesthouses by using different types of viruses, phishing emails, etc., but also able to attack and gain the advantage of Wi-Fi in the hotels. Many hotels even provide free Wi-Fi services to their guests which makes them more vulnerable to hackers. Routers are also hacked by hackers, and they upload the hacking software by updating Flash Players or Adobe Reader to retrieve confidential information of guests from the database or payroll management system. Such updating software contains malware which is used to fetch usernames, passwords, and other confidential information from the system or server.

Many times, the security team might be using the Security Information & Event Management (SIEM) tools for its data security. The tools only checked logs and didn't provide any visibility into internal network traffic. Therefore, SIEM was unable to identify never-before-seen threats in their system. MGM Resorts needed a new security solution that could provide them with total visibility into the mesh of the network and could reduce the burden of the existing lean security team. The Las Vegas Resort would have to concern themselves, particularly for fast-moving attacks like DDoS (Distributed Denial of Service) and ransomware that may have ravaging impacts on the security and privacy of guests' sensitive data in the resort industry. Below are some usual threats that MGM Resorts might face with low cybersecurity maintenance.

DDoS

MGM Resorts have a wide array of systems and therefore, DDos might be the hack of choice for those that are looking for an attack to retrieve the guests' information from the hotel industry. MGM Resorts might also suffer from such a security attack if a perpetrator were to make a malicious attempt to temporarily disrupt the targeted server making the network or machine unavailable to the intended users of MGM management. For example, vulnerable sprinkler systems or security cameras might have been hijacked and the entire computer system of the resorts crashed down.

Ransomware Attacks

Cybersecurity of the hotel industry is also at high risk from ransomware. These malware block the system from accessing their own data or encrypt the files adding extensions to the attacked data in its own systems demanding a ransom to free them. MGM Resorts might also suffer from a ransomware attack where a rogue software holds the data of MGM Resorts systems hostage and threatens to publish the guests' data until a ransom has been paid. It might be possible that a staff member of hotel management clicked on an infected email or visited malicious sites which contained such malicious software. Matters would not be revealed by the management team. If a ransom was not paid, then the guest's data would have been exposed by the hackers. Sometimes, such privacy breaches have been done in terms of other confidential information by management staff. Ransomware can also attack through the VPN to the guests' systems. It might also be a concern for information security directly from the guests' system. WannaCry is one of the ransomware that could be used against the hotel industry by taking their information and certain systems hostage. After gaining the financial statements, the perpetrators free the hotel systems and data.

Phishing Attacks

Phishing is one of the oldest methods of scam over the internet. Untrained employees of MGM Resorts may be trapped by fraudulent emails, sharing all the personal information of their guests in response to an email which appears to be from a genuine source. An employee could even share his password or personal information with such phishing emails. For the last few years, attackers seem to be increasingly targeting an authorized person, convincing them to authorize and share some information over fraudulent emails. Improper training given to the authorized employees of MGM Resorts might have got them trapped by phishing emails and shared guests' information. Clicking on a link or opening an attachment could also initiate the installation of any virus, spyware, or ransomware into the system.

DarkHotel Hacking

DarkHotel hacking is another form of hacking in the hotel industry where hotel Wi-Fi systems are being used by criminals to hack the guests' information. MGM Resorts staff might possibly have been trapped by a forged digital certificate that convinced them to download software. However, this software could contain some malicious code that could be uploaded to the server of hotels to track the guests' information. This is a targeted spear-phishing malware or spyware which selectively targets the guests of the hotels.

Customer Data/Identity Theft – For any successful hotel industry, securing the identity and information of their guests is paramount. Hacking the surrounding guests' information is one of the biggest risks in the hoteliers' report. In the case of MGM Resorts, the data was not secure. Pretending to be a guest, malicious intruders may use other guests' information like social security number and date of birth to access the personal information of the guests, commit fraud, and gain financial benefits via any fraudulent transactions or purchases. Thus, such criminals may commit identity theft or sell the information on the dark web. Such crime is still forever changing all around the world.

Point of Sale/Payment Cards Attack

MGM Resorts might also suffer from point-of-sale attacks which pose one of the biggest threats in the hotel industry. In a third-party crime, the attack on the hotel is indirect, i.e., vendors are being attacked. Somewhere in the system, there is a weakness which has been revealed by human error. Such malware may steal the information of guests of resorts and hotels, leading to bad press for the resort and causing financial implications in the hotel industry.

LESSONS LEARNED

Information Security is determined to protect valuable assets from the disclosure of confidential and personal information and damage to these data assets. Information Security is attained through both non-technological and technological methods such as user identification and authentication, physical security of assets, firewalls, biometrics, and user training. In the case of MGM Las Vegas, one may suppose that their cloud system was hacked to retrieve the personal information of guests of the Resort. Looking at the future concern, a few proactive approaches should be implemented to prevent any further privacy and security breaches such as some of the information and network security tools and techniques discussed below.

User training is the first and foremost line to protect guests' data in the hotel industry. All the concerned employees of MGM Resorts should be sufficiently trained to understand the concerns of data privacy and security. There are various methods to ensure data security. A common-sense safety is to train employees to not share their ID and password with an unauthorized person. Different verticals of guests' data security should be implemented. Front-desk and back-office employees are trained at regular intervals to ensure the confidentiality of data. To avoid any suspicious mail to open or to download any such software into the computer system is strongly emphasized. Employees must report any suspicious activity between the hotel staff to the next level of hotel management.

The hospitality industry needs to encourage the use of digital IDs from the guests for their passports, membership cards, and driving licenses. These digital IDs are often secured with a username and password.

There are Intrusion Detection Systems that analyze the irrelevant or suspicious events happening in the network and computer system to detect any intrusion and attacks. An intrusion tries to evade or bypass the security services that are activated by the system, such as integrity and confidentiality. Sometimes such attacks can make the system unavailable, i.e., DoS attack. Intruders can use various means to attack the system, i.e., the internet and outside networks. A user may try to gain additional privileges for which they are not authorized, or an authorized user may misuse and abuse the given privilege.

All such suspicious activities are detected by the intrusion detection system and must be immediately reported to hotel managers for prompt action.

Firewalls can be software or hardware that are used to monitor the traffic between computers. Many times, a combination of both software and hardware firewalls are used by the hotel industry to stop any malicious packet attack trying to move in a computer network. This combination firewall prevents hacking and thus maintains the confidentiality of guests' data.

Guests' information can also be transformed in a way that is not possible or very hard to interpret. Such encryptions maintain the privacy and confidentiality of data preventing its misuse.

There is Vulnerability Assessment Scan software that examines or scans the system for possible vulnerabilities or weakness and provide reports to the system administrator so that the system can be fortified against these weaknesses. This can be used by MGM Resorts as a proactive security tool to prevent the loss of confidentiality of guests' data from the hotel. During travel and in the hotel, guests should not use online banking on public Wi-Fi and public computers. In an emergency, they can access their email inbox by creating a throwaway email address. Also, technological caution advises users to prevent their smartphones, laptops, or computers from automatically connecting to an unknown Wi-Fi. Instead of saving sensitive and confidential information on smartphones or laptops, they should use remote desktops.

Attackers can be from both – outside and inside of the hospitality industry. In Las Vegas, MGM Resorts have a very high rate of turnover and have a higher possibility of an inside attack in comparison to other small groups of hotels. They must take measures to prevent such attacks. The IS infrastructure must provide the employees with a single sign-on to ensure better security. The system administrator can also restrict the usage of system resources by providing access control to the authorized or relevant person that prevents data leakage and barring access to unauthorized employees of the hotel industry. The Systems Administrator and MGM must ensure that no former employees can access the system once they resign from the hotel or resort responsibilities.

CONCLUSION

MGM Resorts and other hotel industries may lose the competitive advantage and image due to cybersecurity attacks and consequent negative word of mouth. Although customers' comfort and confidence are unique features of the hospitality industry that depend on the reputation of the hotels, the misuse or distortion of customers' information may cause customers to lose their confidence in the security and privacy policies of their preferred accommodation or all of them.

DISCUSSION QUESTIONS

1. How could a guest secure their personal data while booking hotel rooms?
2. How shall the hospitality industry mitigate the risk of loss of personal data of patrons?
3. Should MGM Resorts management have disclosed its lapses in data security to its patrons in the first place?

REFERENCES

Alowibdi, J. (2011). Adopting knowledge based security system for software development life cycle. Athens. https://www.cs.uic.edu/~jalowibd/uploads/SER4014.pdf

Holdsworth, J., & Apeh, E. (2017). An effective immersive cyber security awareness learning platform for businesses in the hospitality sector. In *2017 IEEE 25th International Requirements Engineering Conference Workshops (REW)*.

Kohnke, A., Sigler, K., & Shoemaker, D. (2016). Strategic risk management using the NIST risk management framework. *EDPACS*, *53*(5), 1–6. https://doi.org/10.1080/07366981.2016.1160713

Lee, M. K., Tang, K. P., Forlizzi, J., & Kiesler, S. (2011). Understanding users! Perception of privacy in human-robot interaction. In *2011 6th ACM/IEEE International Conference on Human-Robot Interaction (HRI)*.

Lutz, C., Schöttler, M., & Hoffmann, C. P. (2019). The privacy implications of social robots: Scoping review and expert interviews. *Mobile Media & Communication*, *7*(3), 412–434.

Lutz, C., & Tamó-Larrieux, A. (2020). The robot privacy paradox: Understanding how privacy concerns shape intentions to use social robots. https://search.informit.org/doi/10.3316/INFORMIT.097053479720281

News, G. B. (2011). SBI to ask staff to go on compulsory leave to limit fraud. https://economictimes.indiatimes.com/industry/banking/finance/banking/sbi-makes-offs-compulsory-for-staff-to-clean-out-frauds/articleshow/7315285.cms

Panai, E. (2018). A Cyber Security Framework for Independent Hotels. In *Proceedings 4th EATSA-FRANCE 2018, Challenges of tourism development* (145–152).

Portugal, D., Pereira, S., & Couceiro, M. S. (2017). The role of security in human-robot shared environments: A case study in ROS-based surveillance robots. In *2017 26th IEEE International Symposium on Robot and Human Interactive Communication (RO-MAN)*.

Priyadarshini, I. (2018). Cyber security risks in robotics. In *Cyber security and threats: Concepts, methodologies, tools, and applications* (pp. 1235–1250). IGI Global. https://www.igi-global.com/gateway/chapter/203558

Shabani, N. (2017). A study of cyber security in hospitality industry-threats and countermeasures: Case study in Reno, Nevada. *arXiv preprint arXiv:1705.02749*.

3.7 PIH Health Phishing

Anjali Krishnan
University of Canterbury, Christchurch, New Zealand

Shameer Thottoli
Acclaim Group Limited, Auckland, New Zealand

CASE SUMMARY

A significant data breach was exposed on June 18, 2019, at PIH Health [1], a provider of 10 clinics, emergency medical services, and other healthcare facilities in southern California. Following a phishing scheme that effectively manipulated workplace profiles, the personal identifying details of about 200,000 current and former California healthcare patients could have been jeopardized. On June 18, 2019, PIH Health confirmed that – as a part of a coordinated email phishing program – certain employee email addresses were potentially accessed without permission. PIH Health took action to protect its email service and network after knowing details about the incident which includes reconfiguring the passwords that are used to access potentially compromised employee email accounts. PIH Health has conducted an inquiry promptly and recruited leading, independent cybersecurity experts. As a result of this review, PIH Health found on October 2, 2019, that as a continuation of the above-mentioned phishing program, some employee email addresses were compromised without authorization between June 11, 2019, and June 18, 2019.

After receipt of evidence of unauthorized access to some PIH employee email accounts, PIH Health retained the same leading, independent cybersecurity experts to determine if the employee's email accounts accessed contain any private details and/or secure health data that might eventually lead to unauthorized disclosure. As per the review, PIH Health found some documents relevant to current and former patients were used in the email addresses accessed. PIH Health also worked tirelessly to locate the contact details for all persons directly involved and alert them of the incident. The above-mentioned inappropriate access was confined to email addresses for PIH Health and did not apply to other data systems for PIH Health. However, PIH Health is not aware of any proof that the details used in this accident were misused, and the independent forensic investigation culminated in the discovery of such proof. Nevertheless, PIH Health gave an alert to potentially affected people out of an excess of caution. PIH Health takes all cyber security extremely seriously and introduces new compliance controls to help deter a comparable situation in the future.

PIH Health has set up a toll-free contact center to answer questions about the accident and discuss related issues. Additionally, as a measure, PIH Health provides

DOI: 10.1201/9781003264415-40

certain potentially affected individuals with supportive credit monitoring services through Kroll. PIH Health also reported this incident to the US Health and Human Services for civil rights and consumer reporting agencies.

Although this incident has not impacted any patient's treatment or care, it has caused a 200% increase in insurance claims and multiple insurance claims. In addition, some of this stolen information has been used to forge medical backgrounds that have been sold on the dark web. As per the research carried out by Black Book, over 93% of healthcare providers have encountered some kind of data breach over the past five years and more than 50% of healthcare organizations have suffered at least five occurrences over the same period [2].

CASE ANALYSIS

The Internet of Medical Things (IOMT), Mobile Apps, Information Technology, and Cloud Computing have driven the healthcare sector to digital transformation. Digital medical care has laid the foundation for easier and more accessible diagnoses to make our lives much more comfortable. The healthcare industry has become more digitized, distributed, and mobile in recent years. In this aspect, the IOMT had a critical role. Healthcare agencies receive Personally Identifiable Information (PII) from their patients and archive it on their servers in the network to make it available at any time and to enhance the treatment of patients. Regrettably, the proliferation of smartphones and other mobile gadgets has now been a major cause of violations of privacy [3]. Those accounts are often accessed by unauthorized users due to program bugs, authentication flaws, and human error. This allows confidential information to be revealed through data leaks. Attackers occasionally inflict harm to safe medical records, leading to the destruction, misuse, or release of confidential patient data. On the dark web, the price of a patient's whole data document can be many dollars [4].

As documented by other clinicians [5], the overall number of people affected by cyber-attacks in healthcare was 249.09 million from 2005 to 2019. During the last five years alone, 157.40 million people have been affected. The number of data breaches identified in 2018 was 2,216 from 65 nations. The medical sector was faced with 536 violations of those. This indicates that the healthcare sector confronted the highest number of infringements among all industries. In 2019, more than 2,000 data breaches were reported from 86 countries. The total number of medical documents exposed, robbed, or unlawfully disclosed in 2019 was 41.2 million in 505 security breaches in healthcare. As per an IBM survey, a data breach cost an average of $3.92 million in 2019, whereas a breach in the healthcare sector usually costs $6.45 million. Relative to other nations, the cost was the largest within the US. A data loss will normally cost $8.19 million. The estimate of a data leak in the US, with an estimated violation scale of 25,575 records, is $15 million. The estimated cost of a data breach rose by 12% from 2014 to 2019 and the record violations went up 3.4% during the same period.

The statistics and figures listed above indicate that peoples' and organizations' digital data are at great risk. Rather more disturbingly, hackers are deliberately attacking the healthcare sector, which is making it as most vulnerable. Therefore, the privacy and security of data have become a major issue for both people and organizations. Medical care data are more delicate than other data types because any manipulation of data can lead to incorrect treatment, with devastating and irrecoverable losses. Due to these reasons, healthcare data also requires increased safeguards against attacks.

Healthcare data is a key tool for both scientific and health studies. Clinical data were gathered either through routine hospital treatment or as part of a structured clinical research program. Different categories of healthcare data [6] are Electronic Medical Records (EMR), administrative data, claim data, disease registries, health surveys, and clinical research datasets.

Electronic Medical Records: EMR is the clearest type of medical electronic data. EMRs are collected from the point-of-service, such as a treatment center, hospital, laboratory, or practice. EMR is not commonly open to outside researchers and specific institutions such as hospitals or healthcare networks can decide who gets access to this data and they may provide access to the intern workers. The data gathered contains information on institutional and population, diagnoses, care, prescription medications, clinical testing, statistics on medical examination, hospitalization, health benefits, etc. This is considered confidential data and any breach could lead to serious problems.

Administrative Data: These data are also connected to electronic medical reports, predominantly patient discharge data submitted to a government agency such as AHRQ. This data can be purchased for research from government-secured portals. Hence, the vulnerability of administrative data is not severe.

Claim Data: Claim data explains the billable transactions (insurance claims) between covered individuals and the delivery system for medical services. Data on reports fell into four general categories: hospital stay, rehabilitation, medication, and enrolment. Government (e.g. Medicare) and/or commercial healthcare firms (e.g. United HealthCare) are the sources of claim data.

Disease Registries: A limited set of main data that monitor only certain medical disorders like Alzheimer's disease, cancer, diabetes, heart disease, and asthma are stored in disease registries. Additionally, these registries offer critical information to monitor the health conditions of a patient, and therefore the security concern of disease registries are very crucial.

Health Surveys: To have an accurate assessment of public health, nationwide surveys are generally performed to obtain estimates of the severity of the most prevalent chronic illnesses. National polls are one of the few data forms gathered exclusively for research purposes, making them more available to the general public.

Datasets of Clinical Research: Data on clinical studies can be accessible from different regional or institutional organizations. Access levels are likely limited but datasets are accessible across accepted networks.

From the above-mentioned types of data, the EMR and disease registries are the most vulnerable data sources in healthcare and they need to be protected with proper security measures.

Based on a thorough review [6] of incidents of data violations by HIPAA (Health Insurance Portability and Accountability Act), it has been observed that hacking incidents, unauthorized access, fraud or loss, and improper disposal of unnecessary data were the main disclosure types of protected healthcare information. The different forms of disclosure listed above are discussed in brief as follows.

- *Hacking Incidents*: Hacking crimes cover all cyber-attacks used to obtain unauthorized access to sensitive information. Ransomware and malware are widely used today for exposing protected health information
- *Unauthorized Access (Internal)*: This covers all threats that lead to sensitive health data being revealed with the aid of an internal source from the organization. This may be power misuse, unverified access/disclosure, etc.
- *Fraud or Loss*: This includes all incidents that result in the disclosure of protected health information in the form of fraud or theft, such as theft of hard discs, computers, or any other portable device containing protected healthcare data. This may be attributed to serious damage or the destruction of some of these devices as well.
- *Improper Disposal of Unnecessary Data*: Data that is delicate and confidential, but no longer necessary, should be thrown away promptly so that it cannot be retrieved later. Waste handling of this data may result in the disclosure of protected health information. Incorrect disposal attacks involve all accidents triggered by inappropriate disposal of unwanted, but critical and private health data.

Furthermore, there were a total of 2,860 data breaches between 2010 and 2019 that come under the above-mentioned disclosure:

1. 29.72% of infringement proceedings were due to hacking/IT crimes
2. 29.47% of infringement proceedings were due to unauthorized internal disclosures
3. 37.65% of cases were due to theft/loss
4. 3.14% of cases were due to unauthorized processing of unwanted yet valuable data

The cumulative findings reveal that fraud/loss cases are the largest in number, accompanied by hacking / IT accidents and illegal internal dissemination. There are relatively few cases of improper disposal over 10 years.

Phishing is the method of stealing users' device credentials by manipulation or trickery. The effectiveness of phishing scams is also tied to authenticity and legitimacy – they can claim to be from an agency like a hospital IT department and notify users that their accounts will be shut down when they don't "correct" them by entering their credentials. Phishing domains, where users enter after following phishing email messages, are often designed to imitate official pages with misrepresented logos and have URLs that look like official websites, often supplemented with small spelling mistakes or lowercase letters. Phishing scams have become more advanced nowadays, with e-mails of better quality and more compelling sources for password collection. Therefore, distinguishing between original and phishing emails requires keen observation, more time, and a lot of effort and hackers can easily find their victims in healthcare industries as they misuse the vulnerability of this industry where importance would be given to patient care, and no one put a lot of effort or time on checking mail like this.

Furthermore, there are many factors why phishing attacks are so frequent in healthcare organizations. First, they do have a high rate of success. After all, getting into the corporate network takes just one error – one employee who is tricked into clicking on a malicious page – to get the culprit. In addition, certain phishing attacks, called spear-phishing, are specifically designed to reach individual users, raising the risk of opening the email and clicking on the suspicious sites inside.

Ultimately, spammers are extremely driven to attack healthcare organizations as they contain personal health information (PHI) that is very useful. According to the Centre for Internet Security (CIS), one medical record could sell for as much as $363 in the illegal market, while PII would sell for only $1 or $2. The explanation for this cost difference is simple: offenders can potentially use medical records for fraud, identity theft, or robbery, and victims can not make such information worthless. For example, victims of credit card theft can cancel credit cards, but they cannot invalidate their health history.

LESSONS LEARNED

Data on E-health is extremely vulnerable because it is most commonly exploited by attackers. A long-term analysis of data breaches found that both internal and external threats, such as hacking, compromised patient information, fraud/loss, the unauthentic release of internal data, and inappropriate storage of unwanted but confidential data. Email and Network Servers are the key sites from where personal health information is breached and infringements of healthcare data are much more costly than the average cost of data breaches, particularly in developed countries. In the future, both data breaches and their costs will continue to rise. Preventive steps must therefore be given priority by analysts, security experts, and healthcare organizations.

Healthcare organizations have several ways to strengthen their data breach protections and overall defense posture. Some applicable considerations are mentioned further.

Perform a Risk Analysis: The primary task is to perform an IT services risk assessment by the healthcare organizations. This should be done in compliance with the HIPAA Privacy and Security Rules which regulate the exchange of all electronic patient data. The risk analysis forces medical care providers to evaluate security measures, assess the risks, and identify the security vulnerabilities of their current systems. This is something healthcare firms need to be doing now, but many are not. With compliance and safety being a big concern in the corporate world today, this should be the focus.

Ongoing HIPAA Training: Educate and retrain staff on existing HIPAA laws and regulations. In addition, analyze and exchange state legislation about the protection of patient records. When workers are in the awareness and are aware of the consequences of data violations, the likelihood of abuse can be minimized significantly.

Inspection of Devices and Records: Remind staff to be cautious about electronic devices and/or unsupervised paper documents. Although safeguarding health records is an IT job, staff will always be advised to do their part in keeping information secure. Ensure that the devices are always locked whether it is a laptop, a desktop, or a phone and this must be protected with a password. Multifactor authentication should also be allowed wherever possible.

Encryption of Data: Encryption technology is crucial in preventing data breaches. Although HIPAA doesn't require encryption of data, it also does not consider the loss of encrypted data as a breach. Therefore, it is undoubtedly advisable, to protect data and avoid potential penalties, for healthcare providers to encrypt medical information.

ID and Access Control: Despite too many healthcare system participants regularly accessing patient details – for a variety of purposes – it is critical to closely monitor user identities. For example, make sure that user accounts limit access appropriately, and that procedures for logging in/off on shared devices are simple.

Use of Subnet Wireless Networks: Ensure the confidential health data is not disclosed on networks made accessible for public access. One way to do this is by developing sub-networks devoted to guest operations and dividing more stable networks for medical equipment and software that communicate and hold confidential patient information.

It is high time the healthcare industry needs to have a dedicated team working on information security and risk management. It is only a baby step against the thriving cyber-attacks targeting the vulnerable patient community. Data privacy is a fundamental right and patient care training for healthcare workers must include awareness classes about the same.

DISCUSSION QUESTIONS

1. How aware is the medical community – including patients – of the importance of information security?
2. Are data security and other security compliances a part of organizational culture?
3. What steps can a healthcare center take to ensure they are fully compliant with HIPAA or GDPR?
4. How can the patient community be empowered to act if they face a security incident that compromises their health data?

REFERENCES

1. https://portswigger.net/daily-swig/california-healthcare-data-breach-could-impact-nearly-200-000-patients
2. https://www.prnewswire.com/news-releases/black-books-2020-top-rated-healthcare-analytics-solutions-vendors-and-consultants-announced-300985539.html

3. Smith, T.T. Examining Data Privacy Breaches in Healthcare. Available online: https://scholarworks.waldenu.edu/cgi/viewcontent.cgi?referer=&httpsredir=1&article=3726&context=dissertations
4. Chernyshev, M.; Zeadally, S.; Baig, Z. Healthcare data breaches: Implications for digital forensic Readiness. *J. Med. Syst.* 2019, 43, 7. [CrossRef] [PubMed]
5. Seh, A. H., Zarour, M., Alenezi, M., Sarkar, A. K., Agrawal, A., Kumar, R., & Khan, R. A. (2020, June). Healthcare Data Breaches: Insights and Implications. In *Healthcare* (Vol. 8, No. 2, p. 133). Multidisciplinary Digital Publishing Institute.
6. December 2019 Healthcare Data Breach Report. Available online: https://www.hipaajournal.com/december2019-healthcare-data-breach-report

3.8 Biometric Protection and Security

A Case Study on Clearview AI

Juliet Samandari and Amelia Samandari
University of Canterbury, Christchurch, New Zealand

CASE SUMMARY

Clearview AI provides a level of accuracy and speed that no other facial recognition system has achieved. This is mainly due to the limitations other facial recognition services have in terms of the photos they can use. For example, government-affiliated facial recognition systems can only use photos provided by the government, such as mug shots or driver's license photos. Clearview, as a private organization, has not yet been limited by such regulations and has therefore been using images that have been publicly available at any point since their database started collecting photos (O'Sullivan, 2020). This includes photographs that are currently available on the Internet as well as those that have since been made private (Heilweil, 2020).

Clearview is a secretive company, with a website that provides a minimum of information and a fake address as their place of business. They have only one employee on LinkedIn, which was revealed to be the CEO, Mr. Ton-That, using an alias. They have also been seen to monitor common law enforcement searches (Hill, 2020). This would give Clearview information into what may be occurring in classified investigations by law enforcement. Clearview states that it has been independently investigated, but there has been no proof of this provided. Clearview also says it has been cleared as being compliant with the law, (Clearview AI, 2020) yet the company has been subject to many legal disputes, with many still ongoing. Due to the lack of information shared, Clearview's ability to protect its data is still unclear.

This has raised concerns about how secure Clearview's database is, especially as their list of clients, as well as the number of user accounts and the number of searches each client has made, has been leaked. Could all three billion of the photos in their database be leaked? It may be argued that photos in this database have been scraped from public pages and are therefore already publicly available. However, while this fact has never been explicitly stated, it seems logical to assume that the photos used for searching will be added both to Clearview AI's database as well as to its collection of images. If Clearview AI's system is being used by law enforcement to identify

DOI: 10.1201/9781003264415-41

individuals, the photos used for searching could be sensitive and likely may not be in the public domain. This, naturally, is quite concerning from a privacy standpoint as the Clearview AI database no longer only contains images in the public domain. Even more concerning: when Clearview's lawyer was asked about the leak of their client list, he seemed unworried about the leak of this information, stating, "Unfortunately, data breaches are part of life in the 21st century" (Morrison, 2020).

Clearview AI is also terminating all contracts with groups from Illinois. This is the result of a successful legal case where Clearview was sued for violating the Illinois Biometric Information Privacy Act (BIPA) (Statt, 2020; Robertson, 2022). This law has been in place in Illinois since 2008 with the intention of dealing with the rising worry of identity theft. Especially because when biometric data is jeopardized there is nothing an individual can do to protect their identity, as they do not have a way to change these biometric factors the way they might a password. This law applies to "private entities", which includes everyone from individuals to associations and corporations. BIPA is made up of five elements. First, entities must have a public policy that outlines how long this biometric information will be stored and how this information will be destroyed permanently once the specified storage period has elapsed. Second, biometric data cannot be collected without the informed written consent of the individual whose information is being collected. Third, biometric data cannot be used to gain profit, even if the individual consents for their data to be used for that purpose. Fourth, biometric data cannot be disseminated without consent. The only exception is in special cases, such as the disclosure of biometric data being required by law. Finally, if a private entity has stored biometric data, they must store this as carefully as, or even more carefully than, they would store other confidential information such as Social Security Numbers or passwords. BIPA also allows any private individual to file a lawsuit against private entities that do not comply and can therefore result in significant payouts (De la Torre, et al., 2020).

Due to this act in Illinois, Clearview has stated that they will try to ensure they are not collecting information from Illinois residents. They will do this by banning photos that have metadata indicating that they were taken in Illinois, as well as no longer collecting photos that have been posted from a device with an IP address from Illinois. Clearview AI has also proposed an opt-out tool, but it is yet unclear how this opt-out tool would work (Statt, 2020).

As a result of BIPA, Clearview AI and the American Civil Liberties Union (ACLU) came to a settlement in 2022 that restricts access to Clearview AI's database of images. Comprehensive nationwide measures have been taken to prevent most private individuals or corporations from working with Clearview AI, including government employees who are acting on their own behalf.

The measures outlined by BIPA cannot fully reassure users, as there is no certainty that these steps will be successful at protecting users' information. However, BIPA does hold Clearview AI accountable for how it uses and collects information, although this does only assure a level of privacy and control over the information saved about individuals located in Illinois. Most localities don't have any legislation in place to protect their citizens and BIPA is widely considered to be a standard for measures other states and nations should implement in order to provide protection to their citizens from organizations that collect biometric information.

Clearview AI's facial recognition software has already been used in several police cases to identify individuals and match names to photos of a suspect. The software was also used by banks to try and find individuals involved in fraud cases (Hill, 2020). In response to the COVID-19 pandemic, Clearview is also now offering its database to be used for contact tracing. Clearview CEO, Hoan Ton-That, has stated that they are now in discussions with both state and federal agencies regarding using Clearview's database to efficiently identify, and therefore contact, individuals that have been diagnosed with COVID-19 and those whom they have been in contact with. Clearview says that they are confident in their ability to identify individuals who have been in contact with someone with a confirmed case of COVID-19.

Clearview proposes the use of surveillance videos from locations where people may have encountered an individual with COVID-19 to achieve this, stating that their technology would identify, by name, all individuals in these videos. This could be considered a much better alternative to current invasive contact-tracing methods, which involve constantly tracking an individual's phone signal and therefore tracking their movements. This tracking is sometimes used in conjunction with other information, such as an individual's credit card purchase history (Ward & Sottile, 2020).

Clearview's method could be seen as a better alternative for maintaining a higher level of privacy. Instead of constant location tracking, Clearview will only monitor public spaces that a person with a confirmed case of COVID-19 has visited. This method is also likely to more accurately conduct contact tracing as Clearview would be able to see, using the surveillance footage, individuals who have been in close contact. With location tracking, there is a slight margin of error due to the use of GPS. GPS on a smartphone is accurate within a range of just under 5 m and there are some areas where this would be even less accurate (GPS.gov, 2020). Of course, the main concern with Clearview's implementation is that this method will continue, after contact tracing is no longer needed, for other purposes (Ward & Sottile, 2020).

In 2022, Ton-That offered Ukranian government agencies free access to its database, which by then contained 20 billion faces extracted from various global social media sites. Within one month, five government agencies registered a total of 200 users and had conducted over 5,000 searches (Hill, 2022).

While there are advantages to Clearview AI's proposed methods for contact tracing and law enforcement support, there is concern that the integrity of a database containing a significant volume of biometric data could become compromised.

CASE ANALYSIS

A private company storing biometric data without explicit consent poses a threat to personal privacy, as well as to the integrity of other systems that use the biometric data for security. When Clearview AI was hacked in 2020, the names, user accounts, and searches of law enforcement agencies that use Clearview AI were released. While images reportedly were not compromised, it follows that subsequent attacks could target Clearview AI's image database.

Databases are vulnerable to attacks, such as structured query language (SQL) injection. SQL injection is the misuse of client entry fields to introduce malicious

SQL statements. This abuse of intent is made possible by security vulnerabilities that have been introduced into the application software. As a result, malicious SQL queries are allowed to pass through to the database.

Three methods to secure a database are type validation, allow-list validation, and least-privilege enforcement. Type validation checks all input that is submitted in the SQL database. If the input statement is of a specific (unacceptable) type, it may be rejected. Validation can also be against a specific input allow-list. With allow-list validation, if the input is not within the accepted list, the entire SQL statement is rejected. The effect of SQL injection can be mitigated by enforcing least privilege, where in the event of a successful attack, the attacker's access privilege will be limited. Therefore, the amount of information gained by the attacker is minimal.

A denial of service (DoS) attack can be launched against the database, using connection information to create a program that monopolizes the resources of the database. This is of concern as a DOS attack is one that can be launched by amateurs due to the prevalence of tools and code available online, but attack identification and mitigation techniques are decidedly more difficult and labor-intensive. Access control lists (ACL) can be used to provide some protection to the network on which the database connection is being established. Limiting network traffic rates and implementing session protection are two other mitigation techniques that could be used.

A database can also be compromised through packet sniffing, where attackers listen to the packets being sent from the application to the database (and vice versa). Packet sniffing can provide valuable insight into the database itself, the process of connection establishment, and the queries that are passed to the database. While it is acknowledged that an attacker needs access to the network that facilitates the database connection before they can sniff packets, there is a security risk in making the information contained in packets – transferred from the application to the database – readable in plain text. A protection against this would be to encrypt the packets that are sent between the application and the database.

Clearview AI is being used to support law enforcement agencies, who have cited the usefulness of the image database particularly in criminal cases involving minors. This could pose a security risk if the images used to conduct searches are stored by Clearview AI. For companies like Clearview AI that store biometric data, it is critical for appropriate security measures to be in place to mitigate threats, reduce possible attack vectors, and limit the attack surface.

It has also been recently revealed that the New Zealand Police has been testing this software. While they have ultimately decided not to use this technology, this test has been quite worrying in New Zealand due to the test being undertaken without the knowledge of the public and, arguably more importantly, without the knowledge of either New Zealand's privacy commission or the police commissioner (Smith, 2020). The ability to operate with this level of secrecy reduces accountability for what can be done with the images.

This highlights the importance of having legal protections such as BIPA, where the act clearly states how long information is to be kept and what will be done to destroy it when that time has elapsed. Legal frameworks should also have clear punishments to enforce accountability if companies do not uphold what they claim.

LESSONS LEARNED

Facial Recognition technology is likely only going to become more prevalent moving forward. The New Zealand Privacy Commissioner has stated that facial recognition for criminal investigations is inevitable (Devanesan, 2020). Therefore, legislation being in place to hold companies, such as Clearview AI, accountable for their use of biometric data is critical.

In the US, Clearview AI has been able to operate with minimal accountability thus far. The absence of federal laws prohibiting the extraction of data from public websites, and the prevalence of photo and video sharing on social media and the Internet, makes it far too easy for a company like Clearview AI to build a database with biometric data. The state of Illinois and BIPA have set a precedent that can be followed to hold companies accountable for actions that compromise the integrity of biometric data.

Other states and countries should learn from BIPA and start to put forward their own legislation to protect their citizens. This seems to be the best way to mitigate the many privacy risks that Clearview and other similar companies pose to individuals. Facial recognition technology has been, and will continue to be, a useful tool for law enforcement. The goal is to maximize the benefits of facial recognition technology while minimizing privacy concerns through proper safeguards.

DISCUSSION QUESTIONS

1. What legal protections should be afforded to biometric data?
2. Should biometric data be held by private companies? Is consent necessary if the data is only going to be used by law enforcement agencies?
3. Do you agree with the statement that "data breaches are part of life in the 21st century"?
4. How can we mitigate the risk of a data breach?

REFERENCES

Clearview AI (2020). *Clearview AI - Tested and compliant.* https://clearview.ai

De la Torre, L., Golding, E., & Scarver, I. K. (2020, February 17). *The Illinois biometric information privacy act ("BIPA"): When will companies heed the warning signs?* The National Law Revue. https://www.natlawreview.com/article/illinois-biometric-information-privacy-act-bipa-when-will-companies-heed-warning

Devanesan, J. (2020 May 21). *Police trial controversial Clearview AI facial recognition tech in New Zealand.* Techwire Asia. https://techwireasia.com/2020/05/police-trial-controversial-clearview-ai-facial-recognition-tech-in-new-zealand/

GPS.gov. (2020, April 22). *GPS Accuracy.* https://www.gps.gov/systems/gps/performance/accuracy/

Heilweil, R. (2020, May 8). *The world's scariest facial recognition company, explained.* Recode - Vox. https://www.vox.com/recode/2020/2/11/21131991/clearview-ai-facial-recognition-database-law-enforcement

Hill, K. (2020 January 18). *The secretive company that might end privacy as we know it.* The New York Times. https://www.nytimes.com/2020/01/18/technology/clearview-privacy-facial-recognition.html

Morrison, S. (2020, February 28). *The world's scariest facial recognition company is now linked to everybody from ICE to Macy's.* Recode - Vox. https://www.vox.com/recode/2020/2/26/21154606/clearview-ai-data-breach

O'Sullivan, D. (2020, February 10). *This man says he's stockpiling billions of our photos.* CNN Business. https://edition.cnn.com/2020/02/10/tech/clearview-ai-ceo-hoan-ton-that/index.html

Smith, M. (2020 May 13). *Police trialled facial recognition tech without clearance.* RNZ. https://www.rnz.co.nz/news/national/416483/police-trialled-facial-recognition-tech-without-clearance

Statt, N. (2020, May 7). *Clearview AI to stop selling controversial facial recognition app to private companies.* The Verge. https://www.theverge.com/2020/5/7/21251387/clearview-ai-law-enforcement-police-facial-recognition-illinois-privacy-law

Ward, J. & Sottile, C. (2020, May 1). *A facial recognition company wants to help with contact tracing. A senator has questions.* NBC News. https://www.nbcnews.com/tech/security/facial-recognition-company-wants-help-contact-tracing-senator-has-questions-n1197291

Robertson, A. (2022, May 9). *Clearview AI agrees to permanent ban on selling facial recognition to private companies. The Verge.* https://www.theverge.com/2022/5/9/23063952/clearview-ai-aclu-settlement-illinois-bipa-injunction-private-companies

Hill, K. (2022, April 7). *Facial recognition goes to war.* The New York Times - Breaking News, US News, World News and Videos. https://www.nytimes.com/2022/04/07/technology/facial-recognition-ukraine-clearview.html

3.9 Zoombombing
A Business Process Perspective

Alisha Gilbert
Canterbury District Health Board, Christchurch, New Zealand

Zhigang Gong
University of Canterbury, Christchurch, New Zealand

CASE SUMMARY

Zoom Video Communications, Inc. (Zoom) is a technology company, headquartered in San Jose, California, that provides communication solutions (Wikipedia contributors, 2020). Zoom's business strategy emphasizes offering easier-to-use products than competitors and helping customers gain efficiency. Zoom was founded in April 2011 by Eric Yuan who used to work in Cisco Webex. Stanford University is Zoom's first customer. It launched its service in January 2013. One month after the release, Zoom had 400,000 users. The number of users reached 40 million by February 2015. The maximum number of people allowed to participate in a meeting increased from 5 at the beginning to 1000 in October 2015. From the perspectives of revenue growth and customers' positive feedback on the product, Zoom developed very well. Zoom was valued at $1 billion in 2017 and became a public company in 2019.

In January 2020, a new strain of the SARS-CoV virus, later named COVID-19, was found in Wuhan, China. Despite the Chinese government realizing the severity of the problem and deciding to put Wuhan in lockdown to stop the spread of the disease, COVID-19 spread around the world. The global pandemic that followed enforced quarantine measures worldwide, giving rise to businesses and organizations employing remote work strategies.

To implement the transition from being offline to online, an appropriate IT platform was required. People found Zoom Meetings to be a feasible and suitable platform to implement the transition and used it to achieve many things, such as teaching & learning, gaming, discussions, seminars, parties, and weddings. As a result, Zoom added a huge number of new subscribers. The daily active users increased from 100 million on 1 April to 200 million on 21 April.

Zoom's core product is Zoom Meetings & Chat (hereafter simply referred to as Zoom Meetings). Enterprises and individual consumers use Zoom Meetings to

 DOI: 10.1201/9781003264415-42

conduct video conferences across any device (desktop and mobile) at any time. During a meeting, users can send messages and share content in real-time. Zoom Meetings is built for modern teams. Its main features include HD video and audio, built-in collaboration tools (co-annotate), secure meetings, recording and transcripts, streamlined calendaring, and team chat. In addition, some features like beautification filters and virtual backgrounds make Zoom Meetings rather competitive. Customers switching to Zoom from other platforms claim that their performance, trust, and engagement increased significantly. While Zoom Meetings is free for individual customers, there are limitations to the free service. For example, the length of a meeting cannot exceed 40 minutes.

Individuals who used Zoom Meetings and completed their goals without physically contacting other people felt satisfied with the functionalities provided by Zoom Meetings. Unfortunately, many users of Zoom Meetings fell victim to Zoombombing. For example, the founder and editor in charge of the technology website The Information, Jessica Lessin, tweeted about how her video conference was hijacked by someone who shared pornography. Similarly, The Saint Paulus Lutheran Church in San Francisco fell victim to Zoombombing during a Bible study session. People who encountered Zoombombing felt fury and some even filed lawsuits against Zoom.

CASE ANALYSIS

Wikipedia defines Zoombombing as:

> an unwanted, disruptive intrusion, generally by Internet trolls and hackers, into a video conference call. In a typical Zoombombing incident, a teleconferencing session is hijacked by the insertion of material that is lewd, obscene, racist, or antisemitic in nature, typically resulting in the shutdown of the session.

The attendees of a Zoom meeting have a common interest and want to make the most of it. They want their session to be enjoyable, smooth, and secure. When Zoombombing happens, the host is left with no other option than to shut down the meeting. They then need to reschedule the event later, and all participants in the call will have less interest in joining a Zoom call considering the aftereffects of a disturbing Zoombombing incident. Meeting organizers are therefore likely to consider using other platforms, which could incur a significant time and resource cost.

At present, the people who instigate Zoombombing attacks are hardly being traced, let alone judged in a court of law. Due to the low risk for attackers, there could be more Zoombombing attacks if the security flaws that are being exploited are not fixed. If people tend to have a negative attitude toward Zoom Meetings, it could make the huge market that was gained by Zoom due to the pandemic of COVID-19 a flash in the pan.

It has been almost 7 years since the rollout of Zoom Meetings as a cloud-based communication service. In the beginning, the service was targeted at business users only. This has three implications. First, the host of a Zoom meeting knows all attendees of the meeting. Second, most Zoom Meetings are private. Meeting attendees are invited via the organization's internal communication channels, such as email or calendar invites, and there is no need to publish the meeting link on a public medium.

Third, an employee must provide login credentials to Zoom Meetings when trying to join a meeting. These login credentials are assigned by the organization, an organization's IT department either creating a Zoom Meetings subscription for each employee or linking Zoom Meetings to the organization's existing Single Sign-On (SSO) authentication and authorization platform. Because meeting participants can be identified, they are held accountable for their behavior during a Zoom meeting. These three implications resulted in Zoombombing attacks not being launched during this phase of the development of Zoom Meetings.

The COVID-19 pandemic in 2020 caused the number of daily users of Zoom Meetings to surge, attracting both enterprise and individual customers. Schools and universities can use Zoom Meetings to conduct teaching. Individuals can use it to host happy hour parties. In an era of social networking, individuals tend to publish invites to their events on social media. As a result, many Zoom meeting links started to appear on social media sites like Facebook, Twitter, and Instagram. Individuals who published these Zoom meeting links on social media did not realize the risk. Because the meetings are public and do not have adequate security, anyone who knows the link can join the meeting. In addition, new users of Zoom Meetings are inexperienced. They just want to conduct a meeting and have no idea of the potential dangers of their meeting not being configured properly. In short, a publicly known Zoom Meetings link without any protection is the ideal target for a Zoombombing attack. As such, Zoombombing is a session hijacking attack.

Attackers are always in search of applications' vulnerabilities or weaknesses. If a user is hosting a Zoom meeting and keeps the meeting public by publishing the meeting link online, or if they are not hosting a meeting with a unique ID and password, then a back door is left open, and hackers can exploit this vulnerability to launch an attack on a valid computer session. Hackers who launch Zoombombing attacks frequently change their accounts to avoid being blocked, and therefore a meeting host cannot identify a repeat hacker and is left with no other choice but to shut down a hacked meeting.

The CEO of Zoom Eric Yuan said in a call "We will earn our customers' trust and deliver them happiness with our unwavering focus on providing the most secure platform". When the user base of a service reaches a certain scale, the service provider struggles to meet the security needs. As more security flaws in Zoom Meetings were exposed, Zoom made a 90-day plan to improve its service's security. Zoom has released a software update, Zoom 5.0, to eliminate the Zoombombing security issue (Paul, 2020).

The main features in Zoom 5.0 that enhance security include allowing hosts to "report a user" to Zoom, defaulting users to the waiting room, having passwords enabled by default, and AES 256-bit GCM encryption.

To prevent Zoombombing, users of Zoom Meetings should follow several rules and enable strict configurations for meetings (Vigliarolo, 2020). The Zoombombing starts from searching Zoom meeting links on social media, so rule number one is to stop publishing Zoom meeting links on social media.

Although Zoom allocates a personal meeting ID for each user, users should use a randomly generated Meeting ID instead of their Personal Meeting ID when they are going to organize a meeting.

Every Zoom meeting should be protected by a password. The Zoom meeting links follow a pattern. Hackers could enumerate the possible links for valid meetings. The hacker can't join a valid meeting if he or she doesn't know the password of the meeting.

Users should use the Waiting Room feature in Zoom Meetings. When the Waiting Room feature is enabled, a user who wants to join the meeting gets in line at first, it gives the meeting host a chance to check his or her eligibility. Unapproved users are not allowed to join the meeting.

At last, the autonomy of meeting attendees to play video and audio is the root cause of the destruction. So, turn off any features that can be used to share content with the attendees, such as video, audio, screen sharing, and annotation, for any attendees other than the host.

LESSONS LEARNED

It is not uncommon to see that an enterprise has implemented many measures to assure the security of its assets. The fingerprint locks are installed on doors to prevent unauthorized access. A firewall is installed between the internal network and the Internet. Employees use the VPN to access the internal network. The movable disks are not allowed to be used to transport documents between company computers. Every year the legal department of the company organizes training activities for employees to help them obey the security policies made by the company and industry.

Besides the above security measures, authentication and authorization are must-have components in the security architecture. Every system used by a company requires users to enter credentials (username and password) to log in. To gain stronger security, some companies even use two-factor authentication (2FA). Using 2FA requires users to provide not only a username and password but also another piece of information to prove that they are who they claim to be, such as a code received by SMS. The level of difficulty in managing the authentication and authorization is high when both the number of services used in the company and the number of employees in the company are large. So, big companies usually use Single Sign-On (SSO) to manage authentication and authorization. The accounts of employees are stored in a single place, e.g., A Microsoft Active Directory (AD) server. When an employee quits, the administrator only needs to delete the corresponding account from the AD, the employee can no longer access any system in the company.

A service delivered by software, such as Zoom Meetings, can be used by both enterprise customers and individual customers. The context in which the service is used determines the level of risk that users will face. Enterprises usually won't use the service out of the box. They often consider the impact of the deployment of the service on the security architecture used in the organization. Usually, the enterprise will integrate its authentication system into the service's authentication system. Okta, Inc. (Okta) is a company that provides solutions to facilitate authentication integration. Because of authentication integration, employees can use external services with their existing account credentials. If any issues arise relating to the use of the service, the scope of impact is limited, and the issue can be investigated to determine the responsible parties. Therefore, the level of risk that enterprises face when they use the service is relatively low. Conversely, individual customers often use software services out of the box. Few of them will notice the security issues of the service in the beginning. If the user base of the service is small, problems are unlikely to arise because attackers rarely are interested in attacking a service with low attention and reputation. However, if the user base of the service is large, and the use of the service is ubiquitous and becomes a phenomenon, the service can easily become the prospective target of attack. Attackers can build their reputation by attacking software services with a large user base and their anonymity can make them difficult to trace. If the benefit of an attack vector prevails over its price, more hackers will be encouraged to launch the attack and to launch it repeatedly. As a result, the level of risk that individual customers face is relatively high.

Technology companies who focus on providing a service to the enterprise customer market should plan changes to the security architecture of the service if it has the potential to develop to be a star product in the personal consumer market. Zoom Meetings is a typical example. Zoom didn't prepare well for the significant increase in individual customers or the innovative ways that its service would be used. Fortunately, Zoom identified the security issues and enacted effective response actions.

DISCUSSION QUESTIONS

1. Why are the security problems of Zoom so prominent in 2020?
2. What does a software development company need to prepare if its enterprise-oriented product has the potential to be used by many individual consumers?
3. Will you abandon a productive and ease-of-use software if it has security risks?
4. Do you have any ideas other than Zoom's solutions to solve Zoombombing?
5. What is the social networks' impact on software security?

REFERENCES

Wikipedia contributors. (2020, July 11). Zoom video communications. Retrieved July 11, 2020, from https://en.wikipedia.org/wiki/Zoom_Video_Communications

Paul, K. (2020, July 1). Zoom releases security updates in response to "Zoom-bombings". Retrieved July 11, 2020, from https://www.theguardian.com/technology/2020/apr/23/zoom-update-security-encryption-bombing

Vigliarolo, B. (2020, April 3). How to prevent Zoom bombing: 5 simple tips. Retrieved July 11, 2020, from https://www.techrepublic.com/article/how-to-prevent-zoom-bombing-5-simple-tips/

3.10 Security in the Healthcare Sector

Alisha Gilbert
Canterbury District Health Board, Christchurch, New Zealand

Zhigang Gong
University of Canterbury, Christchurch, New Zealand

CASE SUMMARY

We see changes in almost every sector in this modern world. A rapid change can be witnessed in the healthcare sector, where day by day, new developments and ideas are being expressed and introduced. The healthcare sector plays a huge role in human life, as it is inherently human to desire a healthy body and mind. The healthcare sector is an integration of different sectors in the economy that provides services and goods needed for the treatment of patients. This industry aims to maintain and improve the health of people by offering medical services and by manufacturing and commercializing products needed for those services.

Technology is adopted by every sector to ease the difficulty of work or reduce the time consumed by a given task. Through the adoption of technology, the quality of life can be improved. This motivates the introduction of innovations and techniques in the healthcare sector. Technologies like artificial organs, brain sensors, robotic surgery, and smart inhalers are coming into use and gaining popularity. Thanks to technology, the location of delivery for medical services can be a patient's house, moving ambulance, or office. Furthermore, medical services can sometimes be offered without the physical presence of the doctor in instances where the doctor's virtual presence through video conferencing is deemed adequate.

When considering the administration needs of medical services, technology can be an asset. All data related to patients, hospitals, doctors, management, etc., can be entered and stored in a central digital system. Some hospitals even use Enterprise Resource Planning (ERP) modules for a smooth administration process. In the not-so-distant era where patient details, such as medical records, were entered manually and filed in thick folders and stored in huge file racks, patients needed to request their relevant records and files before entering the doctor's office. Now, this time-consuming process is eliminated from the process of receiving medical services because of the introduction of electronic

DOI: 10.1201/9781003264415-43

patient records. This holds not just for patient records, but every record regarding the hospital, management, and employees.

Access to these records can be done by a single click if the user has the necessary credentials. For example, doctors can access a patient's medical record using their own login credentials, while front office staff can access employee work schedule records. For security purposes, access requests for all electronic records are stored and tracked. These records (in full or part) can be used for other purposes. For example, to determine the health of the community by the government or other organizations like the World Health Organization (WHO). At the same time, data can be transferred to insurance companies to cover expenses incurred by a patient or to check a patient's eligibility for a given insurance scheme. There have also been cases where access has been granted to first responders in an accident, where knowledge of a victim's medical history is necessary for safe treatment.

While there are advantages to adopting technology in the healthcare sector, every coin has two sides and there are also some disadvantages. Recently, it was reported that overseas hackers attempted to target the New Zealand health sector. A District Health Board Information Services Director, Geoff King, indicated that these attacks are increasing daily, with the hackers aiming to ransom money or to enact other inappropriate schemes, such as when a nurse clicked an encrypted mail link and ended up corrupting the computer's desktop (Leaman, 2017).

Similarly, a report was filed by Maryland Health Services and Lorien Health Services that 47,754 patient records were breached. The data from Lorien Health Services was ransomed, displayed on the Lorien Health website, and published on the dark web for sale. The records contained patient names, dates of birth, social security numbers, health-related diagnoses, and treatment plans (Davis, 2020). In February 2019, another ransomware case was reported by the Southeastern Council on Alcoholism and Drug Dependence where 25,148 patient records were stolen from the system (Davis, 2019). According to research on information risk, health data holds second place, after social security numbers, in high-risk information that offers a significant reward for a successful hacker.

Hackers have different reasons for wanting access to health records (Dimov, 2016). Health records are in high demand on the dark web. In ransomware cases, a hacker can sell the information back to the hospital itself. Hospitals are likely to buy back the compromised records so they can preserve the security of the patients and maintain their reputation. On the black market, a person's medical records can be used against them. In some cases, healthcare organizations have even hired hackers to attack competing healthcare providers and defame them.

CASE ANALYSIS

The privacy of the patient and their information (e.g. their social security number, date of birth, name, current and past medical treatments, physical diagnoses, genetic information) are exploited and sold or distributed on the dark web. This risk to patients can be broadly classified into two categories: organizational risk and systemic risk. Organizational risk results from unauthorized exposure to medical records either by internal agents abusing their rights or by foreign agents abusing information system

(IS) instability. Systemic risk occurs when an individual in the information security chain manipulates the details collected outside their actual use. Organizational risk and systemic risk lead to organizational threats and systemic threats, respectively.

Organizational threats can be realized because of a careless employee mistake, such as sending information to an inappropriate recipient, or a deliberate action against the organization, and can be conducted by two types of agents: employees and outsiders. An employee can access the IS and collect data without any valid reason or inappropriately make use of it. An outsider, who can be termed a hacker, can gain unauthorized access to the organization's IS. Both can have a motive to obtain and expose stolen data.

The attack could affect the organization as well as the owner of the compromised data. When it comes to medical organizations, like hospitals, their reputation could be damaged by an attack. This could undermine the trust patients have in that medical organization. This is a non-economic motive for the hacker. An attack can be launched from any location, and experienced hackers have often accumulated the skills and resources necessary to successfully launch attacks anonymously and across multiple attack vectors. For motivated agents who do not have the necessary knowledge to launch their own attack, many hackers provide their services for a fee on the dark web.

Reports indicate that systemic threats are mostly performed by agents who have inside resources, such as valid system login credentials provided by the organization, and are part of the organization's IS. The motive behind an attack can differ between agents. For example, an insurance company may want to reject an application or not pay a customer. Employers could falsify records to wrongly dismiss an employee or prevent a pay raise or promotion. A doctor may wish to use patient records for research purposes and could decide to do so even in the absence of written patient consent.

All healthcare data is saved in a database management system (DBMS). Different attack vectors can be exploited to access the DBMS and steal its data for economic and non-economic purposes. Common attack vectors are phishing, malware, exploits, and dictionary attacks.

LESSONS LEARNED

The healthcare sector uses digital technologies to facilitate both administrative and medical tasks. While advantageous, the adoption of digital technologies can give rise to attack vectors that pose a threat to the organization and its patients. Notably, it's not just hackers that exploit attack vectors, but also agents granted legitimate system access who can commit these crimes knowingly or unknowingly. These crimes can result in economic as well as non-economic benefits.

The sale of medical records on the black market can provide an economic benefit to an attacker while maintaining their anonymity. A stolen medical record can sell for an average of $250 and a projected maximum of $1,000. Compare this to $5 for a credit card number or $1 for a social security number. Medical records can be sold on dark web storefronts and dashboards. Resembling online shopping sites, storefronts provide many anonymous transaction options. Dashboards require entrants to deposit money and prove that they have hacking skills and valuable data that can be sold on the dashboard.

When it comes to the use of technology in any sector, private information stored in an organization's DBMS must be protected. To mitigate threats, it is important to regularly update an organization's IS and implement new security measures (Brown, 2017). While this may necessitate an increase in the budget allocated to IS security, providing a secure IS is a crucial factor in gaining and maintaining consumer trust. Further measures to securely destroy old data, enforce the security of external-facing facets of the IS, as well as annual analyses of compliance with both the Health Insurance Portability and Accountability Act (HIPAA) and the Health Information Technology for Economic and Clinical Health Care Act (HITECH) are valuable additions to an organization's security policy. These analyses could help determine vulnerabilities and threats to the system, allowing the organization to address them before issues arise.

DISCUSSION QUESTIONS

1. How do you respond to a cyberattack in a healthcare organization?
2. Have healthcare organizations implemented sufficient countermeasures against cyberattacks?
3. What are the benefits of HIPAA and HITECH and how can they help prevent cyberattacks?
4. How can you evaluate the attack vector properly to prevent future attacks?
5. How can we prevent cyberattacks in the healthcare sector?

REFERENCES

Brown, L. (2017, May 4). 6 important ways to prevent data breaches in healthcare. Shred-it. https://www.shredit.com/en-us/blog/data-security/ways-to-prevent-data-breaches-in-healthcare

Davis, J. (2019, May 13). Ransomware attack on Connecticut provider impacts 25,148 patients. Health IT Security. https://healthitsecurity.com/news/ransomware-attack-on-connecticut-provider-impacts-25148-patients

Davis, J. (2020, July 21). Lorien health services ransomware attack impacts 48K patients. Health IT Security. https://healthitsecurity.com/news/lorien-health-services-ransomware-attack-impacts-48k-patients

Dimov, D. (2016, May 24). Healthcare hacking. Infosec. https://resources.infosecinstitute.com/healthcare-hacking/

Leaman, A. (2017, April 19). Overseas cyber criminals target New Zealand hospitals, public sector. Stuff. https://www.stuff.co.nz/national/health/91701509/overseas-cyber-crims-target-hospitals-public-sector

3.11 eBay Hack

Tridhawan Choudhary
AFFCO New Zealand Limited, Hamilton, New Zealand

Priyanka Jagre
UNIFY Solutions, Wellington, New Zealand

CASE SUMMARY

In 2014, hackers attacked the e-commerce site eBay and stole user information from a database. This allowed the hackers to obtain personal user details, such as full names, home addresses, mobile numbers, email addresses, dates of birth, and account passwords. In a statement, eBay said that the passwords stored in the database were encrypted and that the hackers would not be able to decrypt them. eBay's spokesperson revealed that a large volume of data was impacted by this hack but did not reveal how many of their 148 million active user accounts were stored in the compromised database. Fortunately, the payment platform used, PayPal, was not affected by the attack because payments are managed on a separate network.

The hackers were able to access a "small number" of login credentials from eBay employees to access the database. The hackers launched their attack undetected because it took eBay weeks to realize that employee credentials had been stolen. Knowledge of the attack prompted eBay to conduct a forensic examination of its systems and devices to determine how it had been launched and what had been compromised.

To eBay buyers, the good news was that the passwords were encrypted and, with a random number added, "salted." If a salt is used during encryption, it makes decryption much harder. The main danger in this case is posed by personal details being shared, such as physical addresses and phone numbers. These personal details, especially if combined with a Social Security number or bank details, could be used for identity theft (Pagilery, 2014).

Electronic commerce (e-commerce) refers to the purchase or sale of goods and services, or the transmission of funds or data through an electronic network. These transactions can take place between businesses (business-to-business), between consumers (consumer-to-consumer), or between businesses and consumers (business-to-consumer). In most cases, the electronic network that is used for e-commerce is the Internet. e-commerce relies on computers, telephones, barcodes, credit cards, order management systems, electronic payment and verification, order fulfillment systems, distribution networks, and customer service teams. e-commerce provides many new opportunities for organizations but also introduces new risks and vulnerabilities such as information systems (IS) security threats from remote attackers. Defense of

DOI: 10.1201/9781003264415-44

e-commerce properties against unauthorized entry, use, modification, or destruction is e-commerce security.

e-commerce security is part of the information security framework and is applied exclusively to e-commerce modules, like computer security and data protection. Stringent protection is required for e-commerce components that impact the end user. User information privacy issues have given rise to a lack of trust in a number of contexts, including the industry, electronic health records, recruitment technologies, social networking, and e-commerce. Providing inadequate information security is one of the main factors impeding customers and organizations from engaging with e-commerce (Panigrahi, Padhy, & Patro, 2016).

CASE ANALYSIS

e-commerce organizations are increasingly solving their information security problems. Since a variety of consumers use online shopping, they have different levels of digital literacy and some may not immediately appreciate the importance of information security. Client systems containing a virus or Trojan horse pose one of the biggest threats to e-commerce sites. Clients must be granted access because this is how consumers can make purchases. Additionally, many e-commerce organizations utilize services from third-party organizations and attacks can originate from those third-party organizations.

All e-commerce transactions require the following security measures:

- Access control
- Integrity checks
- Secure contract identification
- Digital signatures
- Encryption
- Secure tracking and delivery of products

The biggest concerns for e-commerce are attacks that infect systems with a virus, worm, or Trojan horse. These attacks can interrupt the organization's operations. Furthermore, all three attacks can originate from a valid client device and can therefore be difficult to detect. To mitigate this risk, the authentication and encryption methods used need to be sufficiently advanced (Panigrahi, Padhy, & Patro, 2016).

e-commerce security has various dimensions, such as the following:

- Integrity
  - prevention of unauthorized data change
- Nonrepudiation
  - prevention of reneging on an agreement after the event against any party
- Authenticity
  - data source protection
- Confidentiality
  - Protection from unauthorized knowledge disclosure
- Privacy
  - data control and divulgence provisions

Some of the security measures to be taken for an e-commerce site are as follows:

- Firewalls
- Public Key Infrastructure (PKI)
- Encryption
- Security certificates
- Digital Signatures
- Passwords

None of the aforementioned security measures are airtight, and they work best in combination. For example, the advantage of PKI is that each protected transaction does not require a centralized, highly open intermediary. It also makes it difficult to know whether the key of another party has been stolen or otherwise compromised. In asymmetric key systems, a public key and private key pair are used to maintain security. The public key can be widely distributed without compromising security. Digital signatures, provide digital identity authentication and require transmission of a public key. Using PKI, a receiver can validate the identity of the public key sender before creating the digital signature.

LESSONS LEARNED

Security threats to e-commerce are posed by clients, communication channels, and servers. The openness required for e-commerce makes securing systems challenging. e-commerce platforms must follow best practices to maintain security. For example, HTTP is largely outdated and should not be used because it does not provide security. Similarly, it is now known that sensitive data, such as passwords, must be stored in an encrypted form because leaving it in plain text does not provide adequate protection against attacks.

Multi-layer security makes it more difficult for a hacker to access a system. To achieve multi-layer security, many security measures must be put in place concurrently. For example, antivirus and anti-malware software, firewalls, and two-factor authentication.

DISCUSSION QUESTIONS

1. Why do people save their credit card as the default card while shopping online?
2. Why don't people install antivirus software on their personal computers?
3. Why is HTTP still used?

REFERENCES

Pagilery, J. (2014, May 21). *CW BUSINESS.* Retrieved from EBay customers must reset passwords after major hack: https://money.cnn.com/2014/05/21/technology/security/ebay-passwords

Panigrahi, R., Padhy, N., & Patro, S. P. (2016). Security Issues over E-Commerce and their Solutions. *International Journal of Advanced Research in Computer and Communication Engineering ISO 3297:2007 Certified Vol. 5, Issue 12*, 81–85.

3.12 Using AI to Maintain Security of Healthcare Systems

Alastair Blackett
M2X, Auckland, New Zealand

Cherie Chun Mei Mak
LivePerson, Singapore, Singapore

CASE SUMMARY

In 2009, US President Barack Obama signed the Health Information Technology for Economic and Clinical Health Care Act (HITECH Act) to encourage the adoption of health records and support the advancement of technology within the sector. In support of the HITECH Act, the Affordable Care Act (ACA) of 2010 was enacted as a tool to force the healthcare sector to shift from hard copies to electronic documentation for all medical records of Medicare patients (Ghosh, 2012). The focus of implementing federal government strategies and legislation to automate patient healthcare records and hold electronic records was the overall improvement of quality, safety, efficiency, and cost-effectiveness of care delivery (Ghosh, 2012). Perhaps what the government did not have in mind was that the transitioning from hard copy to soft copy documentation of patients' records would expose the sector to immense cyber threats, since hackers can steal medical records remotely from their computers. Researchers have seen the backlash of these federal government regulations, requiring healthcare providers to implement electronic healthcare records and healthcare exchange systems – where the healthcare industry has become a major target for cyber attackers. For instance, since the introduction of the two acts, HITECH and ACA, from 2010 to 2015, there has been a 125% increase in security breaches (Ghosh, 2012). The first half of 2016 also saw an increase of 11 million electronic medical records, demonstrating rapid growth in the digitalization of patient records. As the healthcare industry becomes more technologically sophisticated, it faces greater exposure to data breaches.

Throughout the US, healthcare centers are becoming more highly integrated, including the use of communication technology to facilitate data exchange. However, due to IT integration and the lack of robust data security systems and strategies, this leads to inherited vulnerability dependencies (Dwivedi, 2019). These vulnerabilities

 DOI: 10.1201/9781003264415-45

include disruptive attacks that have shut down hospital systems such as the 2017 WannaCry attack that affected many hospitals in the United Kingdom (UK), almost causing the loss of many lives (Vicars, 2019). Other similar examples include identity theft for insurance fraud, loss of historical data that is crucial for treating patients, gaining access to control of laboratories affecting their research, and unauthorized modification or injection of false data that can impact medical decisions. The Inspector General of the Head of Investigations at the UK's National Health Service (NHS) office, Gary Cantrell, said hackers are attracted to stealing a "treasure trove of all this information about you". The hackers gain enough information to take out a loan or set up a line of credit under the patient's name (Vicars, 2019). A victim of identity theft, Brandon Reagin's identity was stolen in 2004, when hackers used his personal information to steal cars and incur $20,000 worth of medical procedures (Vicars, 2019).

CASE ANALYSIS

According to researchers, a security or data breach is described as the compromise of the security, confidentiality, integrity, or loss of electronic data resulting in the unauthorized attainment of sensitive personally identifiable information (Abeyratne, 2019). The healthcare sector is one of the main areas of life that is experiencing these breaches. Today, the healthcare sector sees many big data breaches due to cybercrimes, which have now been identified as the new healthcare crisis. Table 3.12.1

TABLE 3.12.1
The Different Methods Hackers Use for Attacks on the Healthcare Sector

Network packet detection makes use of critical system information, like user account information, passwords, and network topology. It can insert new information or change existing information in the packet.

Spoofing attack (incl. IP spoofing) is done to impersonate an authorized user, without having to use its password, by falsifying data to gain access to the system.

Password attacks include password guessing, capturing, cracking, and sniffing and can be facilitated using dictionary, brute force, phishing, etc., strategies to gain access to accounts and services.

SQL injections to bypass authentication and tamper with (insert, update, delete) database data.

Deception attacks or stealthy deception attacks are another form of false data injection that introduce into the communication signal an incorrect sensor measurement or control input, an incorrect time stamp, or a falsified identity of the sending device.

Intrusion attacks compromise the confidentiality of a system to illegally access the cyber system, such as using malware to introduce a backdoor.

Eavesdropping or sniffing is a passive attack in which the attacker listens to the communication medium and captures the network traffic packets.

Application layer attack targets the application servers to cause a fault by reading, deleting, or modifying data and enables the attacker to launch further attacks such as placing a packet sniffer in the internal network to gain confidential information or introducing viruses to spread and cause a system failure.

demonstrates the various methods used by hackers to exploit vulnerabilities. There are many reasons why this sector faces the most challenges compared to others, and the difficulties of maintaining cybersecurity are compounded by limited resources and lack of experience due to being late adopters of electronic data exchange (Abeyratne, 2019).

With the external pressures from federal governments and hackers, the healthcare sector is required to implement AI to ensure compliance while preserving the safety of their patients. AI has been implemented to support cybersecurity in safeguarding electronic resources (Madnick, 2019). Continuous performance improvements in both software and hardware, along with new paradigms such as big data and cloud computing, have led to the development and deployment of many AI applications with a wide variety of capabilities. In terms of AI applications supporting cybersecurity, detection systems are the most common. Detection system cybersecurity solutions will perform traffic analysis and will classify Internet traffic as either legitimate or malicious (Madnick, 2019). As medical centers get more integrated, so do their Internet-connected devices and applications. This is observable in the large amounts of network traffic being generated in real-time. The downside is that providing security using a detection system involves the time consumption of creating rules for effective traffic analysis; hence the introduction of Machine Learning (ML) techniques for better cybersecurity management. ML techniques work autonomously to collect data and distinguish malicious traffic from legitimate traffic (Madnick, 2019). Researchers have found that ML deep learning algorithms can detect Denial of Service (DoS) attacks with a 96.65% accuracy. This technique is said to be ideal in a Software-Defined Networking (SDN) environment, which is important given the relevance of SDN considering the rapid growth in Internet devices connected to the healthcare sector network (Madnick, 2019). When using ML, AI has been highly successful in identifying some of the most sophisticated cyberattacks in the last five years. This is due to ML knowingly luring the hacker into the system, while at the same time being agile and quick enough to catch them in the act before the hacker can do any damage. AI supports businesses by being able to investigate incidents that have been detected and take necessary action. However, the wider discussion surrounding AI's role in solving cybersecurity issues requires further investigation, and some concerns have arisen concerning where and how AI should be regulated (Madnick, 2019). Furthermore, the extended use of AI to support the needs of the healthcare sector, outside of cybersecurity, has caused some debate (Wired Insider, 2020).

Intelligent machines like AI are becoming more integral solutions for humanity and these machines will consume fundamental resources for life (Vedllido, 2019). The use of AI in the healthcare sector speaks to how "AI is improving diagnostics, patient care and clinical decision support across the medical field". Though we mainly focus on the link between the use of AI and its effect on cybersecurity, it is equally important to see how the healthcare sector uses AI in other ways because it will affect society's perceptions of how the industry uses this technology. AI can overcome the computationally intensive, intellectual, and perhaps creative limitations of humans (Vedllido, 2019). The benefits of AI support strategic elements in the plans for many sectors of business, medicine, and government on a global scale, where it has become an integral component of the business model in the design and improvement of algorithms that help computers crunch enormous amounts of data very little expressivity programming (Vedllido, 2019). Despite the success and glorification of AI around the world, some questions have arisen concerning its use and the effect of reducing the subjective relationship between a healthcare provider and their patient. Furthermore, AI is so powerful that medical professionals feel their sense of being in control is taken away and replaced with a false sense of security. Additionally, deep learning for AI is not transparent in how it decides its many-to-many data points, and an operator can blindly trust an AI decision without giving any justification as to why this decision was made for the patient (Resnik, 2017). This allows offloading responsibility to the system and limits the accountability of medical professionals. The frustration built by the patients may lead to limitations in medical research in the future, as there seems to be an agreement that each patient should be able to decide if and how their information can be used (Resnik, 2017). These privacy issues will continue to be a major concern, even though new computer programs can readily remove names and other personal information from records that are transported into large databases (Resnik, 2017). Given public knowledge of cyberattacks for exploitation of patients' personal information, and the high personal risk if the healthcare system is breached or treatment is miscalculated, there is not much public confidence that their information is truly safe and free from hackers or technological programming bias. The healthcare sector needs to face the level of legitimacy that is expected by society when using AI. It is immensely important for the healthcare sector to operate within the spirit of legitimacy theory, which describes the case where generalized perceptions that the actions of an entity are desirable and appropriate within some socially constructed norms, values, beliefs, and definitions determine the legitimacy of those actions (Idowu & Capaldi, 2013). If not, the sector will be faced with a plethora of ethical issues when combining cybersecurity and AI in a risk management approach to a strategic approach (Idowu & Capaldi, 2013). In other words, society and government may subsequently implement regulations to hinder the sector's ability to use AI that is beneficial for medical practice and cybersecurity. Therefore, the healthcare sector will need to focus more on its ability to show its legitimacy in using AI, and this may necessitate a broadening in focus away from simply treating patients.

The healthcare sector is still exposed to many online hacking threats, even with the implementation of strong AI technology. Threats that increase the risk of falling victim to cybercrimes are not only hacker attacks but also staff negligence, third parties such as insurance companies and physicians, and loss of unsecured devices such as laptops, tablets, and cell phones (Karasaridis, 2018). Indeed, the significantly increased security risk to the healthcare sector is attributed to employee negligence from critical electronic devices getting stolen or lost. This has become one of the leading causes of data breaches in the healthcare industry. Cybercrime is growing rapidly from these stolen and lost devices, as thieves have noted ease in hacking such devices versus healthcare databases (Karasaridis, 2018). As mentioned before, the large quantity of Internet-connected devices within the healthcare sector shows an easier route for hackers to gain access to private patient information. Research has found that often these devices do not possess the appropriate safeguards or encryptions necessary to mitigate attacks. This issue requires serious attention, and healthcare organizations need to look introspectively at their medical employees' existing careless security practices as history has shown these are factors that cause data breaches to occur (Karasaridis, 2018). These poor security practices may imply that the healthcare sector is naïve to the innovative thinking of hackers and believes that a single safeguard, their AI application, will protect them. Healthcare organizations should thoroughly reassess and reevaluate their cybersecurity programs and formulate plans to prevent costly data breaches (Karasaridis, 2018). These practices and programs could include: improving encryption, implementing equipment custody policies and procedures, employee training on effective privacy and security measures, installation of remote wipe software on portable devices, effective data security programs, and an urge for stronger passwords (Karasaridis, 2018). The implementation of better security practices and procedures will further enhance the chosen AI's capability. This shows that AI and humans are required to work together to ensure medical records are safe, therefore protecting the lives of many.

LESSONS LEARNED

The CIA Triad of Information Security

This case is exemplary of the information security CIA triad of Confidentiality, Integrity, and Availability of information in practice. In the US, the healthcare industry is under legislative pressure to digitalize patient medical records. This has created industrial challenges when ensuring the confidentiality of these electronic health records. Viewing of data or information by unauthorized people creates issues with confidentiality (Nweke, 2017). The healthcare sector has started to adopt AI as a technical cybersecurity safeguard to protect its electronic assets against data breaches. For instance, AI algorithms can be deployed to study and identify deviations in healthcare employees' security behavior to inform appropriate organizational measures to improve conscious security practices (Yeng, Nweke, Woldaregay, Yang, & Snekkenes, 2020). Integrity is concerned with the 'trustworthiness' of data – whether data is accurate, relevant, and complete – and can be interpreted to imply protection against unauthorized modification or destruction of data (Maconachy, Schou, Ragsdale, & Welch, 2001). Vulnerabilities in data security systems that create loopholes for unauthorized modification or insertion of inaccurate data can impact medical decision-making. In literature, patients' privacy concerns often impede the use of health information systems. However, healthcare providers can mitigate such privacy and security concerns by incorporating robust security safeguards, where trust in health information and security beliefs contribute to improved quality of patient care (Kisekka & Giboney, 2018). The goal of cybersecurity concerning availability ensures that authorized users have access to the necessary resources to perform their work with ease (Nweke, 2017). Clinicians require timely access to patient electronic health records, and the availability of relevant information is crucial to facilitate clinical decision-making and enhance patient care (Jacob, 2020).

Government Versus Governance

There is a clear distinction between governments and governance, with the former referring to political organizations and the latter describing the informal and formal rules that shape social and economic interactions within a system (Abimbola, Negin, Martiniuk, & Jan, 2017). Governmental legislation can impact governance in the healthcare industry. For instance, legislation to drive the adoption of electronic health records requires healthcare organizations to adopt rules that steer the proper implementation of such technologies. It has been suggested that one of the factors influencing cyberattack vulnerability in healthcare is due to fragmented governance, which results in a lack of clarity and responsibility regarding data and system security (Martin, Martin, Hankin, Darzi, & Kinross, 2017). This case briefly touched on the use of AI in healthcare diagnostics, an area that encompasses ethical considerations. On a governance level, the deployment of AI systems is likely to involve the emergence of problems ranging from liability to disaster insurance, where challenges exist in the development of legal and regulatory frameworks for responsible governance of AI systems (Maas, 2018). Opportunities for AI usage are tremendous, but the technology comes with substantial risks that can pose extreme governance challenges (Dafoe, 2018).

Risk Management and the Identify Function in the NIST Cybersecurity Framework

In research, organizations continue to face new risks with new technologies – however, some traditional socio-technical cybersecurity risks continue to exist, e.g., theft of unattended assets containing confidential data, password sharing, and sending confidential emails to the wrong recipients (van Deursen, Buchanan, & Duff, 2013). Regarding the case, transitioning from physical to electronic patient health records creates a new risk and threat to patient data privacy. Despite the potential of AI as a cybersecurity safeguard, the healthcare sector continues to be plagued by cybersecurity threats resulting from more traditional risks such as staff negligence and loss of devices containing confidential data. This highlights the importance of risk management to safeguard against not just external threats but internal organizational threats. In healthcare literature, the application of machine learning in intrusion detection and prevention systems, to counter internal threats, is an emerging area of interest (Walker-Roberts, Hammoudeh, & Dehghantanha, 2018). The National Institute of Standards and Technology (NIST) Cybersecurity Framework consists of the Framework Core, Implementation Tiers, and Profiles, where the Core comprises five high-level functions that apply to risk management: Identify, Protect, Detect, Respond, and Recover (NIST, 2020). The Identify Function aims to assist organizational understanding to manage their cybersecurity risks – such as the identification of established cybersecurity policies to define governance programs, identification of organizational assets to establish asset management programs, and identification of asset vulnerabilities and threats as part of organizational risk assessment and management (NIST, 2018).

DISCUSSION QUESTIONS

1. Why is it important to secure confidentiality, integrity, and availability of healthcare information?
2. How do governments and governance impact the use of AI within the healthcare industry?
3. What are some cybersecurity threats that plague the healthcare industry?

REFERENCES

Abeyratne, R. (2019). Regulating cyber security. *Legal Priorities in Air Transport*, 157–194. https://doi.org/10.1007/978-3-030-18391-2_11

Abimbola, S., Negin, J., Martiniuk, A. L., & Jan, S. (2017). Institutional analysis of health system governance. *Health Policy and Planning*, 32(9), 1337–1344. Retrieved from https://doi.org/10.1093/heapol/czx083

Dafoe, A. (2018). *AI Governance: A Research Agenda.* Retrieved from https://www.fhi.ox.ac.uk/wp-content/uploads/GovAI-Agenda.pdf

Dwivedi, Y. K. (2019). Artificial Intelligence (AI): Multidisciplinary perspectives on emerging challenges, opportunities, and agenda for research, practice and policy. *International Journal of Information Management*, 12(1), 777–802. https://doi.org/10.1016/j.ijinfomgt.2019.08.002

Ghosh, B. (2012). Healthcare systems using clinical data. *Cases on Healthcare Information Technology for Patient Care Management*, 208–223. https://doi.org/10.4018/978-1-4666-2671-3.ch013

Idowu, S. O., & Capaldi, N. (2013). Corporate social and environmental responsibility. *Encyclopedia of Corporate Social Responsibility*, 17(4), 544–544. https://doi.org/10.1007/978-3-642-28036-8_100402

Jacob, P. D. (2020). Management of patient healthcare information. In *Fundamentals of Telemedicine and Telehealth* (pp. 35–57). Elsevier. Retrieved from https://doi.org/10.1016/B978-0-12-814309-4.00003-3

Karasaridis, A. (2018). Artificial intelligence for cybersecurity. *Artificial Intelligence for Autonomous Networks*, 231–262. https://doi.org/10.1201/9781351130165-10

Kisekka, V., & Giboney, J. S. (2018). The effectiveness of health care information technologies: Evaluation of trust, security beliefs, and privacy as determinants of health care outcomes. *Journal of Medical Internet Research*, 20(4), e107. Retrieved from https://doi.org/10.2196/jmir.9014

Maas, M. M. (2018). Regulating for 'Normal AI Accidents'. In *Proceedings of the 2018 AAAI/ACM Conference on AI, Ethics, and Society* (pp. 223–228). New York, NY: ACM. Retrieved from https://doi.org/10.1145/3278721.3278766

Maconachy, W. V., Schou, C. D., Ragsdale, D., & Welch, D. (2001). A model for information assurance:an integrated approach. In *2001 IEEE Workshop on Information Assurance and Security United States Military Academy*. West Point, NY. Retrieved from https://www.researchgate.net/publication/235470635_A_Model_for_Information_AssuranceAn_Integrated_Approach

Madnick, S. E. (2019). The ethics of AI: What does AI do when humans cannot agree what is 'Right'? *SSRN Electronic Journal*, 10(14), 16–24. https://doi.org/10.2139/ssrn.3542524

Martin, G., Martin, P., Hankin, C., Darzi, A., & Kinross, J. (2017). Cybersecurity and healthcare: how safe are we? *BMJ*, 358, j3179. Retrieved from https://doi.org/10.1136/bmj.j3179

National Institute of Standards and Technology (NIST). (2018). The five functions. Retrieved from https://www.nist.gov/cyberframework/online-learning/five-functions

National Institute of Standards and Technology (NIST). (2020). An introduction to the components of the framework. Retrieved from https://www.nist.gov/cyberframework/online-learning/components-framework

Nweke, L. O. (2017). Using the CIA and AAA models to explain cybersecurity activities. *PM World Journal*, VI(XII). Retrieved from https://pmworldlibrary.net/wp-content/uploads/2017/05/171126-Nweke-Using-CIA-and-AAA-Models-to-explain-Cybersecurity.pdf

Resnik, D. B. (2017). Ethical issues in human genomic databases in addiction research. *Genetic Research on Addiction*, 108–121. https://doi.org/10.1017/cbo9781139058971.009

van Deursen, N., Buchanan, W. J., & Duff, A. (2013). Monitoring information security risks within health care. *Computers & Security*, 37, 31–45. Retrieved from https://doi.org/10.1016/j.cose.2013.04.005

Vedllido, A. (2019). Societal issues concerning the application of Artificial Intelligence in Medicine. *Intelligent Data Science and Artificial Intelligence (IDEAI)*, 5(17), 88–112. https://www.karger.com/Article/Pdf/492428

Vicars, A. (2019, March 1). What hackers actually do with your stolen medical records. Advisory Board. https://www.advisory.com/daily-briefing/2019/03/01/hackers#:~:text=Gary%20Cantrell%2C%20head%20of%20

Walker-Roberts, S., Hammoudeh, M., & Dehghantanha, A. (2018). A systematic review of the availability and efficacy of countermeasures to internal threats in healthcare critical infrastructure. *IEEE Access*, 6, 25167–25177. Retrieved from https://doi.org/10.1109/ACCESS.2018.2817560

Wired Insider. (2020). AI in Healthcare: Protecting the Systems that Protect Us. Retrieved from https://www.wired.com/brandlab/2020/04/ai-healthcare-protecting-systems-protect-us/

Yeng, P., Nweke, L. O., Woldaregay, A. Z., Yang, B., & Snekkenes, E. A. (2020). Data-driven and Artificial Intelligence (AI) approach for modelling and analyzing healthcare security practice: A systematic review. *Advances in Intelligent Systems and Computing, 2020*. Retrieved from https://www.researchgate.net/profile/Livinus_Nweke/publication/339071164_Data-Driven_and_Artificial_Intelligence_AI_Approach_for_Modelling_and_Analyzing_Healthcare_Security_Practice_A_Systematic_Review/links/5e5cc21892851cefa1d4e209/Data-Driven-and-Artificial-Intelligence-AI-Approach-for-Modelling-and-Analyzing-Healthcare-Security-Practice-A-Systematic-Review.pdf

Research Cases

3.13 Digital Identity Theft Using Deepfakes

Alisha Gilbert
Canterbury District Health Board, Christchurch, New Zealand

Zhigang Gong
University of Canterbury, Christchurch, New Zealand

CASE SUMMARY

It was reported by The Wall Street Journal on 30 August 2019 that the CEO of a UK-based energy firm was deceived by criminals who called him, impersonated his boss, and requested the transfer of €220,000 ($243,000) to an account. This voice-spoofing attack was described by cybercrime experts as an unusual case because artificial intelligence (AI) was used in the attack. Although it had been predicted that AI could be used as a weapon to conduct cyberattacks, scams using AI are a novel challenge for companies because their existing cybersecurity tools can't recognize spoofed voices (Stupp, 2019). In the above case, the identity of the CEO's boss, namely his voice characteristics (accent and melody), has been stolen by criminals using deepfake technology.

In our daily lives, we visit websites on the Internet to do a lot of things such as browsing the news, checking emails, paying bills, and learning. Most websites need to remember the identities of the users they serve. For example, an online learning website needs to remember users so that their learning schedules can be kept, and they can be automatically prompted to continue from their last lesson every time they visit the website. Websites provide sign-up functionality for a user to create an account to represent their presence on the website. Users often need to use a combination of a username and a password as credentials to log in to the website. The website confirms login credentials that are used to validate that a user is eligible to access a given account. A user account is a form of digital identity (hereafter identity). Essentially, identity is a collection of attributes that is used as a digital representation of an individual or entity in an online environment. The attributes of identity can range from straightforward data (e.g. name, birthday, identification number) to more complex data (e.g. medical history, behavior, and social graph data). Various credentials could be used to access an identity, such as a username, password, card, device, biometric data, or a combination. Identities

DOI: 10.1201/9781003264415-47

make identity-based authentication, digital signatures, and form-population easier, but identities are at risk of theft from a data breach.

Identity theft is the intentional use of another person's identity (e.g. their name, identification number, cell phone number, or credit card number) as a method to gain benefit from, or cause loss to, that person (Wikipedia contributors, 2020a). For example, in China, the Cainiao logistics platform provided by Alibaba is used by merchants to deliver goods to customers. To make the delivery process efficient, packages processed by Cainiao won't be delivered to a house but instead to a shop located in the residential community where the customer lives. When a customer comes to the shop and tells the clerk that they want to claim a package, the clerk will ask them for their cell phone number. If there is a package in the system linked to the cell phone number given by the customer, the clerk will locate the package and give it to the customer. In this situation, identity authentication by matching cell phone numbers is simple and efficient but risky. An attacker can collect cell phone numbers and use them as the identity to impersonate a package's receiver.

"Deepfakes (a portmanteau of 'deep learning' and 'fake') are synthetic media in which a person in an existing image or video is replaced with someone else's likeness" (Wikipedia contributors, 2020b). The process of creating deepfakes was labor-intensive and time-consuming in the past. Today, deepfakes can be created quickly through a mobile app. Computer vision and computer graphics researchers don't like the term deepfake because people use it to refer to not only state-of-the-art videos created by AI but also seemingly fraudulent videos and images. Machine learning (ML), the main technical advance that has given rise to an increase in deepfakes, makes it possible to create deepfakes quicker and at a lower cost (Tolosana et al., 2020). Although at present most deepfakes are not convincing, we are worried about deepfakes because technology always develops. The research communities have yet to reach a consensus on when deepfakes will be a true threat, but experts agree that, eventually, anyone will be able to create realistic deepfakes of any person simply by using an app on a smartphone.

There are many communities on the dark web for deepfakes. Their main aim is the creation of pornographic deepfake content, and their members are hard to trace. Aside from this, some people make deepfakes to increase their number of online followers and for these individuals, their deepfake videos or memes are likely to be for entertainment. Some hobbyists use deepfakes to display their skills and get job proposals or collaboration offers from media or television companies. They will get paid for this and no one gets mentally harmed or shamed. Deepfakes can also be used for voting campaigns in politics. Political figures, campaign agencies, or activists can spread false information to manipulate public opinion (Pantserev, 2020). Misinformation can spread, and in severe cases sow civil unrest or even prompt war. Further, deepfakes can be used by fraudsters for currency and asset manipulation or for other financial crimes. Fraudsters have already used deepfake audio and in the future, it will be possible to fake video calls and video conferences in real-time. For example, the large volume of online video content featuring C-level executives, such as TED Talks published on YouTube, can be used as the input for ML that generates deepfakes.

The effects of deepfakes are varied, and they can affect many people and organizations. Primarily, Women are vulnerable to deepfakes, with 96% of deepfakes currently present on the Internet having female victims. Deepfakes can also be a method of bullying in schools or workplaces. Anyone has the risk of having their deepfake placed into annoying, ridiculous, frightening, and dangerous scenarios. Deepfakes can also make scams more powerful. People are worried about identity theft using deepfakes, which could lead to fraudulent online payments and hacking into personal banking services. Additionally, deepfakes could be used as a cover for people's wrongdoing. Governments also fear the danger posed to democracy by deepfakes. A politician running for election could star in a deepfake where they denounce their campaign promises. An unwell governor could surface on a video telling people that they are, in fact, healthy.

Malicious deepfakes must be stopped. Enforceable laws criminalizing non-consensual deepfakes need to be introduced. Fortunately, technologies used to identify and detect manipulated videos are being developed. A company named Deeptrace works on an API that pre-screens incoming media and puts obviously manipulated content into a quarantine zone. A platform named Reality Defender, developed by a company named AI Foundation, aims to tag and bag obviously manipulated videos and images before they start to do any damage.

When people receive a phone call or a voice chat, they can identify who the caller is by recognizing the tone, accent, speed, and other characteristics of the caller's voice. Unfortunately, the voice that people hear in the phone call or voice chat could not be real, but instead, be a deepfake. An attacker uses collected data relating to a victim's voice as material and then uses AI technology to synthesize voices to sound like the voice spoken by the victim in real-time. The victim's voice-based identity has thereby been stolen by a deepfake.

Biometric facial recognition is widely used to authenticate users. It is assumed to be a secure method of authentication because an individual's face is unique. Unfortunately, the face that appears in front of a camera used to collect biometric facial information might not be a real one. It could be a deepfake in which the victim's face has been superimposed on a recorded person. An attacker uses collected data relating to a victim's face as material and then uses AI technology to synthesize videos to make it look like the victim appeared in them. The victim's face-based identity has thereby been stolen by deepfakes. At present, biometric facial recognition can detect deepfakes. However, as technology advances, deepfakes can become more convincing, and sophisticated deepfakes could compromise the security of biometric facial recognition.

The Emergence of Deepfake Technology

Distinguishing between real and fake media becomes increasingly difficult because of new digital technologies. The emergence of deepfakes makes this problem more prominent. Deepfakes' impact on society can be extended by social media. Although deepfakes are a serious threat to our society, economy, and political system, they can be defeated by legislation, regulation, policy, education, training, detection technologies, and content authentication (Westerlund, 2019). Therefore, the need for tools that can check the authenticity of media content is urgent. To defeat deepfakes, we need to understand what deepfakes are, who produces them, why they exist, and what technology they use (Westerlund, 2019).

The producers of deepfakes include hobbyists, political rivals, fraudsters, and legitimate players like movie studios (Westerlund, 2019). Deepfakes impose a threat to cybersecurity because they can be used to create fraudulent identities (Westerlund, 2019). Real-time digital impersonation of an executive can be implemented by deepfakes. Fraudsters can use this kind of deepfakes to ask employees to do things that will cause loss to the company, such as transferring money and exposing confidential information. High-performance, low-cost hardware, and open-source software for manipulating videos are prerequisites for the emergence of deepfakes. As technology develops, the number of photos required to create a deepfake can be reduced to only one. It means that a deepfake can be created by just using a selfie.

The technologies used to create deepfakes are not exclusive to certain groups. For example, mobile apps for creating deepfakes are widely available to download. Tensorflow, released by Google as an artificial intelligence tool, is famous for its machine learning and image processing capabilities. However, the open-source backend has been used to create malicious deepfakes. The introduction of AI to classic face-morphing software programs makes them capable of creating convincing videos. If the original video used by a deepfake is of lower quality (e.g. CCTV footage), the deepfake is almost impossible to be recognized as being fake. As technology develops, deepfakes won't be distinguishable from real videos by the naked eye. Eventually, individuals will be capable of fabricating information, manipulating existing data, and disseminating misinformation including fake video and audio recordings (Maras & Alexandrou, 2018).

In some courts of law, an image or video, submitted as evidence, that has been obtained reliably doesn't need a witness to verify its authenticity. The difficulty is establishing that the method used to create an image or video is reliable (i.e. accurate and authentic). To answer this question, we first need to understand the process of creating a digital image taken by a digital camera. First, light goes into the digital sensor through the aperture. Second, the sensor captures the intensity of light on the sensor and allocate a color filter, which is Red, Green, or Blue. The pixels in a digital image are too small for human eyes to see. Therefore, the brain interprets the light emitted from the pixels as a single image. The way to create a digital video is similar to the way to create a digital image. A digital video consists of frames. Each frame is a still image. A video with 30 frames per second (30 fps) plays 30 pictures in a second.

Photographic images could be manipulated long before digital image technology was invented. At that time, the authenticity of the image could be verified by comparing the photo with its negative. In the era of digital imaging, the camera uncompressed (.RAW) file format plays the same role as a film negative. The .RAW file format produced by digital cameras stores the original and unaffected pixel information of an image. Therefore, the footprints left by the post-processing of an image can be used to verify its authenticity (Maras & Alexandrou, 2018).

With the use of editing tools, it is easy to manipulate photos and videos to create a deepfake. Anyone can create or swap their face with that of an actor or high-profile person with the help of software and editing tools. At present, the research on face manipulation detection is sparse and the number of available datasets to assist in the detection of face manipulation and video alternation is small. Another problem is that the advance in deepfake detection technology can also be used to increase the quality of the deepfake. In the future machine learning and AI have the potential to make fake videos difficult to be identified as being fake (Maras & Alexandrou, 2018).

The broader area of face manipulation can be divided into four categories according to the level of image manipulation: Entire Face Synthesis, Identity Swap, Attribute Manipulation, and Expression Swap.

With Entire Face Synthesis, the entire face of an individual can be swapped with the face of another individual. This type of image can be developed using an ML framework known as Generative Adversarial Networks (GAN). StyleGAN, created by Nvidia Corporation, is an extension of the progressive growing GAN that is an approach for training generator models capable of synthesizing very large high-quality images via the incremental expansion of both discriminator and generator models from small to large images during the training process (Brownlee, 2019). As a result, high-quality images with a real impact can be created. These are used mainly for video games and modeling of 3D industries. Entire Face Synthesis can be used by cybercriminals to create fake profiles on social media and spread false information.

Examples of Identity Swaps are face swap applications that use deepfake technologies. Face swap applications are an easy entry point to making deepfakes because they package the technical expertise that is necessary for an otherwise challenging task.

Attribute manipulation is a face-editing process that changes elements of an identity, such as changing hair color, eye color, complexion, gender, or age. Attribute manipulation can also be achieved using a GAN framework. Some types of attribute manipulation are also referred to as face editing or retouching.

With expression swap, the expression of an identity's face is changed. This can be done using an image of another identity that contains the desired expression, for example, the widely circulated video of Mark Zuckerburg seemingly giving a press release detailing the level of control Facebook has. More advanced techniques are being developed for expression swap, particularly to achieve convincing video manipulation.

CASE ANALYSIS

Phone calls are often used as a measure to deceive people. In a classical phone call deception, the caller can speak familiarly to the receiver and spin a story that refers to a person the receiver knows. The caller creates a lie on top of the truth of the relationship between the receiver and the known person. The truth in the lie increases the receiver's trust. Some inexperienced receivers could be deceived in this way.

The situation described at the opening of this case study is far more sophisticated than the common phone call deceptions. In this situation, the fraudster assumes the identity of a person by speaking in their voice. There are a few things the fraudster needs to do to conduct such a fraud. First, the fraudster needs to figure out the organizational structure of the company. They need to know who the boss of the company is and who can execute the boss's orders and their phone numbers. Second, the fraudster needs to know what business activities are in progress, so they can make use of them to fabricate requirements. Third, the fraudster needs to collect the person's voice and use it as material to create deepfakes. Fourth, the fraudster needs to be proficient in deepfake technology. In these types of situations, high-quality and responsive deepfakes are required. The deepfake created is not simply predetermined words or phrases in the person's voice that need to be played in a scheduled order in a phone call. Instead, the deepfake will generate responses during the conversation, making the deepfake more convincing for the receiver of the call.

The fraud in the case appears to be unavoidable because it is normal for an employee to execute orders from their supervisors. However, fraud can be avoided through information security management to protect confidential assets and enhance risk management processes. The confidential information assets, in this case, include the profiles of the boss and CEO and the information relating to current business activities. Access to confidential information should, at least, be limited to the organization. In some cases, more strict access control mechanisms will be required. Employees should be trained not to disclose confidential information to the public. The information security department in the organization should also pay attention to technology trends. By doing so, they can recognize what emerging technologies, such as deepfakes, could be a threat to information security.

Deepfakes include two categories: voice and video. The information security department should identify parts of the organization's information system that could be impacted by malicious voice and video. For instance, an access control system based on facial recognition could be impacted by video deepfakes. Because the fraud in the case is conducted through a phone call, we focus on an organization's phone system. The information security department should conduct a security test against voice deepfakes for the phone system to determine if an employee identifies calls controlled by a deepfake, and whether they are willing to follow the instructions given by the deepfake. The fact that employees could be manipulated by fake identities is a security vulnerability. The information security department needs to assess a vulnerability's impact on all business processes and the whole information system.

There are four main entities behind fake images and videos:

1. Deepfake hobbyist groups
2. Political figures or activists
3. Fraudsters
4. Television companies

Deepfakes are going to pose a massive threat and burden on the communications and media industries. If deepfakes are used, it is going to be very hard to distinguish between real and fake news. Apart from journalism, the government is concerned about the threat deepfakes pose to national security. Deepfakes could be a powerful technique for spreading false information during a war or altering the opinions and beliefs of voters during an election. Cybersecurity problems are another danger that deepfakes bring. The business sector has also demonstrated an interest in shielding itself from viral fraud, as deepfakes may be used for consumer and asset manipulation.

There are four main methods to fight against deepfake harm:

1. Legislation and regulation
2. Corporate policies and voluntary action
3. Education and training
4. Anti-deepfake technology, including deepfake detection, content authentication, and deepfake prevention

LESSONS LEARNED

As technologies develop, harmless things could be a threat to information security. Identity theft by deepfakes exemplifies this idea, where voice and video can be used in a way that compromises information system security. Information security development should pay attention to technology trends and use the risk management process to identify, analyze, evaluate, and treat novel risks caused by new threats and vulnerabilities.

With editing software, the handling of images and videos, such as facial manipulation, is now quite simple. Facial manipulation can be classified into four groups, based on the degree of distortion of the images as follows:

1. Entire face synthesis
2. Identity swap
3. Attribute manipulation
4. Expression swap

Identification is mostly used in access control systems. Because of deepfakes, facial recognition becomes less reliable as an access control mechanism. For example, in China, when older people remotely apply for an annual retirement pension via a mobile app, they need to prove to the government that they are still alive through facial recognition. To make facial recognition reliable, liveness detection methods are employed asking users to blink, smile, turn/nod, watch colored flashing lights, make random faces, and speak random numbers. Most liveness detection technologies claim that they can detect deepfakes, making liveness detection a viable method of multi-factor authentication. Ordinal facial recognition systems should be upgraded to include liveness detection.

There are four main categories of people who create deepfakes:

1. Deepfake hobbyist groups
2. Political figures, activists, and government organizations
3. Fraudsters
4. Media companies

Deepfake technology provides the tools to create false images (e.g. by swapping faces in images or videos). Anti-deepfake technology provides tools to detect false images, authenticate images, and prevent content from being used in a deepfake

(Westerlund, 2019). Social engineering cybercriminals use deepfakes, so organizations and companies should develop good security protocols and cyber-resilience plans.

Deepfakes are a significant danger to our culture, democratic system, and industry as they place tension on reporters trying to separate legitimate and false news, endanger national security by distributing misinformation and intervening in elections, make citizens doubt policy intelligence and government, and give rise to cybercrimes against individuals, organizations, and governments. While it is important to introduce appropriate policies and legal restrictions, deepfakes should not be banned outright by laws and regulations, as this would be unethical.

DISCUSSION QUESTIONS

1. How can the creators of deepfakes be identified and brought to justice?
2. What methods can be used to stop deepfake news and false information?
3. Is it ethical to ban deepfake technology by just considering its harmful effects?
4. What kinds of tools should be developed to determine the authenticity of images?

REFERENCES

Brownlee, J. (2019, August 19). A gentle introduction to StyleGAN the style generative adversarial network. Machine Learning Mastery. https://machinelearningmastery.com/introduction-to-style-generative-adversarial-network-stylegan

Maras, M.-H., & Alexandrou, A. (2018). Determining authenticity of video evidence in the age of artificial intelligence and in the wake of Deepfake videos. *The International Journal of Evidence & Proof*, 23(3), 255–262. https://doi.org/10.1177/1365712718807226

Pantserev, K. A. (2020). The Malicious Use of AI-Based Deepfake Technology as the New Threat to Psychological Security and Political Stability. In *Cyber Defence in the Age of AI, Smart Societies and Augmented Humanity* (pp. 37–55). Springer, Cham. https://link.springer.com/chapter/10.1007/978-3-030-35746-7_3

Stupp, C. (2019, August 30). Fraudsters used AI to Mimic CEO's voice in unusual cybercrime case. Retrieved August 10, 2020, from https://www.wsj.com/articles/fraudsters-use-ai-to-mimic-ceos-voice-in-unusual-cybercrime-case-11567157402

Tolosana, R., Vera-Rodriguez, R., Fierrez, J., Morales, A., & Ortega-Garcia, J. (2020). Deepfakes and beyond: A survey of face manipulation and fake detection. arXiv preprint arXiv:2001.00179.

Westerlund, M. (2019). The emergence of Deepfake technology: A review. *Technology Innovation Management Review*, 9(11), 39–52. https://doi.org/10.22215/timreview/1282

Wikipedia contributors. (2020a, August 1). Identity theft. Retrieved from https://en.wikipedia.org/wiki/Identity_theft

Wikipedia contributors. (2020b, August 7). Deepfake. Retrieved August 10, 2020, from https://en.wikipedia.org/wiki/Deepfake

Index

Pages in *italics* refer to figures and pages in **bold** refer to tables.

E

F

G

H

I

J

K

L